VOLUME ONE HUNDRED AND THIRTY

ADVANCES IN IMMUNOLOGY

Tumor Immunology

ASSOCIATE EDITORS

VOLUME ONE HUNDRED AND THIRTY

ADVANCES IN IMMUNOLOGY

Tumor Immunology

Edited by

ROBERT D. SCHREIBER

Washington University School of Medicine,
St. Louis, Missouri, USA

ELSEVIER

AMSTERDAM • BOSTON • HEIDELBERG • LONDON
NEW YORK • OXFORD • PARIS • SAN DIEGO
SAN FRANCISCO • SINGAPORE • SYDNEY • TOKYO
Academic Press is an imprint of Elsevier

Academic Press is an imprint of Elsevier
50 Hampshire Street, 5th Floor, Cambridge, MA 02139, USA
525 B Street, Suite 1800, San Diego, CA 92101-4495, USA
The Boulevard, Langford Lane, Kidlington, Oxford OX5 1GB, UK
125 London Wall, London, EC2Y 5AS, UK

First edition 2016

ISBN: 978-0-12-805156-6
ISSN: 0065-2776

For information on all Academic Press publications visit our website at http://store.elsevier.com/

CONTENTS

CONTRIBUTORS

Etienne Becht
INSERM UMR_S 1138, Cancer, Immune Control and Escape; Université Paris-Descartes, Sorbonne Paris Cité, UMR_S 1338; Sorbonne Universités, UPMC Univ Paris 06, UMR_S 1138, Cordeliers Research Centre, Paris, France

Leticia Corrales
The University of Chicago, Chicago, IL, United States

Aurélien de Reyniès
Programme Cartes d'Identité des Tumeurs, Ligue Nationale Contre le Cancer, Paris, France

Marie-Caroline Dieu-Nosjean
INSERM UMR_S 1138, Cancer, Immune Control and Escape; Université Paris-Descartes, Sorbonne Paris Cité, UMR_S 1338; Sorbonne Universités, UPMC Univ Paris 06, UMR_S 1138, Cordeliers Research Centre, Paris, France

Glenn Dranoff
Novartis Institutes for Biomedical Research, Cambridge, MA, United States

Wolf H. Fridman
INSERM UMR_S 1138, Cancer, Immune Control and Escape; Université Paris-Descartes, Sorbonne Paris Cité, UMR_S 1338; Sorbonne Universités, UPMC Univ Paris 06, UMR_S 1138, Cordeliers Research Centre, Paris, France

Claire F. Friedman
Memorial Sloan Kettering Cancer Center; Weill Cornell Medical College, New York, NY, United States

Thomas F. Gajewski
The University of Chicago, Chicago, IL, United States

Claire Germain
INSERM UMR_S 1138, Cancer, Immune Control and Escape; Université Paris-Descartes, Sorbonne Paris Cité, UMR_S 1338; Sorbonne Universités, UPMC Univ Paris 06, UMR_S 1138, Cordeliers Research Centre, Paris, France

Nicolas A. Giraldo
INSERM UMR_S 1138, Cancer, Immune Control and Escape; Université Paris-Descartes, Sorbonne Paris Cité, UMR_S 1338; Sorbonne Universités, UPMC Univ Paris 06, UMR_S 1138, Cordeliers Research Centre, Paris, France

Matthew M. Gubin
Washington University School of Medicine, St. Louis, MO, United States

Matthew D. Hellmann
Memorial Sloan Kettering Cancer Center; Weill Cornell Medical College, New York, NY, United States

Pierre Laurent-Puig
Université Paris Descartes, Sorbonne Paris Cite; INSERM, UMR-S1147, Paris, France

WeiWei Aileen Li
Wyss Institute for Bioinspired Engineering, Harvard University, Boston; School of Engineering and Applied Sciences, Harvard University, Cambridge, MA, United States

Sherene Loi
Peter MacCallum Cancer Centre, East Melbourne; University of Melbourne, Parkville, VIC, Australia

David J. Mooney
Wyss Institute for Bioinspired Engineering, Harvard University, Boston; School of Engineering and Applied Sciences, Harvard University, Cambridge, MA, United States

Shin Foong Ngiow
QIMR Berghofer Medical Research Institute; University of Queensland, Herston, QLD, Australia

Steven A. Rosenberg
Center for Clinical Research, National Cancer Institute, Bethesda, MD, United States

Catherine Sautès-Fridman
INSERM UMR_S 1138, Cancer, Immune Control and Escape; Université Paris-Descartes, Sorbonne Paris Cité, UMR_S 1338; Sorbonne Universités, UPMC Univ Paris 06, UMR_S 1138, Cordeliers Research Centre, Paris, France

Robert D. Schreiber
Washington University School of Medicine, St. Louis, MO, United States

Ayelet Sivan
The University of Chicago, Chicago, IL, United States

Mark J. Smyth
QIMR Berghofer Medical Research Institute; University of Queensland, Herston, QLD, Australia; Peter MacCallum Cancer Centre, East Melbourne; University of Melbourne, Parkville, VIC, Australia

Stefani Spranger
The University of Chicago, Chicago, IL, United States

David Thomas
Garvan Institute of Medical Research, Darlinghurst, NSW, Australia

Jeffrey P. Ward
Washington University School of Medicine, St. Louis, MO, United States

Jedd D. Wolchok
Memorial Sloan Kettering Cancer Center; Weill Cornell Medical College, New York, NY, United States

Karrie K. Wong
Novartis Institutes for Biomedical Research, Cambridge, MA, United States

James C. Yang
Center for Clinical Research, National Cancer Institute, Bethesda, MD, United States

Jessica Zucman-Rossi
INSERM, UMR-1162, Génomique fonctionnelle des tumeurs solides, Paris, France

PREFACE

The fields of cancer immunology and cancer immunotherapy have undergone explosive growth over the last 15 years. In part, this growth has come about as a natural consequence of our enhanced understanding of the molecular and cellular basis of immunologic reactions and the highly sophisticated manner by which we now view the biologic consequences of these interactions. However, additional discoveries made during this time led to the recognition that immunity plays important roles throughout the entire cancer cycle—from unresolved inflammation promoting cellular transformation, to the roles of innate and adaptive immunity in the cancer immunoediting process that both protects the host against cancer outgrowth and shapes cancer immunogenicity, to the observation that cancer can co-opt the complex machinery that evolved to prevent autoimmunity and use it to its own advantage by inducing, in otherwise immunocompetent hosts, a state of tumor-specific immune unresponsiveness that favors cancer expansion over cancer patient survival.

Thus, the field has undergone a major paradigm shift in its thinking about the interaction of the immune system with developing and established tumors. Just two decades ago, most would have said that immunity had no role in promoting or controlling cancer. However, today the results are unequivocal and the pendulum has swung in the complete opposite direction. The immune system is seen as a key player in the complex interactions that lead to the fate decisions of cancer development and cancer control. These fundamental observations have set the stage for the generation of novel and effective cancer immunotherapies that in some cases can promote extraordinary, durable responses in cancer patients that have never been seen before. Some of these treatments have now received FDA approval and many more are in the pipeline. Thus, cancer immunotherapy has indeed become a reality and to many has become the fourth pillar of cancer therapy (together with surgery, chemotherapy, and radiation therapy).

However, many critical questions about cancer immunotherapy remain unanswered and these are the focus of this volume of *Advances in Immunology*. In Chapter 1, Nglow, Loi, Thomas, and Smyth deal with the question "What are the best experimental models to translate basic findings in cancer immunology to the therapeutic situation." They discuss the major classes of mouse models that are currently being used to assess the potential for different immunotherapies, document the utility of carcinogen-induced cancer models for assessing novel immunotherapies, and discuss the use of xenograft models using mice with reconstituted human immune systems. In Chapter 2, we (Ward, Gubin, and Schreiber) address the question of whether there are a set of tumor antigens that are particularly well suited for use as targets for cancer immunotherapy. We further discuss the possibility of using cancer-specific mutant neoantigens in cancer-specific vaccines and especially as the basis for personalized therapeutic cancer vaccines. Chapters 3 and 4 focus on the dilemma of why some patients respond to cancer immunotherapy, while others do not. In Chapter 3, Spranger, Sivan, Corrales, and Gajewski discuss their observation that patients who tend to respond positively to immunotherapy display T cells that infiltrate the tumor microenvironment, while those that do not respond the cancer immunotherapy display tumors that contain very few T cells. They describe some of the elements they have discovered that play important roles in determining the T cell inflamed phenotype and discuss how one might go about converting tumors that display a noninflamed tumor microenvironment to ones that are inflamed. In Chapter 4, Becht, Giraldo, Germain, de Reyniès, Laurent-Puig, Zucman-Rossi, Dieu-Nosjean, Sautès-Fridman, and Fridman continue the discussion about the tumor microenvironment and specifically on the immune contexture of this environment. They specifically describe their fundamentally important observation that by defining the immune and inflammatory landscapes of human tumors, key insights can be obtained that facilitate determining the course of treatment that the patient should undergo. Their work has made important connections between the bi-directional nature of the dialog between tumor cells and cells of the host and has helped establish prognostic and theranostic classifications to different cancers. Chapters 5-7 deal with different aspects of the question whether principles can be established that will help one select between specific types of immunotherapies to identify the one(s) best suited to treat a particular type of cancer. In Chapter 5, Wong, Li, Mooney, and Dranoff deal with this question in the context of therapeutic cancer vaccines. They discuss the relative lack of success in past attempts at developing cancer vaccines and then go on to discuss recent advances and next-generation cancer vaccines that perhaps will be more successful. In Chapter 6, Hellmann, Friedman, and Wolchok argue that the most successful cancer immunotherapies will most likely come as a result of combinatorial therapeutic approaches. They raise the interesting possibility of combining checkpoint blockade that target T cells with therapies that target other co-inhibitory or co-stimulatory molecules as well as other anti-cancer approaches such as cancer vaccines, chemotherapy, and radiation therapy. Finally, Yang and Rosenberg discuss their experiences, past and present, with adoptive T cell therapy for cancer and the refinement of their approach to focus on T cells that are specific for cancer neoantigens.

ROBERT D. SCHREIBER

CHAPTER ONE

Mouse Models of Tumor Immunotherapy

Shin Foong Ngiow*,†, Sherene Loi‡,§, David Thomas¶, Mark J. Smyth*,†,‡,§,1

*QIMR Berghofer Medical Research Institute, Herston, QLD, Australia
†University of Queensland, Herston, QLD, Australia
‡Peter MacCallum Cancer Centre, East Melbourne, VIC, Australia
§University of Melbourne, Parkville, VIC, Australia
¶Garvan Institute of Medical Research, Darlinghurst, NSW, Australia
[1]Corresponding author: e-mail address: mark.smyth@qimrberghofer.edu.au

Contents

Abstract

Immunotherapy is now evolving into a major therapeutic option for cancer patients. Such clinical advances also promote massive interest in the search for novel immunotherapy targets, and to understand the mechanism of action of current drugs. It is projected that a series of novel immunotherapy agents will be developed and assessed for their therapeutic activity. In light of this, in vivo experimental mouse models that recapitulate human malignancies serve as valuable tools to validate the efficacy and safety profile of immunotherapy agents, before their transition into clinical trials. In this review, we will discuss the major classes of experimental mouse models of cancer commonly used for immunotherapy assessment and provide examples to guide the selection of appropriate models. We present some new data concerning the utility of a carcinogen-induced tumor model for comparing immunotherapies and combining immunotherapy with chemotherapy. We will also highlight some recent advances in experimental modeling of human malignancies in mice that are leading towards personalized therapy in patients.

Advances in Immunology, Volume 130
ISSN 0065-2776
http://dx.doi.org/10.1016/bs.ai.2015.12.004

1. INTRODUCTION

Immunotherapy has a longstanding history in cancer treatment. Recently, this approach to cancer therapy reinvigorates a massive interest from the academic investigators and pharmaceutical companies. There has been a revolutionizing success of T cell checkpoint inhibitors and chimeric antigen receptor (CAR) T cells in treating various malignancies (Ansell et al., 2015; Hamid et al., 2013; Herbst et al., 2014; Hodi et al., 2010; Maude et al., 2014; Porter, Levine, Kalos, Bagg, & June, 2011; Powles et al., 2014; Robert et al., 2015; Topalian et al., 2012; Wolchok et al., 2013). Beyond these advances in the clinic, it is foreseen that a broad spectrum of promising preclinical immunotherapeutic agents (alone or in combination) will soon be enlisted for clinical trials (Galluzzi et al., 2014). The bench-to-bedside transition of these immunotherapeutic agents has largely stemmed from preclinical assessment in experimental cancer models in mice.

However, the tumor microenvironment (TME) is a dynamic milieu. It is now clear that the interaction between tumor (or early transformed cells) and immune cells plays a significant role in determining disease outcome. Assisted by intrinsic genomic instability and mutations, and multiple dysregulated cellular machineries, tumor cells continue to adapt and survive in their host body (Hanahan & Weinberg, 2011). Human malignancies demonstrate regional heterogeneity in their tumor–immune cellular makeup and variation in the effectiveness of natural tumor immunity. This heterogeneity is perhaps reflected by the absence of an effective clinical immune characterization (immune contexture and/or immunoscore at different tumor stages) for most types of cancer (Fridman, Pagès, Sautès-Fridman, & Galon, 2012). More importantly, for preclinical research, these complexities limit the ideal mimicry of human malignancies by experimental mouse models of cancer. Despite these limitations, in vivo experimental tumor models are valuable tools for the assessment of therapeutic agents before their transition into clinical trials. These experimental studies in mice have provided an understanding of the role for tumor immunity in shaping the TME at different stages of tumor growth, and a proof-of-principle for the therapeutic activity and the associated overt toxicities of numerous immunotherapeutic agents (Gyorki, Callahan, Wolchok, & Ariyan, 2013; Liu, Blake, Smyth, & Teng, 2014). It is not the major purpose of this review to compare the strengths and weaknesses of various experimental mouse tumor models, since this has been done before (Dranoff, 2012;

DuPage & Jacks, 2013; Gould, Junttila, & de Sauvage, 2015), but to highlight some of the advances in experimental mouse tumor models, and to discuss the use of appropriate models to explore antitumor activity of the therapeutic agents of interest.

2. TRANSPLANTABLE TUMOR MODELS

For decades, in vivo mouse tumor cell line transplants have been the major approach to study tumor immunology and immunotherapy. Following injection of in vitro-cultured tumor cells into syngeneic mice, the mice must be sacrificed within weeks due to a large and rapidly growing tumor burden. The relatively reproducible and rapid in vivo tumor growth of transplantable mouse tumor lines when injected subcutaneously or intravenously are major advantages of this experimental model in studies of immunotherapy. It is notable that most of the experimental tumor cell lines are tumor cells that have been derived in immunocompetent mice and thus have evaded immunological pressure and/or been immunologically tolerated (Vesely, Kershaw, Schreiber, & Smyth, 2011). Hence, the tumorigenesis of transplantable tumor models is likely to be different from de novo tumors. The local innate inflammatory response triggered by an injection and presence of a large number of tumor cells may also condition the TME of transplantable tumors (Apetoh et al., 2007; Kroemer, Galluzzi, Kepp, & Zitvogel, 2013; Ladoire et al., 2013; Obeid et al., 2007). To a certain extent, the presence of dead tumor cells might provide a vaccination effect, affecting the therapeutic response of an immunotherapeutic agent (Kroemer et al., 2013; Ladoire et al., 2013). Despite these limitations, transplantable tumor models remain valuable tools for immunotherapy assessment.

An emerging area of importance concerns the epitopes driving tumor-suppressor responses. To track and study tumor-specific T cells, tumor cell lines are often engineered with model antigen, such as ovalbumin or hemagglutinin (Gilboa, 1999). These engineered tumor cells can be subsequently transplanted into transgenic mice expressing T cell receptor specific for the model antigen (eg, OT-I or OT-II transgenic mouse), transgenic mice expressing the model antigen (eg, RIP-OVA mouse), or syngeneic mice; where antigen-specific T cells can be monitored and assessed (Gilboa, 1999). However, it is important for the researcher to note that these modified tumor cells could change the tumor cell immunogenicity and trigger different antitumor immune responses (Schreiber, Old, & Smyth, 2011). In light of this, a careful interpretation of the therapeutic outcome in these models is required. This is particularly essential in an assessment of antitumor immunity and potential adverse immune responses provoked by immunotherapy agents (John et al., 2013; Klages et al., 2010). In contrast, the assessment of endogenous tumor antigen-specific T cells could provide a less biased readout (Bloom et al., 1997; Overwijk et al., 1999). For example, the detection of melanoma and prostate antigen-specific CD4 and CD8 T cells in B16 melanomas and TRAMP-C1 prostate adenocarcinoma, respectively, can be performed by staining using antigen–tetramer complexes or antibodies to known αβ TCR chains (Hanson et al., 2000; Malchow et al., 2013; Muranski et al., 2008; Overwijk et al., 2003; Savage et al., 2008). As an additional example, in a genomic (exome) analysis of MC38 colon adenocarcinoma, this tumor model was shown to harbor seven mutated antigens, and prophylactic vaccination against these antigens was shown to impair the growth of MC38 subcutaneous tumors (Yadav et al., 2014). Indeed, vaccination against tumor neoantigens has recently been shown to increase the breadth and diversity of antigen-specific T cells in advanced melanoma patients (Carreno et al., 2015). Clinically, better understanding the antigens that drive cell-mediated tumor rejection will lead to the development of tumor-specific, peptide-based vaccination strategies to complement checkpoint inhibition.

To ease the monitoring of tumor growth, tumor cell lines are generally injected subcutaneously in mice. It is however important to understand that unlike de novo formed primary tumors, subcutaneous implanted tumors are not growing within a relevant TME. To mimic the physiological environment of a tumor, some tumor cell lines can be orthotopically injected into the relevant organs (Westwood, Darcy, & Kershaw, 2014). These include intravenous injection for a series of leukemia and lymphoma cell lines (Mattarollo et al., 2012; Minard-Colin et al., 2008; Zhou et al., 2011), mammary fat pad injection for mammary tumor cell lines (Verbrugge et al., 2012), intrapancreatic injection for pancreatic ductal adenocarcinoma lines (Zhu et al., 2014), and intracranial injection for glioma cell lines (Wainwright et al., 2014). This variation in site of tumor formation gives rise to a different tumor–immune cellular makeup, impacting on the progression of tumor and therapeutic outcome of a number of immunotherapy agents (Devaud et al., 2014; DuPage et al., 2011; Knight et al., 2013; Westwood et al., 2014). We have previously shown that Tri-mAb (a combination of monoclonal antibodies targeting DR5, CD40, and CD137) was less effective to treat orthotopically injected Renca renal cell carcinoma, RM-1 prostate carcinoma, and CT26 colon adenocarcinoma in comparison to their subcutaneously injected counterparts (Devaud et al., 2014). Despite the advantages of mimicking tumor growth in a relevant microenvironment, orthotopic tumor model approaches sometimes require labor-intensive techniques to perform tumor cell implantation (such as surgery techniques) and to monitor/image tumor growth (such as engineered bioluminescence cell lines and imaging technology to monitor visceral tumor sites), limiting its application in less-equipped laboratories.

Metastasis of tumor is the primary obstacle in various cancer treatments. An experimental surrogate of the formation of tumor metastases in mice is commonly performed by intravenous injection of tumor cells. Tumor cells are experimentally disseminated to organs like lung, liver, spleen, or bones (Sathe et al., 2014). Experimental lung metastases models that are relatively well characterized are the B16 melanoma, 3LL Lewis lung carcinoma, and RM-1 prostate carcinoma (Chow, Sceneay, et al., 2012; Teng, von Scheidt, Duret, Towne, & Smyth, 2011). Given the dependency on myeloid and natural killer (NK) cells to control the dissemination of tumor cells (Chow, Sceneay, et al., 2012; Sceneay et al., 2012; Teng et al., 2011), these metastatic tumor models serve as valuable tool for evaluating immunotherapy that could target these immune subsets. In some cases, to model brain metastases, tumor cell lines may also be intracerebral/intracardially injected into mice (Connell et al., 2013; O'Brien et al., 2014). In addition, under specific experimental conditions, a small number of tumor cell lines are capable of forming spontaneous metastases. One of the most commonly used conditions is to surgically remove primary 4T1 or EO771 mammary adenocarcinomas (either subcutaneously or orthotopically injected) for the induction of spontaneous metastases to distant organs (Chow, Sceneay, et al., 2012; Mittal et al., 2014; Stagg et al., 2010). The formation of metastases under these conditions is likely driven by factors like prolonged time span to obtain overt metastases upon primary tumor challenge, and surgery-induced NK cell dysfunction (Tai et al., 2013). Other approaches to model spontaneous metastases include intrapinna injection of B16 melanoma to generate lymph node metastases (Bobek et al., 2010), intracutaneous injection of HCmel12 and HCmel31 melanomas to generate spontaneous lung metastases (Bald et al., 2014), and intrakidney injection of Renca renal cell carcinoma to generate lung and abdominal metastases (Devaud et al., 2014). In light of the different biology of metastatic disease, these tumor models may be utilized to preclinically assesses the ability of immunotherapy to inhibit for metastasis.

3. GENETICALLY ENGINEERED TUMOR MODELS

For most malignancies, the initiation of transforming process is driven by genomic instability and mutations. These genetic changes could lead to the activation of oncogenes (such as *Kras* and *Braf*) and/or suppression of tumor suppressor genes (TSG) (such as *Tp53*), and eventually progress to form cancer (Hanahan & Weinberg, 2000, 2011). Before the recognition of a role for endogenous immune response in affecting tumor progression, most cancer studies were focused on understanding the contribution of cell intrinsic defects to cancer formation (Hanahan & Weinberg, 2000, 2011). This is evidenced by the previous lack of recognition of immunoevasion as a feature characterizing cancer (Hanahan & Weinberg, 2000, 2011), and the massive effort invested in constructing cancer genetic databases and genetically engineered (GE) tumor models for cancer biology studies (DuPage & Jacks, 2013). The development of a diverse collection of GE tumor models allows researchers to study in situ cellular transformation events under defined genetic changes, and to visualize histopathological features at different stages of tumor progression. In contrast to transplantable tumor models, GE tumor models provide a physiological relevant environment for the study of interplay between tumor and immune cells.

GE tumor models are generally developed using transgenic technologies to enforce a systemic or tissue-specific oncogene expression and/or tumor suppressor gene deletion (DuPage & Jacks, 2013; Frese & Tuveson, 2007). These transgenic models can be further categorized into germline GE models and conditional GE models (DuPage & Jacks, 2013; Frese & Tuveson, 2007). For the former, transgenic mice would develop malignancies in a spontaneous manner. As an example, a series of *Tp53*-mutated mice have been shown to induce a wide spectrum of solid and hematological malignancies (Frese & Tuveson, 2007; Zender, Zuber, & Lowe, 2007). As an additional example, an mouse mammary tumor virus-polyoma middle T (MMTV-PyMT) transgenic mouse was shown to closely recapitulate the progression of human breast carcinoma (Fantozzi & Christofori, 2006). A series of mouse models of this class that are useful for preclinical therapy (monotherapy or in combination with targeted therapy) assessment include the MMTV-driven ErbB2/neuT transgenic mouse (mammary adenocarcinoma; anti-Her2 mAb-sensitive) (Guy et al., 1992; Stagg et al., 2008), $Kit^{V558\Delta/+}$ transgenic mouse (gastrointestinal stromal tumor; Imatinib-sensitive) (Balachandran et al., 2011; Sommer et al., 2003), and

$Kras^{LSL-G12D/+}Trp53^{LSL-R172H/+}$Pdx1-Cre (KPC) (pancreatic ductal adenocarcinoma; anti-CD40 mAb-sensitive) (Beatty et al., 2011). We have used the MMTV-driven ErbB2/neuT transgenic mouse to demonstrate the combination activity of anti-PD-1 and anti-erbB2 therapy (Fig. 1). This study and others with transplanted tumors have led to a phase Ib/II trial evaluating the efficacy of MK-3475 (anti-PD-1) and trastuzumab in patients with trastuzumab-resistant, HER2-positive metastatic breast cancers (International Breast Cancer Study Group—IBCSG 45-13/BIG 4–13). On the other hand, conditional GE models allow spatiotemporal control of the transformation onset. The initiation of tumorigenesis in conditional GE model is generally tissue-specific and triggered by a specific ligand (such as Tet-on/off and tamoxifen-inducible Cre recombinase systems) (DuPage & Jacks, 2013; Frese & Tuveson, 2007). In the context of experimentation, the tumor stages are relatively well controlled in comparison to germline GE model. For example, we and others have recently assessed the efficacy of a series of immunotherapeutic agents in a mouse model of conditionally $Braf^{V600E}$-mutated melanoma (Ho et al., 2014; Hooijkaas et al., 2012; Knight et al., 2013). In addition, a mouse model of conditionally

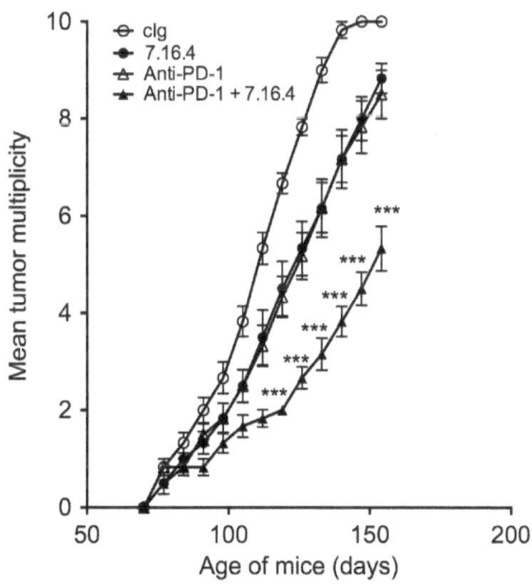

Figure 1 Combined anti-PD-1 and anti-ErbB-2 mAb therapy prolonged the latency of spontaneous mammary carcinomas. Groups of six female ErbB-2/neuT transgenic mice were treated for 6 wk starting at 77 d of age with control Ig (cIg, Mac4, 100 μg i.p.) and/or anti-PD-1 (RMP1-14, 100 μg i.p.) and/or anti-erbB2 (7.16.4, 100 μg i.p.). Mean tumor multiplicity ± standard errors of mean versus the age of mice (days). *** $p < 0.001$ combination compared to any single treatment by Mann–Whitney test.

Kras/Tp53-mutated lung carcinoma developed by Tyler Jacks group was shown to be an ideal system to study the role of T cell immunity in controlling lung adenocarcinoma (DuPage et al., 2011).

Gene targeting via homologous recombination in embryonic stem (ES) cells has been a widely used method for genetic modification in mice (Capecchi, 2005). This construction of a GE mouse model of tumor is labor-intensive, time-consuming, and expensive. In many cases, homozygous germline mutations could result in embryonic lethality, off-target developmental defects, and/or nonspecific genetic defects. These flaws limit the study of a gene of interest in adult tissues. More recently, the clustered, regularly interspace, palindromic repeats (CRISPR)-Cas9 (CRISPR-associated protein) technology is being explored to generate GE tumor models in a more efficient manner (Hsu, Lander, & Zhang, 2014; Pelletier, Gingras, & Green, 2015; Sander & Joung, 2014). CRISPR-Cas9 technology utilizes single guide RNA (sgRNA; a small artificial RNA molecule) to direct mutagenesis in zygotes, accelerating the generation of GE models (Jinek et al., 2012). However, similar to ES gene-targeting approach (such as transgenic mouse model), genetic mosaicism is also present in CRISPR-Cas9-generated GE models. Hence, the production of a large cohort of experimental mice requires further intercrossing of the founder mice. Regardless, CRISPR-Cas9 technology is a very efficient new platform to develop GE tumor models. Excitingly, a number of independent laboratories have recently reported the generation of GE mouse models of liver cancer, lung cancer, and intestinal hyperplasia using this technology (Blasco et al., 2014; Dow et al., 2015; Maddalo et al., 2014; Platt et al., 2014; Sanchez-Rivera et al., 2014; Xue et al., 2014). Like conventional GE tumor models, we foresee this class of tumor model will be interrogated in tumor immunology studies in the near future.

However, GE models have limitations. Perhaps the most fundamental of these is the assumption of the order of genetic events in real world tumor development. A specific concern for recapitulation of immune-mediated tumor suppression is initiation of tumor development by inactivation of genes required for induction of senescence—for example, *Trp53*, *Rb1*, and *Cdkn2a*. Senescence is an important tumor-suppressor process that engages the immune system by expression of a slew of signaling molecules—including IL-1, IL-6, and IL-8 (Kansara et al., 2013). Models that abrogate this response ab initio cannot account for a plausible early phase of tumor suppression that eradicates neoplastic clones that sustain an increased mutation burden in other genes. More practically, in stark contrast to transplantable tumor models, most

GE tumor models require a longer period of time for tumors to progress. Perhaps such temporal variation in GE tumor models resembles the chronic inflammatory features of malignancies in humans and allows the development of immune tolerance, immunoediting, and/or immunosuppressive processes (Kim & Ahmed, 2010; Vesely et al., 2011). In concordance, GE tumor models have relatively heterogeneous tumor formation, potentially dictated by the tumor–immune interaction. The constant expression of a transgene may vary in cohorts of transgenic mice, hence producing different levels of immunosurveillance in controlling tumor outgrowth. We now appreciate that certain mutations in cancer may give rise to the development of T cell-sensitive neoantigens (Linnemann et al., 2015; Rizvi et al., 2015; Robbins et al., 2013; Snyder et al., 2014; van Rooij et al., 2013). Of note, the presence of these neoantigens was shown to correlate with immunotherapy outcome (Rizvi et al., 2015; Snyder et al., 2014). In line with this, the assessment of immunotherapy using GE tumor models is particularly important in an examination of antitumor T cell immunity. However, it remains unclear if most of the currently available GE tumor models fulfill such features of cancer (ie, the impact of oncogene/TSG mutation(s) on neoantigen generation). Most of the current GE tumor models demonstrated an inability to fully recapitulate the endogenous T cell response toward tumor antigens. Although there has been effort to construct model antigens in GE tumor models (Cheung, DuPage, Dong, Chen, & Jacks, 2008; DuPage, Mazumdar, Schmidt, Cheung, & Jacks, 2012), it remains challenging to model T cell responses in malignancies. Despite these drawbacks, it is likely that with the appreciation of common ground between tumor geneticists, tumor biologists, and tumor immunologists, more human cancer-like GE models might be developed in the near future to guide preclinical immunotherapy testing.

4. CARCINOGEN-INDUCED TUMOR MODELS

Another class of experimental mouse model that are widely used in the study of tumor immunology is the carcinogen-induced tumor model. A number of carcinogen-induced (CI) tumor models have been used to study tumor immunosurveillance and immunoediting mechanisms (Schreiber et al., 2011; Swann & Smyth, 2007; Vesely et al., 2011). Some of the relatively well-studied CI models include: methylcholanthrene (MCA)-induced fibrosarcomas, 7,12-dimethylbenz[*a*]anthracene (DMBA)/12-*O*-tetradecanoylphorbol-13-acetate (TPA)-induced skin papillomas, azoxymethane/dextran sodium sulfate-induced colon carcinoma, and ultraviolet B-induced skin cancers (Schreiber et al., 2011; Swann & Smyth, 2007; Vesely et al., 2011). In contrast to the transplantable and GE tumor models, a carcinogen is used to induce de novo tumor formation. Similar to GE tumor models, the in situ induction of genomic instability and mutation using carcinogen allows the natural formation of a tumor–immune interaction in its relevant microenvironment (Hanahan & Weinberg, 2011). For a limited number of tumor types, CI tumor models also provide histopathological features resembling human malignancies (Chow, Tschopp, Moller, & Smyth, 2012; Kansara et al., 2013; Mossman & Churg, 1998). Although both the GE and CI tumor models require a relatively longer period of time for tumor progression, CI tumor models have much complexity in their genetic makeup, depending on the types and doses of carcinogen used, perhaps more accurately reflecting most sporadic carcinogenesis in man. Such variables in carcinogen are potentially generating different levels of immunogenic neoantigens, hence affecting the penetrance and/or latency of disease.

Given the absence of defined genetic manipulation in CI tumor models, it has been technically challenging to effectively track tumor antigen T cell responses in primary tumors, despite the likely presence of a vast number of neoantigens. It is however exciting that in recent studies reported by Schreiber and colleagues, in silico prediction algorithms of high affinity epitopes for major histocompatibility complex class I (H-2D^b and H-2K^b) presentation and exome analyses were used to identify T cell-reactive neoantigens in a series of MCA-induced fibrosarcoma cell lines (Gubin et al., 2014; Matsushita et al., 2012). In this context, together with the use of DNA sequencing technologies, we may be able to identify and predict tumor-derived peptides for the stimulation of endogenous antitumor T cell response in CI tumor models. Another useful aspect of CI tumor models is their application in most mice strains. The construction of a GE tumor model is a labor-intensive and time-consuming process. Unlike GE tumor models, carcinogens can be applied into a wide range of gene-deficient mice, to help the identification of novel immunotherapy targets. Examples of carcinogen-induced models include radiation-induced osteosarcomas and de novo MCA-induced fibrosarcomas and DMBA/TPA-induced skin papillomas (Kansara et al., 2009, 2013; Teng et al., 2010, 2011). The latter models were used to study the resistance to tumor development in IL-23p19-deficient mice, prompting the use of anti-IL23p19 mAb as an immunotherapy agent (Langowski et al., 2006; Teng et al., 2010, 2011). Similarly, we are currently developing strategies to inhibit CD73 and

CD96 signaling, driven by the tumor formation resistance phenotypes seen in CD73- and CD96-deficient mice (Chan et al., 2014; Stagg et al., 2012), respectively.

The MCA-induced fibrosarcoma model in particular, has also been used as a model to test therapies, given its relative reproducibility and comparatively short latency to palpable tumor size. We first described an important synergy for the Tri-mab (as mentioned above) in collectively triggering tumor cell apoptosis, antigen-presenting cell activation and T cell activation and survival (Uno et al., 2006). More recently, this MCA induction of sarcoma model has been used to demonstrate the role of host immune elements in the efficacy of the chemotherapeutic, doxorubicin (Mattarollo et al., 2011). As with human sarcomas, doxorubicin has limited single agent activity in the MCA model. Compellingly, we show that anti-CD40, anti-CD137, or their combination augments the antitumor efficacy of doxorubicin, preempting clinical scenarios for their use (Fig. 2).

5. HUMANIZED MOUSE TUMOR MODELS

Due to ethical and economical constraints, the use of models more closely related to humans (such as nonhuman primate models) for experimental purposes remains limited. Hence, experimental mouse models are generally recognized as the major surrogate models for biology research. However, it is well known that humans and mice have considerable differences in a number of their biological machineries. Therefore, a complete recapitulation of human physiological or pathological processes in mice is very challenging. In concordance, a huge number of preclinical drug candidates that demonstrated promising outcomes in mice failed to show efficacy when tested in humans (Arrowsmith & Miller, 2013; Kamb, 2005). This issue is particularly obvious when modeling diseases where the hematological system is involved, such as infections, autoimmune conditions, and malignancies. To overcome this problem, there has been a great effort in the field to "humanize" experimental mice for a better modeling of human pathophysiological conditions (Holzapfel, Wagner, Thibaudeau, Levesque, & Hutmacher, 2015; Legrand et al., 2009; Rongvaux et al., 2013; Scheer, Snaith, Wolf, & Seibler, 2013).

One promising approach is the xenotransplantation of functional human tissues (such as immune cells and/or tumor tissues) into immunodeficient mice, for studies of cancer biology. Immunodeficient mouse models, like athymic nude mice, recombination-activating gene (Rag) × common

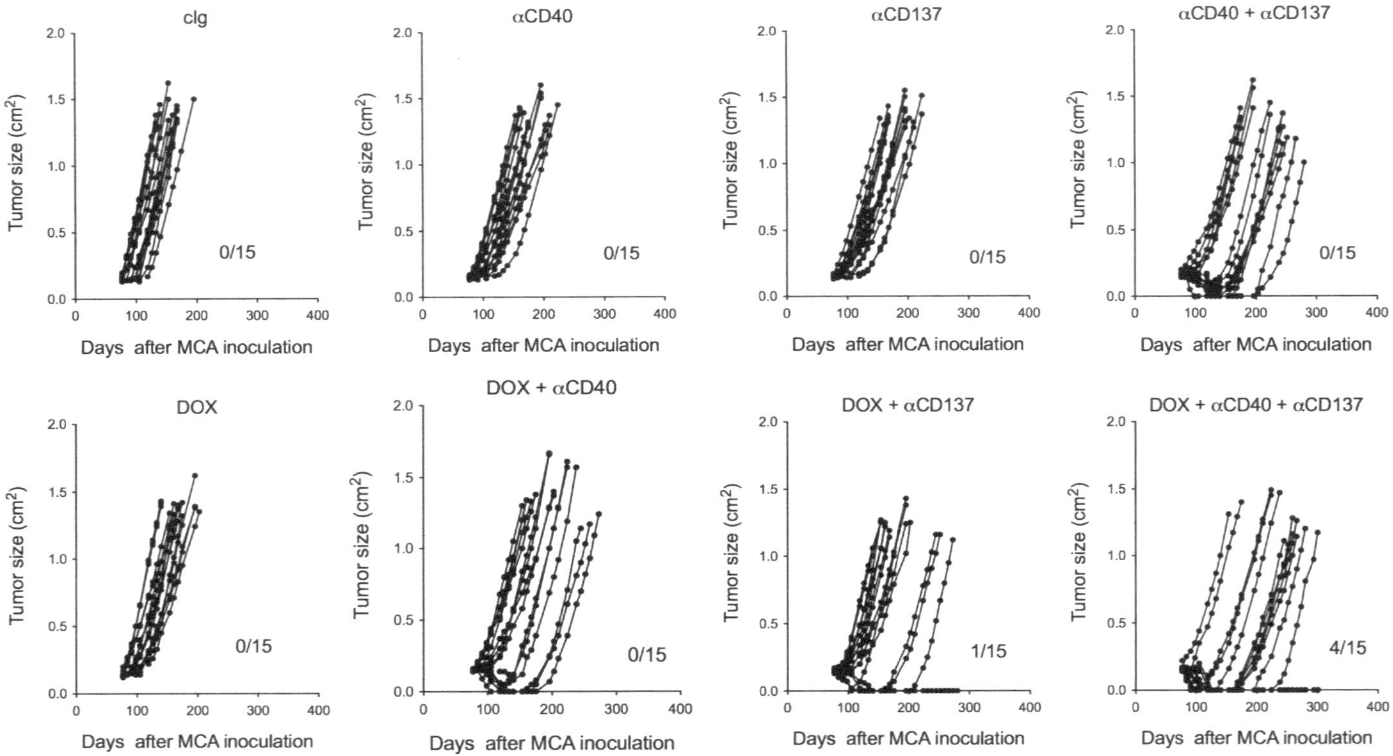

Figure 2 Combination efficacy of doxorubicin (DOX) alone and in combination with mAbs against tumors established de novo in host. Groups of 15 male BALB/c WT mice were inoculated s.c. in the hind flank with 400 μg of MCA in 0.1 mL of corn oil as described in Section 7. Mice were treated with PBS or DOX (2 mg/kg i.v.) from the second palpable tumor measurement (0.1–0.3 cm^2, days 77–119 relative to MCA inoculation). Some groups of mice also received cIg (Mac4), anti-αCD40 (FGK45), αCD137 (3H3), or their combination (100 μg i.p., four times/3 d apart) from the first palpable tumor measurement. Mice were then monitored for fibrosarcoma development over 300 d, with measurements made with a caliper square as the product of two perpendicular diameters (cm^2). Data were recorded as tumor size in cm^2 of individual mice with tumor in each group.

gamma chain (γc)-deficient mice, nonobese diabetic (NOD)-severed combined immunodeficiency (SCID)-IL-2 receptor gamma (IL2rg)-deficient mice are commonly used host for xenograft implantation (Aparicio, Hidalgo, & Kung, 2015; Holzapfel et al., 2015; Legrand et al., 2009; Rongvaux et al., 2013). Mouse models xenografted with human tumor tissues are also widely used for immune-based targeted therapy studies, such as anti-VEGF and anti-EGFR mAbs, and adoptive-transferred engineered T cells (Holzapfel et al., 2015; Zhou, Facciponte, Jin, Shen, & Lin, 2014). However, it is notable that an assessment of these classes of therapy is flawed by the lack of intact adaptive immunity in the recipients, including important regulatory T and B cells. For instance, the antibody-dependent cell-mediated cytotoxicity mechanism triggered by a therapeutic mAb might be masked in a host deficient for NK cells and/or lacking the appropriate Fc receptors (Barok et al., 2007; Ito et al., 2009; Kohrt et al., 2012; Park et al., 2010; Pincetic et al., 2014; Weiner, 2010; Zhang et al., 2004).

To overcome this problem, human immune cells are often cotransplanted (prior or concurrently with the tumor xenograft) into recipient mice. The recipient mice can be reconstituted with human hematopoietic progenitor cells (HPCs; generally based on CD34+) or sorted immune subsets (Legrand et al., 2009; Rongvaux et al., 2014). Substantial evidence supports functional properties of the human immune cells in these immunodeficient hosts and tumor–immune interactions were evidenced. Using this approach, Palucka and colleagues demonstrated the role for thymic stromal lymphopoietin (TSLP) in driving OX40L+ dendritic cell (DC)-mediated Th2 inflammatory responses in breast cancer (Pedroza-Gonzalez et al., 2011). More importantly, in this study, the authors also demonstrated suppression of tumor (xenografted Hs578T breast cancer cells) growth in mice treated with anti-OX40L, anti-TSLP, and anti-TSLP receptor (Pedroza-Gonzalez et al., 2011). As an additional example, in human breast cancer (MDA-MB-453) and γδ cotransplanted with Tregs, Ye and colleagues demonstrated that the administration of anti-IP10 (CXCL10) could suppress tumor growth in vivo (Ye et al., 2013). In a more recent study, Rongvaux and colleagues reported a development of two strains of mice, MITRG and MISTRG, which are capable of supporting the development of human innate immune cells (Rongvaux et al., 2014). MITRG were generated in an immunodeficient *Rag2*$^{-/-}$*IL2rg*$^{-/-}$ background, and they additionally encode human M-CSF (macrophage colony-stimulating factor), human IL-3, human GM-CSF (granulocyte macrophage colony-stimulating factor), and human thrombopoietin. A bacterial artificial chromosome transgene encoding for human SIRPα was inserted into the MITRG strain, to generate the MISTRG strain (Rongvaux et al., 2014). In these models, the administration of Avastin (human vascular endothelial growth factor (VEGF) inhibitor) was shown to suppress the growth of the xenografted melanoma (Me290 melanoma cells) (Rongvaux et al., 2014).

These examples (Pedroza-Gonzalez et al., 2011; Rongvaux et al., 2014; Ye et al., 2013) and others (Aspord, Leccia, Charles, & Plumas, 2013; Bankert et al., 2011; Liddy et al., 2012; Obenaus et al., 2015; Wege et al., 2011), indicate the potential of humanized mouse tumor models in clinical investigations. This is particularly important when assessing a combination therapy involving immunomodulatory agents. However, additional studies are required to interrogate and validate the general translational value of humanized mouse tumor models for such purposes. The origin of immune cells (HPC or sorted immune subsets; patient PBMC or tumor infiltrating immune cells), dynamics of immune cells in the recipient (tumor and periphery), and tumor specimens (regional heterogeneity) are important variables to be considered in the context of personalized therapy. With advances that will further improve humanized mice, we might expect application of this class of tumor model to inform biomedical research.

6. PERSPECTIVES

It is worth noting that many immunotherapy agents target preexisting host immunity. The basal immune system of a tumor-bearing host can be conditioned via multiple tumor intrinsic factors (such as genetics, expression of receptors/ligands, and secretion of soluble factors; Facciabene et al., 2011; Motz et al., 2014; Qian et al., 2011; Ruffell et al., 2013) and extrinsic factors (such as stromal cells, microbiota, and housing environment; Engels, Rowley, & Schreiber, 2012; Grivennikov et al., 2012; Kokolus et al., 2013; Zitvogel et al., 2015). In this light, these variables should be carefully considered in an in vivo experiment. Despite the value of individual experimental mouse tumor models to inform clinical investigation, their findings need to be replicated across multiple tumor models, and ideally more than one strain of mouse. In addition, studies need to be corroborated with the use of appropriate statistical principles, experimental reproducibility, group sizes, endpoint readouts, and careful interpretation, to guide the translation of a new agent into the clinic (Couzin-Frankel, 2014; Kohrt et al., 2012; Peers, South, Ceuppens, Bright, & Pilling, 2014).

Although currently available mouse tumor models may individually recapitulate limited aspects of human cancer, together they allow the functional dissection of phenomena observed in and relevant to human tumor development and therapy. We speculate that with advances in genetic manipulation and imaging technologies, the study of tumor immunology and immunotherapy will eventually tested in more life-like and human-like experimental tumor models. These models should have the capacity to model the physiological environment typical for the tumor's growth, should demonstrate specific cancer-causing genetic changes, should identify relevant neoantigens, and/or should predict immune-related adverse events.

The capacity of tumors to secrete soluble factors (such as cytokines and chemokines) and immunoregulatory receptors/ligands is highly variable between tumor models, hence affecting the recruitment of immune subsets and determining the immunosuppressive mechanisms being deployed. Nonetheless, the immunogenicity of different tumor models might also dictate their tumor–immune contexture (Vesely et al., 2011). Undoubtedly, a better understanding of every tumor model helps the design of therapy. A novel immunotherapeutic agent might first be tested in a series of transplantable tumor models with varied immunogenicity, known immune contexture, and characterized immunosuppressive mechanisms. For example, the B16 melanoma and TRAMP-C1 prostate adenocarcinoma are considered poorly immunogenic compared to MCA-induced fibrosarcoma cell lines or MC38, and so these latter models can be used to validate immunotherapy that targets T cell effector function. Their variations in tumor immunogenicity are likely driven by the presence (and levels) of mutated antigens (ie, neoantigens). By contrast, 4T1 mammary adenocarcinomas have a higher myeloid:T cell ratio compared to CT26 colon adenocarcinoma, and hence these models are useful tools to validate immunotherapy that targets myeloid cell function. These cell lines are also useful tools to assess Treg-targeted therapy, given that CT26 has a higher dependency on Treg number and function to support its growth in vivo, when compared to 4T1 (unpublished data). As an additional example, Rosenberg and colleagues have utilized the pmel transgenic models to uncover a number fundamental principles of adoptive cellular transfer therapy in melanoma (Restifo, Dudley, & Rosenberg, 2012). The efficacy, scheduling and dosing of the immunotherapy of interest might subsequently be assessed in other relevant GI, CI, and/or humanized mouse tumor models.

The in vivo mechanism of action of a therapeutic agent is generally discovered in mouse tumor models appreciably sensitive to the agent (ie, "a therapeutic window of opportunity"). We suggest that therapy-resistance results in tumor models are less often reported. While substantial effort is seen in assessing novel immunotherapy agents, one should consider to revisit clinically established immunotherapy agents in tumor models of resistance. These studies have the potential to yield key information leading to better clinical practice in patients. For example, by studying mouse melanoma that was resistant to a combination of radiotherapy and anti-CTLA-4 therapy, Twyman-Saint Victor and colleagues reported that the PDL1 upregulation on tumor cells and associated T cell exhaustion are the major drivers of therapy resistance. In concordance, a coblockade of PD-1/PD-L1 pathways using anti-PD-1/PD-L1 mAbs resulted in tumor clearance (Twyman-Saint Victor et al., 2015). This example highlights the importance of mouse tumor models of resistance in the preclinical assessment of immunotherapy agents. Of relevance, with advances in genetic manipulation in experimental tumor models (such as CRISPR), we might foresee a significant interest in the field to further study tumor–immune interactions in therapy-sensitive/resistance models of clinically useful agents (completing the virtuous "bed-to-bench side" cycle). Many future tumor model studies in mice will continue to inspire new investigation in the clinic.

7. MATERIALS AND METHODS

7.1 *BALB/c* MMTV-Her2 Mammary Carcinoma

BALB/c MMTV-ErbB-2/neuT mice were bred and maintained in house. Treatments were scheduled as indicated. Development of mammary tumors was monitored by palpation of mammary glands.

7.2 MCA-Induced Fibrosarcoma

BALB/c wild-type mice were obtained from the Walter and Eliza Hall Institute for Medical Research and were maintained at the Peter MacCallum Cancer Centre. All experiments were approved by the Peter MacCallum Cancer Centre Animal Ethics Committee. Groups of 15–20 male BALB/c WT mice were inoculated s.c. in the hind flank with 400 μg of 3-methylcholanthrene (MCA; Sigma-Aldrich, St. Louis, MO) in 0.1 mL of corn oil as described (Swann et al., 2008). Mice were treated as indicated. Mice were then monitored for fibrosarcoma development over 300 d, with measurements made with a caliper square as the product of two perpendicular diameters (cm^2). Data were recorded as tumor size in cm^2 of individual mice with tumor in each group.

ACKNOWLEDGMENTS

The authors thank the members of the Immunology in Cancer and Infection Laboratory for useful discussion.

Financial Support: S.N. and M.J.S. were supported by a National Health and Medical Research Council of Australia Senior Principal Research Fellowship (1078671), a Cancer Council of Queensland Project Grant (1083776), and a QIMR Berghofer Rio Tinto Ride to Conquer Cancer Research Grant 2013 (RTCC130004). D.T. is supported by a National Health and Medical Research Council of Australia Senior Research Fellowship (1003929).

Conflict of Interest: Mark Smyth declares a research agreement grant with Bristol Myers Squibb. The remaining authors of this manuscript have declared that no conflict of interest exists.

REFERENCES

Ansell, S. M., Lesokhin, A. M., Borrello, I., Halwani, A., Scott, E. C., Gutierrez, M., … Armand, P. (2015). PD-1 blockade with nivolumab in relapsed or refractory Hodgkin's lymphoma. *New England Journal of Medicine*, *372*, 311.

Aparicio, S., Hidalgo, M., & Kung, A. L. (2015). Examining the utility of patient-derived xenograft mouse models. *Nature Reviews. Cancer*, *15*, 311.

Apetoh, L., Ghiringhelli, F., Tesniere, A., Obeid, M., Ortiz, C., Criollo, A., … Zitvogel, L. (2007). Toll-like receptor 4-dependent contribution of the immune system to anticancer chemotherapy and radiotherapy. *Nature Medicine*, *13*, 1050.

Arrowsmith, J., & Miller, P. (2013). Trial watch: Phase II and phase III attrition rates 2011–2012. *Nature Reviews. Drug Discovery*, *12*, 569.

Aspord, C., Leccia, M. T., Charles, J., & Plumas, J. (2013). Plasmacytoid dendritic cells support melanoma progression by promoting Th2 and regulatory immunity through OX40L and ICOSL. *Cancer Immunology Research*, *1*, 402.

Balachandran, V. P., Cavnar, M. J., Zeng, S., Bamboat, Z. M., Ocuin, L. M., Obaid, H., … DeMatteo, R. P. (2011). Imatinib potentiates antitumor T cell responses in gastrointestinal stromal tumor through the inhibition of Ido. *Nature Medicine*, *17*, 1094.

Bald, T., Quast, T., Landsberg, J., Rogava, M., Glodde, N., Lopez-Ramos, D., … Tuting, T. (2014). Ultraviolet-radiation-induced inflammation promotes angiotropism and metastasis in melanoma. *Nature*, *507*, 109.

Bankert, R. B., Balu-Iyer, S. V., Odunsi, K., Shultz, L. D., Kelleher, R. J., Jr., Barnas, J. L., … Yokota, S. J. (2011). Humanized mouse model of ovarian cancer recapitulates patient solid tumor progression, ascites formation, and metastasis. *PLoS One*, *6*, e24420.

Barok, M., Isola, J., Palyi-Krekk, Z., Nagy, P., Juhasz, I., Vereb, G., … Szollosi, J. (2007). Trastuzumab causes antibody-dependent cellular cytotoxicity-mediated growth inhibition of submacroscopic JIMT-1 breast cancer xenografts despite intrinsic drug resistance. *Molecular Cancer Therapeutics*, *6*, 2065.

Beatty, G. L., Chiorean, E. G., Fishman, M. P., Saboury, B., Teitelbaum, U. R., Sun, W., … Vonderheide, R. H. (2011). CD40 agonists alter tumor stroma and show efficacy against pancreatic carcinoma in mice and humans. *Science*, *331*, 1612.

Blasco, R. B., Karaca, E., Ambrogio, C., Cheong, T. C., Karayol, E., Minero, V. G., … Chiarle, R. (2014). Simple and rapid in vivo generation of chromosomal rearrangements using CRISPR/Cas9 technology. *Cell Reports*, *9*, 1219.

Bloom, M. B., Perry-Lalley, D., Robbins, P. F., Li, Y., El-Gamil, M., Rosenberg, S. A., & Yang, J. C. (1997). Identification of tyrosinase-related protein 2 as a tumor rejection antigen for the B16 melanoma. *The Journal of Experimental Medicine*, *185*, 453.
Bobek, V., Kolostova, K., Pinterova, D., Kacprzak, G., Adamiak, J., Kolodziej, J., … Hoffman, R. M. (2010). A clinically relevant, syngeneic model of spontaneous, highly metastatic B16 mouse melanoma. *Anticancer Research*, *30*, 4799.
Capecchi, M. R. (2005). Gene targeting in mice: Functional analysis of the mammalian genome for the twenty-first century. *Nature Reviews. Genetics*, *6*, 507.
Carreno, B. M., Magrini, V., Becker-Hapak, M., Kaabinejadian, S., Hundal, J., Petti, A. A., … Linette, G. P. (2015). A dendritic cell vaccine increases the breadth and diversity of melanoma neoantigen-specific T cells. *Science*, *348*, 803.
Chan, C. J., Martinet, L., Gilfillan, S., Souza-Fonseca-Guimaraes, F., Chow, M. T., Town, L., … Smyth, M. J. (2014). The receptors CD96 and CD226 oppose each other in the regulation of natural killer cell functions. *Nature Immunology*, *15*, 431.
Cheung, A. F., DuPage, M. J. P., Dong, H. K., Chen, J., & Jacks, T. (2008). Regulated expression of a tumor-associated antigen reveals multiple levels of T-cell tolerance in a mouse model of lung cancer. *Cancer Research*, *68*, 9459.
Chow, M. T., Sceneay, J., Paget, C., Wong, C. S. F., Duret, H., Tschopp, J., … Smyth, M. J. (2012). NLRP3 suppresses NK cell-mediated responses to carcinogen-induced tumors and metastases. *Cancer Research*, *72*, 5721.
Chow, M. T., Tschopp, J., Moller, A., & Smyth, M. J. (2012). NLRP3 promotes inflammation-induced skin cancer but is dispensable for asbestos-induced mesothelioma. *Immunology & Cell Biology*, *90*, 983.
Connell, J. J., Chatain, G., Cornelissen, B., Vallis, K. A., Hamilton, A., Seymour, L., … Sibson, N. R. (2013). Selective permeabilization of the blood–brain barrier at sites of metastasis. *Journal of the National Cancer Institute*, *105*, 1634.
Couzin-Frankel, J. (2014). The littlest patient. *Science*, *346*, 24.
Devaud, C., Westwood, J. A., John, L. B., Flynn, J. K., Paquet-Fifield, S., Duong, C. P. M., … Kershaw, M. H. (2014). Tissues in different anatomical sites can sculpt and vary the tumor microenvironment to affect responses to therapy. *Molecular Therapy*, *22*, 18.
Dow, L. E., Fisher, J., O'Rourke, K. P., Muley, A., Kastenhuber, E. R., Livshits, G., … Lowe, S. W. (2015). Inducible in vivo genome editing with CRISPR-Cas9. *Nature Biotechnology*, *33*, 390–394.
Dranoff, G. (2012). Experimental mouse tumour models: What can be learnt about human cancer immunology? *Nature Reviews. Immunology*, *12*, 61.
DuPage, M., Cheung, A. F., Mazumdar, C., Winslow, M. M., Bronson, R., Schmidt, L. M., … Jacks, T. (2011). Endogenous T cell responses to antigens expressed in lung adenocarcinomas delay malignant tumor progression. *Cancer Cell*, *19*, 72.
DuPage, M., & Jacks, T. (2013). Genetically engineered mouse models of cancer reveal new insights about the antitumor immune response. *Current Opinion in Immunology*, *25*, 192.
DuPage, M., Mazumdar, C., Schmidt, L. M., Cheung, A. F., & Jacks, T. (2012). Expression of tumour-specific antigens underlies cancer immunoediting. *Nature*, *482*, 405.
Engels, B., Rowley, D. A., & Schreiber, H. (2012). Targeting stroma to treat cancers. *Seminars in Cancer Biology*, *22*, 41.
Facciabene, A., Peng, X., Hagemann, I. S., Balint, K., Barchetti, A., Wang, L.-P., … Coukos, G. (2011). Tumour hypoxia promotes tolerance and angiogenesis via CCL28 and Treg cells. *Nature*, *475*, 226.
Fantozzi, A., & Christofori, G. (2006). Mouse models of breast cancer metastasis. *Breast Cancer Research: BCR*, *8*, 212.
Frese, K. K., & Tuveson, D. A. (2007). Maximizing mouse cancer models. *Nature Reviews. Cancer*, *7*, 654.

Fridman, W. H., Pagès, F., Sautès-Fridman, C., & Galon, J. (2012). The immune contexture in human tumours: Impact on clinical outcome. *Nature Reviews. Cancer*, *12*, 298.

Galluzzi, L., Vacchelli, E., Bravo-San Pedro, J. M., Buque, A., Senovilla, L., Baracco, E. E., ... Kroemer, G. (2014). Classification of current anticancer immunotherapies. *Oncotarget*, *5*, 12472.

Gilboa, E. (1999). The makings of a tumor rejection antigen. *Immunity*, *11*, 263.

Gould, S. E., Junttila, M. R., & de Sauvage, F. J. (2015). Translational value of mouse models in oncology drug development. *Nature Medicine*, *21*, 431.

Grivennikov, S. I., Wang, K., Mucida, D., Stewart, C. A., Schnabl, B., Jauch, D., ... Karin, M. (2012). Adenoma-linked barrier defects and microbial products drive IL-23/IL-17-mediated tumour growth. *Nature*, *491*, 254.

Gubin, M. M., Zhang, X., Schuster, H., Caron, E., Ward, J. P., Noguchi, T., ... Schreiber, R. D. (2014). Checkpoint blockade cancer immunotherapy targets tumour-specific mutant antigens. *Nature*, *515*, 577.

Guy, C. T., Webster, M. A., Schaller, M., Parsons, T. J., Cardiff, R. D., & Muller, W. J. (1992). Expression of the neu protooncogene in the mammary epithelium of transgenic mice induces metastatic disease. *Proceedings of the National Academy of Sciences of the United States of America*, *89*, 10578.

Gyorki, D. E., Callahan, M., Wolchok, J. D., & Ariyan, C. E. (2013). The delicate balance of melanoma immunotherapy. *Clinical & Translational Immunology*, *2*(8), e5.

Hamid, O., Robert, C., Daud, A., Hodi, F. S., Hwu, W.-J., Kefford, R., ... Ribas, A. (2013). Safety and tumor responses with lambrolizumab (anti-PD-1) in melanoma. *New England Journal of Medicine*, *369*, 134.

Hanahan, D., & Weinberg, R. A. (2000). The hallmarks of cancer. *Cell*, *100*, 57.

Hanahan, D., & Weinberg, R. A. (2011). Hallmarks of cancer: The next generation. *Cell*, *144*, 646.

Hanson, H. L., Donermeyer, D. L., Ikeda, H., White, J. M., Shankaran, V., Old, L. J., ... Allen, P. M. (2000). Eradication of established tumors by CD8+ T cell adoptive immunotherapy. *Immunity*, *13*, 265.

Herbst, R. S., Soria, J.-C., Kowanetz, M., Fine, G. D., Hamid, O., Gordon, M. S., ... Hodi, F. S. (2014). Predictive correlates of response to the anti-PD-L1 antibody MPDL3280A in cancer patients. *Nature*, *515*, 563.

Ho, P.-C., Meeth, K. M., Tsui, Y.-C., Srivastava, B., Bosenberg, M. W., & Kaech, S. M. (2014). Immune-based antitumor effects of BRAF inhibitors rely on signaling by CD40L and IFNγ. *Cancer Research*, *74*, 3205.

Hodi, F. S., O'Day, S. J., McDermott, D. F., Weber, R. W., Sosman, J. A., Haanen, J. B., ... Urba, W. J. (2010). Improved survival with ipilimumab in patients with metastatic melanoma. *The New England Journal of Medicine*, *363*, 711.

Holzapfel, B. M., Wagner, F., Thibaudeau, L., Levesque, J.-P., & Hutmacher, D. W. (2015). Humanized models of tumour immunology in the 21st century: Convergence of cancer research and tissue engineering. *Stem Cells*, *33*, 1696–1704.

Hooijkaas, A., Gadiot, J., Morrow, M., Stewart, R., Schumacher, T., & Blank, C. U. (2012). Selective BRAF inhibition decreases tumor-resident lymphocyte frequencies in a mouse model of human melanoma. *Oncoimmunology*, *1*, 609.

Hsu, P. D., Lander, E. S., & Zhang, F. (2014). Development and applications of CRISPR-Cas9 for genome engineering. *Cell*, *157*, 1262.

Ito, A., Ishida, T., Yano, H., Inagaki, A., Suzuki, S., Sato, F., ... Ueda, R. (2009). Defucosylated anti-CCR4 monoclonal antibody exercises potent ADCC-mediated antitumor effect in the novel tumor-bearing humanized NOD/Shi-scid, IL-2Rγnull mouse model. *Cancer Immunology, Immunotherapy*, *58*, 1195.

Jinek, M., Chylinski, K., Fonfara, I., Hauer, M., Doudna, J. A., & Charpentier, E. (2012). A programmable dual-RNA-guided DNA endonuclease in adaptive bacterial immunity. *Science, 337*, 816.

John, L. B., Devaud, C., Duong, C. P. M., Yong, C. S., Beavis, P. A., Haynes, N. M., … Darcy, P. K. (2013). Anti-PD-1 antibody therapy potently enhances the eradication of established tumors by gene-modified T cells. *Clinical Cancer Research, 19*, 5636.

Kamb, A. (2005). What's wrong with our cancer models? *Nature Reviews. Drug Discovery, 4*, 161.

Kansara, M., Leong, H. S., Lin, D. M., Popkiss, S., Pang, P., Garsed, D. W., … Thomas, D. M. (2013). Immune response to RB1-regulated senescence limits radiation-induced osteosarcoma formation. *The Journal of Clinical Investigation, 123*, 5351.

Kansara, M., Tsang, M., Kodjabachian, L., Sims, N. A., Trivett, M. K., Ehrich, M., … Thomas, D. M. (2009). Wnt inhibitory factor 1 is epigenetically silenced in human osteosarcoma, and targeted disruption accelerates osteosarcomagenesis in mice. *The Journal of Clinical Investigation, 119*(4), 837–851. http://dx.doi.org/10.1172/JCI37175. Epub 2009 Mar 23.

Kim, P. S., & Ahmed, R. (2010). Features of responding T cells in cancer and chronic infection. *Current Opinion in Immunology, 22*, 223.

Klages, K., Mayer, C. T., Lahl, K., Loddenkemper, C., Teng, M. W. L., Ngiow, S. F., … Sparwasser, T. (2010). Selective depletion of Foxp3+ regulatory T cells improves effective therapeutic vaccination against established melanoma. *Cancer Research, 70*, 7788.

Knight, D. A., Ngiow, S. F., Li, M., Parmenter, T., Mok, S., Cass, A., … Smyth, M. J. (2013). Host immunity contributes to the anti-melanoma activity of BRAF inhibitors. *The Journal of Clinical Investigation, 123*, 1371.

Kohrt, H. E., Houot, R., Weiskopf, K., Goldstein, M. J., Scheeren, F., Czerwinski, D., … Levy, R. (2012). Stimulation of natural killer cells with a CD137-specific antibody enhances trastuzumab efficacy in xenotransplant models of breast cancer. *The Journal of Clinical Investigation, 122*, 1066.

Kokolus, K. M., Capitano, M. L., Lee, C.-T., Eng, J. W.-L., Waight, J. D., Hylander, B. L., … Repasky, E. A. (2013). Baseline tumor growth and immune control in laboratory mice are significantly influenced by subthermoneutral housing temperature. *Proceedings of the National Academy of Sciences of the United States of America, 110*, 20176.

Kroemer, G., Galluzzi, L., Kepp, O., & Zitvogel, L. (2013). Immunogenic cell death in cancer therapy. *Annual Review of Immunology, 31*, 51.

Ladoire, S., Hannani, D., Vetizou, M., Locher, C., Aymeric, L., Apetoh, L., … Zitvogel, L. (2013). Cell-death-associated molecular patterns as determinants of cancer immunogenicity. *Antioxidants & Redox Signaling, 20*, 1098.

Langowski, J. L., Zhang, X., Wu, L., Mattson, J. D., Chen, T., Smith, K., … Oft, M. (2006). IL-23 promotes tumour incidence and growth. *Nature, 442*, 461.

Legrand, N., Ploss, A., Balling, R., Becker, P. D., Borsotti, C., Brezillon, N., … Ziegler, P. (2009). Humanized mice for modeling human infectious disease: Challenges, progress, and outlook. *Cell Host & Microbe, 6*, 5.

Liddy, N., Bossi, G., Adams, K. J., Lissina, A., Mahon, T. M., Hassan, N. J., … Jakobsen, B. K. (2012). Monoclonal TCR-redirected tumor cell killing. *Nature Medicine, 18*, 980.

Linnemann, C., van Buuren, M. M., Bies, L., Verdegaal, E. M. E., Schotte, R., Calis, J. J. A., … Schumacher, T. N. M. (2015). High-throughput epitope discovery reveals frequent recognition of neo-antigens by CD4+ T cells in human melanoma. *Nature Medicine, 21*, 81.

Liu, J., Blake, S. J., Smyth, M. J., & Teng, M. W. (2014). Improved mouse models to assess tumour immunity and irAEs after combination cancer immunotherapies. *Clinical & Translational Immunology, 3*, e22.

Maddalo, D., Manchado, E., Concepcion, C. P., Bonetti, C., Vidigal, J. A., Han, Y. C., ... Ventura, A. (2014). In vivo engineering of oncogenic chromosomal rearrangements with the CRISPR/Cas9 system. *Nature*, *516*, 423.

Malchow, S., Leventhal, D. S., Nishi, S., Fischer, B. I., Shen, L., Paner, G. P., ... Savage, P. A. (2013). Aire-dependent thymic development of tumor-associated regulatory T cells. *Science*, *339*, 1219.

Matsushita, H., Vesely, M. D., Koboldt, D. C., Rickert, C. G., Uppaluri, R., Magrini, V. J., ... Schreiber, R. D. (2012). Cancer exome analysis reveals a T-cell-dependent mechanism of cancer immunoediting. *Nature*, *482*, 400.

Mattarollo, S. R., Loi, S., Duret, H., Ma, Y., Zitvogel, L., & Smyth, M. J. (2011). Pivotal role of innate and adaptive immunity in anthracycline chemotherapy of established tumors. *Cancer Research*, *71*, 4809.

Mattarollo, S. R., West, A. C., Steegh, K., Duret, H., Paget, C., Martin, B., ... Smyth, M. J. (2012). NKT cell adjuvant-based tumor vaccine for treatment of myc oncogene-driven mouse B-cell lymphoma. *Blood*, *120*, 3019.

Maude, S. L., Frey, N., Shaw, P. A., Aplenc, R., Barrett, D. M., Bunin, N. J., ... Grupp, S. A. (2014). Chimeric antigen receptor T cells for sustained remissions in leukemia. *The New England Journal of Medicine*, *371*, 1507.

Minard-Colin, V., Xiu, Y., Poe, J. C., Horikawa, M., Magro, C. M., Hamaguchi, Y., ... Tedder, T. F. (2008). Lymphoma depletion during CD20 immunotherapy in mice is mediated by macrophage FcgammaRI, FcgammaRIII, and FcgammaRIV. *Blood*, *112*, 1205.

Mittal, D., Young, A., Stannard, K., Yong, M., Teng, M. W. L., Allard, B., ... Smyth, M. J. (2014). Antimetastatic effects of blocking PD-1 and the adenosine A2A receptor. *Cancer Research*, *74*, 3652.

Mossman, B. T., & Churg, A. (1998). Mechanisms in the pathogenesis of asbestosis and silicosis. *American Journal of Respiratory and Critical Care Medicine*, *157*, 1666.

Motz, G. T., Santoro, S. P., Wang, L.-P., Garrabrant, T., Lastra, R. R., Hagemann, I. S., ... Coukos, G. (2014). Tumor endothelium FasL establishes a selective immune barrier promoting tolerance in tumors. *Nature Medicine*, *20*, 607.

Muranski, P., Boni, A., Antony, P. A., Cassard, L., Irvine, K. R., Kaiser, A., ... Restifo, N. P. (2008). Tumor-specific Th17-polarized cells eradicate large established melanoma. *Blood*, *112*, 362.

O'Brien, E. R., Kersemans, V., Tredwell, M., Checa, B., Serres, S., Soto, M. S., ... Sibson, N. R. (2014). Glial activation in the early stages of brain metastasis: TSPO as a diagnostic biomarker. *Journal of Nuclear Medicine*, *55*, 275.

Obeid, M., Tesniere, A., Ghiringhelli, F., Fimia, G. M., Apetoh, L., Perfettini, J.-L., ... Kroemer, G. (2007). Calreticulin exposure dictates the immunogenicity of cancer cell death. *Nature Medicine*, *13*, 54.

Obenaus, M., Leitao, C., Leisegang, M., Chen, X., Gavvovidis, I., van der Bruggen, P., ... Blankenstein, T. (2015). Identification of human T-cell receptors with optimal affinity to cancer antigens using antigen-negative humanized mice. *Nature Biotechnology*, *33*, 402.

Overwijk, W. W., Lee, D. S., Surman, D. R., Irvine, K. R., Touloukian, C. E., Chan, C.-C., ... Restifo, N. P. (1999). Vaccination with a recombinant vaccinia virus encoding a "self" antigen induces autoimmune vitiligo and tumor cell destruction in mice: Requirement for CD4(+) T lymphocytes. *Proceedings of the National Academy of Sciences of the United States of America*, *96*, 2982.

Overwijk, W. W., Theoret, M. R., Finkelstein, S. E., Surman, D. R., de Jong, L. A., Vyth-Dreese, F. A., ... Restifo, N. P. (2003). Tumor regression and autoimmunity after reversal of a functionally tolerant state of self-reactive CD8 + T cells. *The Journal of Experimental Medicine*, *198*, 569.

Park, S., Jiang, Z., Mortenson, E. D., Deng, L., Radkevich-Brown, O., Yang, X., … Fu, Y. X. (2010). The therapeutic effect of anti-HER2/neu antibody depends on both innate and adaptive immunity. *Cancer Cell*, *18*, 160.

Pedroza-Gonzalez, A., Xu, K., Wu, T. C., Aspord, C., Tindle, S., Marches, F., … Palucka, A. K. (2011). Thymic stromal lymphopoietin fosters human breast tumor growth by promoting type 2 inflammation. *The Journal of Experimental Medicine*, *208*, 479.

Peers, I. S., South, M. C., Ceuppens, P. R., Bright, J. D., & Pilling, E. (2014). Can you trust your animal study data? *Nature Reviews. Drug Discovery*, *13*, 560.

Pelletier, S., Gingras, S., & Green, D. R. (2015). Mouse genome engineering via CRISPR-Cas9 for study of immune function. *Immunity*, *42*, 18.

Pincetic, A., Bournazos, S., DiLillo, D. J., Maamary, J., Wang, T. T., Dahan, R., … Ravetch, J. V. (2014). Type I and type II Fc receptors regulate innate and adaptive immunity. *Nature Immunology*, *15*, 707.

Platt, R. J., Chen, S., Zhou, Y., Yim, M. J., Swiech, L., Kempton, H. R., … Zhang, F. (2014). CRISPR-Cas9 knockin mice for genome editing and cancer modeling. *Cell*, *159*, 440.

Porter, D. L., Levine, B. L., Kalos, M., Bagg, A., & June, C. H. (2011). Chimeric antigen receptor-modified T cells in chronic lymphoid leukemia. *The New England Journal of Medicine*, *365*, 725.

Powles, T., Eder, J. P., Fine, G. D., Braiteh, F. S., Loriot, Y., Cruz, C., … Vogelzang, N. J. (2014). MPDL3280A (anti-PD-L1) treatment leads to clinical activity in metastatic bladder cancer. *Nature*, *515*, 558.

Qian, B.-Z., Li, J., Zhang, H., Kitamura, T., Zhang, J., Campion, L. R., … Pollard, J. W. (2011). CCL2 recruits inflammatory monocytes to facilitate breast-tumour metastasis. *Nature*, *475*, 222.

Restifo, N. P., Dudley, M. E., & Rosenberg, S. A. (2012). Adoptive immunotherapy for cancer: Harnessing the T cell response. *Nature Reviews. Immunology*, *12*, 269.

Rizvi, N. A., Hellmann, M. D., Snyder, A., Kvistborg, P., Makarov, V., Havel, J. J., … Chan, T. A. (2015). Mutational landscape determines sensitivity to PD-1 blockade in non-small cell lung cancer. *Science*, *348*(6230), 124–128.

Robbins, P. F., Lu, Y.-C., El-Gamil, M., Li, Y. F., Gross, C., Gartner, J., … Rosenberg, S. A. (2013). Mining exomic sequencing data to identify mutated antigens recognized by adoptively transferred tumor-reactive T cells. *Nature Medicine*, *19*, 747.

Robert, C., Long, G. V., Brady, B., Dutriaux, C., Maio, M., Mortier, L., … Ascierto, P. A. (2015). Nivolumab in previously untreated melanoma without BRAF mutation. *New England Journal of Medicine*, *372*, 320.

Rongvaux, A., Takizawa, H., Strowig, T., Willinger, T., Eynon, E. E., Flavell, R. A., & Manz, M. G. (2013). Human hemato-lymphoid system mice: Current use and future potential for medicine. *Annual Review of Immunology*, *31*, 635.

Rongvaux, A., Willinger, T., Martinek, J., Strowig, T., Gearty, S. V., Teichmann, L. L., … Flavell, R. A. (2014). Development and function of human innate immune cells in a humanized mouse model. *Nature Biotechnology*, *32*, 364.

Ruffell, B., Affara, N. I., Cottone, L., Junankar, S., Johansson, M., DeNardo, D. G., … Coussens, L. M. (2013). Cathepsin C is a tissue-specific regulator of squamous carcinogenesis. *Genes & Development*, *27*, 2086.

Sanchez-Rivera, F. J., Papagiannakopoulos, T., Romero, R., Tammela, T., Bauer, M. R., Bhutkar, A., … Jacks, T. (2014). Rapid modelling of cooperating genetic events in cancer through somatic genome editing. *Nature*, *516*, 428.

Sander, J. D., & Joung, J. K. (2014). CRISPR-Cas systems for editing, regulating and targeting genomes. *Nature Biotechnology*, *32*, 347.

Sathe, P., Delconte, R. B., Souza-Fonseca-Guimaraes, F., Seillet, C., Chopin, M., Vandenberg, C. J., … Huntington, N. D. (2014). Innate immunodeficiency following genetic ablation of Mcl1 in natural killer cells. *Nature Communications*, *5*, 4539.

Savage, P. A., Vosseller, K., Kang, C., Larimore, K., Riedel, E., Wojnoonski, K., … Allison, J. P. (2008). Recognition of a ubiquitous self antigen by prostate cancer-infiltrating CD8+ T lymphocytes. *Science*, *319*, 215.

Sceneay, J., Chow, M. T., Chen, A., Halse, H. M., Wong, C. S. F., Andrews, D. M., … Möller, A. (2012). Primary tumor hypoxia recruits CD11b+/Ly6Cmed/Ly6G+ immune suppressor cells and compromises NK cell cytotoxicity in the premetastatic niche. *Cancer Research*, *72*, 3906.

Scheer, N., Snaith, M., Wolf, C. R., & Seibler, J. (2013). Generation and utility of genetically humanized mouse models. *Drug Discovery Today*, *18*, 1200.

Schreiber, R. D., Old, L. J., & Smyth, M. J. (2011). Cancer immunoediting: Integrating immunity's roles in cancer suppression and promotion. *Science*, *331*, 1565.

Snyder, A., Makarov, V., Merghoub, T., Yuan, J., Zaretsky, J. M., Desrichard, A., … Chan, T. A. (2014). Genetic basis for clinical response to CTLA-4 blockade in melanoma. *New England Journal of Medicine*, *371*, 2189.

Sommer, G., Agosti, V., Ehlers, I., Rossi, F., Corbacioglu, S., Farkas, J., … Besmer, P. (2003). Gastrointestinal stromal tumors in a mouse model by targeted mutation of the Kit receptor tyrosine kinase. *Proceedings of the National Academy of Sciences of the United States of America*, *100*, 6706.

Stagg, J., Beavis, P. A., Divisekera, U., Liu, M. C. P., Möller, A., Darcy, P. K., & Smyth, M. J. (2012). CD73-deficient mice are resistant to carcinogenesis. *Cancer Research*, *72*, 2190.

Stagg, J., Divisekera, U., McLaughlin, N., Sharkey, J., Pommey, S., Denoyer, D., … Smyth, M. J. (2010). Anti-CD73 antibody therapy inhibits breast tumor growth and metastasis. *Proceedings of the National Academy of Sciences of the United States of America*, *107*, 1547.

Stagg, J., Sharkey, J., Pommey, S., Young, R., Takeda, K., Yagita, H., … Smyth, M. J. (2008). Antibodies targeted to TRAIL receptor-2 and ErbB-2 synergize in vivo and induce an antitumor immune response. *Proceedings of the National Academy of Sciences of the United States of America*, *105*, 16254.

Swann, J. B., & Smyth, M. J. (2007). Immune surveillance of tumors. *The Journal of Clinical Investigation*, *117*, 1137.

Swann, J. B., Vesely, M. D., Silva, A., Sharkey, J., Akira, S., Schreiber, R. D., & Smyth, M. J. (2008). Demonstration of inflammation-induced cancer and cancer immunoediting during primary tumorigenesis. *Proceedings of the National Academy of Sciences of the United States of America*, *105*, 652.

Tai, L.-H., Zhang, J., Scott, K. J., de Souza, C. T., Alkayyal, A. A., Ananth, A. A., … Auer, R. C. (2013). Perioperative influenza vaccination reduces postoperative metastatic disease by reversing surgery-induced dysfunction in natural killer cells. *Clinical Cancer Research*, *19*, 5104.

Teng, M. W. L., Andrews, D. M., McLaughlin, N., von Scheidt, B., Ngiow, S. F., Möller, A., … Smyth, M. J. (2010). IL-23 suppresses innate immune response independently of IL-17A during carcinogenesis and metastasis. *Proceedings of the National Academy of Sciences of the United States of America*, *107*, 8328.

Teng, M. W. L., von Scheidt, B., Duret, H., Towne, J. E., & Smyth, M. J. (2011). Anti-IL-23 monoclonal antibody synergizes in combination with targeted therapies or IL-2 to suppress tumor growth and metastases. *Cancer Research*, *71*, 2077.

Topalian, S. L., Hodi, F. S., Brahmer, J. R., Gettinger, S. N., Smith, D. C., McDermott, D. F., … Sznol, M. (2012). Safety, activity, and immune correlates of anti-PD-1 antibody in cancer. *New England Journal of Medicine*, *366*, 2443.

Twyman-Saint Victor, C., Rech, A. J., Maity, A., Rengan, R., Pauken, K. E., Stelekati, E., … Minn, A. J. (2015). Radiation and dual checkpoint blockade activate non-redundant immune mechanisms in cancer. *Nature*, *520*, 373.

Uno, T., Takeda, K., Kojima, Y., Yoshizawa, H., Akiba, H., Mittler, R. S., … Smyth, M. J. (2006). Eradication of established tumors in mice by a combination antibody-based therapy. *Nature Medicine*, *12*, 693.

van Rooij, N., van Buuren, M. M., Philips, D., Velds, A., Toebes, M., Heemskerk, B., … Schumacher, T. N. (2013). Tumor exome analysis reveals neoantigen-specific T-cell reactivity in an ipilimumab-responsive melanoma. *Journal of Clinical Oncology*, *31*, e439.

Verbrugge, I., Hagekyriakou, J., Sharp, L. L., Galli, M., West, A., McLaughlin, N. M., … Haynes, N. M. (2012). Radiotherapy increases the permissiveness of established mammary tumors to rejection by immunomodulatory antibodies. *Cancer Research*, *72*, 3163.

Vesely, M. D., Kershaw, M. H., Schreiber, R. D., & Smyth, M. J. (2011). Natural innate and adaptive immunity to cancer. *Annual Review of Immunology*, *29*, 235.

Wainwright, D. A., Chang, A. L., Dey, M., Balyasnikova, I. V., Kim, C. K., Tobias, A., … Lesniak, M. S. (2014). Durable therapeutic efficacy utilizing combinatorial blockade against IDO, CTLA-4, and PD-L1 in mice with brain tumors. *Clinical Cancer Research*, *20*, 5290.

Wege, A. K., Ernst, W., Eckl, J., Frankenberger, B., Vollmann-Zwerenz, A., Mannel, D. N., … Brockhoff, G. (2011). Humanized tumor mice—A new model to study and manipulate the immune response in advanced cancer therapy. *International Journal of Cancer*, *129*, 2194.

Weiner, G. J. (2010). Rituximab: Mechanism of action. *Seminars in Hematology*, *47*, 115.

Westwood, J. A., Darcy, P. K., & Kershaw, M. H. (2014). The potential impact of mouse model selection in preclinical evaluation of cancer immunotherapy. *Oncoimmunology*, *3*, e946361.

Wolchok, J. D., Kluger, H., Callahan, M. K., Postow, M. A., Rizvi, N. A., Lesokhin, A. M., … Sznol, M. (2013). Nivolumab plus ipilimumab in advanced melanoma. *New England Journal of Medicine*, *369*, 122.

Xue, W., Chen, S., Yin, H., Tammela, T., Papagiannakopoulos, T., Joshi, N. S., … Jacks, T. (2014). CRISPR-mediated direct mutation of cancer genes in the mouse liver. *Nature*, *514*, 380.

Yadav, M., Jhunjhunwala, S., Phung, Q. T., Lupardus, P., Tanguay, J., Bumbaca, S., … Delamarre, L. (2014). Predicting immunogenic tumour mutations by combining mass spectrometry and exome sequencing. *Nature*, *515*, 572.

Ye, J., Ma, C., Wang, F., Hsueh, E. C., Toth, K., Huang, Y., … Peng, G. (2013). Specific recruitment of γδ regulatory T cells in human breast cancer. *Cancer Research*, *73*, 6137.

Zender, L., Zuber, J., & Lowe, S. W. (2007). SnapShot: Genetic mouse models of cancer. *Cell*, *129*, 838.e1.

Zhang, M., Zhang, Z., Garmestani, K., Goldman, C. K., Ravetch, J. V., Brechbiel, M. W., … Waldmann, T. A. (2004). Activating Fc receptors are required for antitumor efficacy of the antibodies directed toward CD25 in a murine model of adult t-cell leukemia. *Cancer Research*, *64*, 5825.

Zhou, Q., Facciponte, J., Jin, M., Shen, Q., & Lin, Q. (2014). Humanized NOD-SCID IL2rg−/− mice as a preclinical model for cancer research and its potential use for individualized cancer therapies. *Cancer Letters*, *344*, 13.

Zhou, Q., Munger, M. E., Veenstra, R. G., Weigel, B. J., Hirashima, M., Munn, D. H., … Blazar, B. R. (2011). Coexpression of Tim-3 and PD-1 identifies a CD8+ T-cell exhaustion phenotype in mice with disseminated acute myelogenous leukemia. *Blood*, *117*, 4501.

Zhu, Y., Knolhoff, B. L., Meyer, M. A., Nywening, T. M., West, B. L., Luo, J., … DeNardo, D. G. (2014). CSF1/CSF1R blockade reprograms tumor-infiltrating macrophages and improves response to T-cell checkpoint immunotherapy in pancreatic cancer models. *Cancer Research*, *74*, 5057.

Zitvogel, L., Galluzzi, L., Viaud, S., Vétizou, M., Daillère, R., Merad, M., & Kroemer, G. (2015). Cancer and the gut microbiota: An unexpected link. *Science Translational Medicine*, 7, 271ps1.

CHAPTER TWO

The Role of Neoantigens in Naturally Occurring and Therapeutically Induced Immune Responses to Cancer

Jeffrey P. Ward[1], Matthew M. Gubin[1], Robert D. Schreiber[2]
Washington University School of Medicine, St. Louis, MO, United States
[2]Corresponding author: e-mail address: schreiber@immunology.wustl.edu

Contents

[1] These authors contributed equally to this work.

Advances in Immunology, Volume 130
ISSN 0065-2776
http://dx.doi.org/10.1016/bs.ai.2016.01.001

Abstract

Definitive experimental evidence from mouse cancer models and strong correlative clinical data gave rise to the Cancer Immunoediting concept that explains the dual host-protective and tumor-promoting actions of immunity on developing cancers. Tumor-specific neoantigens can serve as targets of spontaneously arising adaptive immunity to cancer and thereby determine the ultimate fate of developing tumors. Tumor-specific neoantigens can also function as optimal targets of cancer immunotherapy against established tumors. These antigens are derived from nonsynonymous mutations that occur during cellular transformation and, because they are foreign to the host genome, are not subject to central tolerance. In this review, we summarize the experimental evidence indicating that cancer neoantigens are the source of both spontaneously occurring and therapeutically induced immune responses against cancer. We also review the advances in genomics, bioinformatics, and cancer immunotherapy that have facilitated identification of neoantigens and have moved personalized cancer immunotherapies into clinical trials, with the promise of providing more specific, safer, more effective, and perhaps even more generalizable treatments to cancer patients than current immunotherapies.

1. INTRODUCTION

After decades of controversy, the ability of the immune system to influence cancer development and progression has now become apparent (Grivennikov, Greten, & Karin, 2010; Mantovani, Allavena, Sica, & Balkwill, 2008; Schreiber, Old, & Smyth, 2011; Shankaran et al., 2001). Two parallel lines of investigation, one focused on assessing naturally occurring immune responses to developing cancers and the other focused on immunotherapy-induced durable responses to established tumors have ultimately led to unequivocal resolution of this long-standing argument. These independent approaches have demonstrated the importance of tumor-specific neoantigens as critical targets of antitumor immune responses (Schumacher & Schreiber, 2015). Immune recognition of neoantigens has the potential to destroy developing cancers before they become clinically apparent, shape the immunogenicities of cancer cells rendering them more fit to grow progressively in an immunocompetent environment, and ultimately to facilitate the immune elimination of growing tumors when manipulated in the appropriate therapeutic manner. The concept that neoantigens may be optimal targets for cancer immunotherapy is a very old one dating back to the 1940s and steadily evolving since that time (Table 1). The evolution of this idea has undergone a dramatic acceleration

Table 1 Pioneering Studies Revealing the Importance of Cancer Neoantigens

Year	Discovery	Reference
1943	Mice with carcinogen-induced tumors are protected against rechallenge with the same tumor line, indicating tumors have tumor-specific antigens	Gross (1943)
1977	Generation of CTL clones against tumor antigen of nonviral origin	Gillis and Smith (1977)
1985	Antigens recognized by T cells are presented on MHC	Babbitt, Allen, Matsueda, Haber, and Unanue (1985) and Bjorkman et al. (1987)
1987	T cells from human melanoma patients react with autologous tumor but not normal tissue	Herin et al. (1987) and Van den Eynde et al. (1989)
1988	Identification of a tumor-specific mutant antigen in an in vitro mutagenized mouse tumor	De Plaen et al. (1988)
	Use of ACT therapy in patients with metastatic melanoma	Rosenberg et al. (1988)
1991	Identification of the first human tumor antigen, the CTA antigen MAGEA1	van der Bruggen et al. (1991)
1994	Identification of melanoma antigens using mass spectrometry	Cox et al. (1994)
1995	First tumor-specific mutant antigen in human tumors identified	Coulie et al. (1995) and Wolfel et al. (1995)
1996	Use of peptide–MHC tetramers to analyze antigen-specific T cells	Altman et al. (1996)
	Antitumor activity of anti-CTLA-4 demonstrated in mice	Leach, Krummel, and Allison (1996)
1997	Use of SEREX to identify the CTA antigen NY-ESO-1	Chen et al. (1997)
1999	Development of the MHC class I epitope database and prediction algorithm SYFPEITHI	Rammensee, Bachmann, Emmerich, Bachor, and Stevanovic (1999)

Continued

Table 1 Pioneering Studies Revealing the Importance of Cancer Neoantigens—cont'd

Year	Discovery	Reference
2001	Demonstration that the immune system can protect against cancer and shape tumor immunogenicity Proposal of Cancer Immunoediting, thus unifying the dual host-protective and tumor promoting and sculpting ability of the immune system	Shankaran et al. (2001)
2003	Development of the NetMHC epitope prediction algorithm	Nielsen et al. (2003)
2004	Establishment of IEDB	Vita et al. (2015)
2005	T cells specific for tumor-specific mutant antigens persist in the blood and tumor of a melanoma patient after ACT	Zhou, Dudley, Rosenberg, and Robbins (2005)
	Autologous T cells to a human melanoma is dominated by responses to tumor-specific mutant antigens	Lennerz et al. (2005)
2007	Experimental demonstration of cancer immune equilibrium	Koebel et al. (2007)
	First cancer whole exome sequencing	Wood et al. (2007)
2008	First cancer whole genome sequencing	Ley et al. (2008)
	Burt Vogelstein and Jim Allison propose that all cancers have mutations that could form neoantigens	Segal et al. (2008)
2009	Detection of antigen-specific T cells by combinatorial encoding of MHC multimers	Hadrup et al. (2009)
2011	FDA approval of the immune checkpoint inhibitor ipilumumab (anti-CTLA-4)	Hodi et al. (2010)

Table 1 Pioneering Studies Revealing the Importance of Cancer Neoantigens—cont'd

Year	Discovery	Reference
2012	First use of genomic sequencing and epitope prediction algorithms to identify tumor-specific mutant antigens	Castle et al. (2012) and Matsushita et al. (2012)
	Demonstration that tumor-specific mutant antigens can drive Cancer Immunoediting	DuPage, Mazumdar, Schmidt, Cheung, and Jacks (2012) and Matsushita et al. (2012)
2013	Use of genome sequencing and epitope prediction to identify human mutant neoantigens recognized by adoptively transferred T cells	Robbins et al. (2013)
	In vivo expansion of mutant antigen-specific T cells in a human melanoma patient following anti-CTLA-4 treatment	van Rooij et al. (2013)
2014	Predicted mutant neoantigens correlate with increased CTL cytotoxicity and patient survival	Brown et al. (2014)
	Autologous $CD4^+$ T cells largely specific for a tumor-specific mutant antigen leads to tumor regression when adoptively transferred into a cancer patient	Tran et al. (2014)
	Demonstration that tumor-specific mutant antigens are targets of checkpoint blockade cancer immunotherapy	Gubin et al. (2014)
	Tumor-specific mutant antigen SLP vaccines provide therapeutic tumor protection in preclinical models	Gubin et al. (2014) and Yadav et al. (2014)
	Mutational load and neoantigen landscape may predict patients who benefit from checkpoint blockade cancer immunotherapy	Snyder et al. (2014)

Continued

Table 1 Pioneering Studies Revealing the Importance of Cancer Neoantigens—cont'd

Year	Discovery	Reference
2015	Genetic analysis reveals CTL activity correlates with mutant neoantigens load and provides evidence of immunoediting for some human tumors	Rooney, Shukla, Wu, Getz, and Hacohen (2015)
	Identification of neoantigen-specific $CD4^+$ T cells that infiltrate melanoma metastases	Linnemann et al. (2015)
	Demonstration that vaccination with MHC II epitopes induces therapeutic antitumor responses in preclinical models	Kreiter et al. (2015)

with the advent and employment of next generation sequencing and computational approaches which have made it possible to predict cancer specific mutations that function as neoantigens for adaptive immunity (Gubin, Artyomov, Mardis, & Schreiber, 2015). The analyses of therapeutically active neoantigens has also led to the realization that both major histocompatibility complex (MHC) class I (MHC I) and MHC class II (MHC II) epitopes are required for effective antitumor immune responses. These developments now leave cancer immunologists and clinical oncologists poised to develop truly personalized treatment approaches against established cancers with the goal of increasing specificity and eliminating toxicity compared to the current therapies. The focus of this review is to summarize the key experimental evidence that has led to a paradigm shift in thinking about immune system–cancer interactions resulting in the current excitement over using neoantigens as tumor-specific targets for immune control of cancer.

2. CANCER IMMUNOEDITING AS AN ENCOMPASSING MODEL OF IMMUNE SYSTEM–TUMOR INTERACTIONS

The dual host-protective and tumor-promoting actions of the immune system on developing cancers have been codified as a process termed "Cancer Immunoediting" (Fig. 1; Schreiber et al., 2011; Shankaran et al., 2001). Cancer Immunoediting initiates after cellular transformation has occurred and intrinsic tumor suppressor mechanisms have been

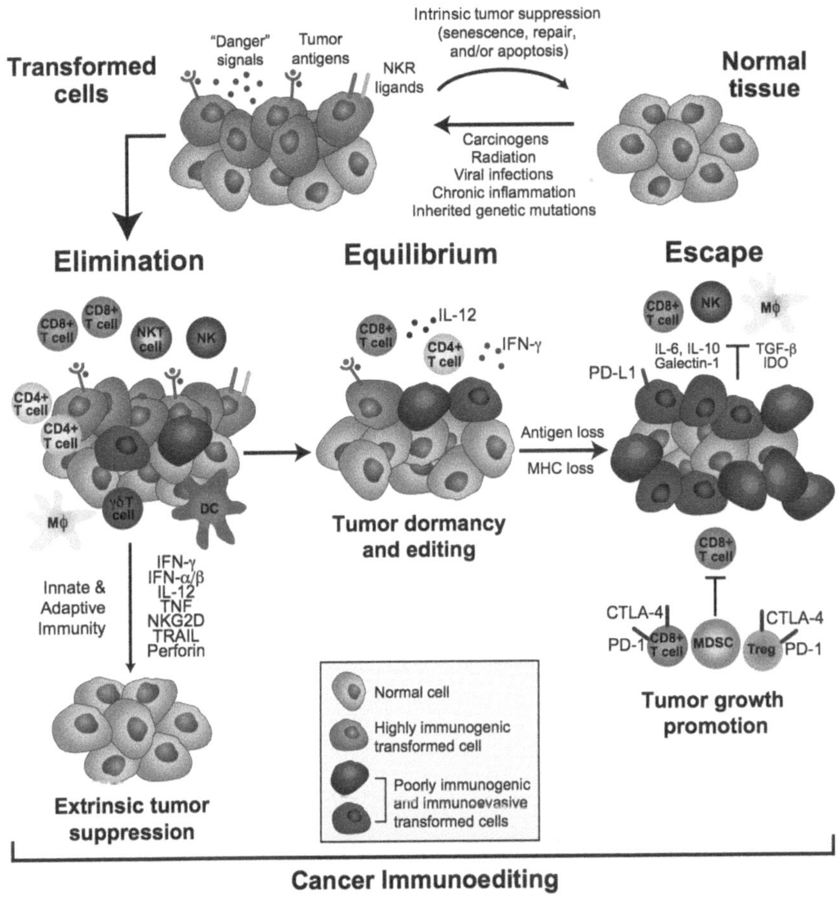

Figure 1 Cancer Immunoediting is an extrinsic tumor-suppressor mechanism that engages after cellular transformation has occurred and intrinsic tumor-suppressor mechanisms have failed. In its most complex form, Cancer Immunoediting consists of three phases: Elimination, Equilibrium, and Escape. In the Elimination phase, innate and adaptive immunity work in concert to destroy emerging tumors before they become clinically apparent. This phase may represent the full extent of the process upon complete tumor elimination, whereby the host remains cancer free. If, however, a cancer cell variant resists elimination, it may then enter the Equilibrium phase, in which its outgrowth is immunologically constrained. Editing of tumor immunogenicity occurs in the Equilibrium phase. Equilibrium may curb outgrowth of occult cancers for the lifetime of the host. However, as a consequence of immune selection pressure, tumor cell variants may arise that are no longer recognized by adaptive immunity, become insensitive to immune effector mechanisms, and/or induce an immunosuppressive tumor microenvironment. These tumor cells may then enter the Escape phase, in which their outgrowth is no longer impeded by immunity and thus manifest as clinically apparent cancer. *Figure adapted from Vesely, M. D., Kershaw, M. H., Schreiber, R. D., & Smyth, M. J. (2011). Natural innate and adaptive immunity to cancer.* Annual Review of Immunology, 29, *235–271. http://dx.doi.org/10.1146/annurev-immunol-031210-101324 and Schreiber, R. D., Old, L. J., & Smyth, M. J. (2011). Cancer immunoediting: Integrating immunity's roles in cancer suppression and promotion.* Science, 331*(6024), 1565–1570. http://dx.doi.org/10.1126/science.1203486.* (See the color plate.)

circumvented. In its most complex form, Cancer Immunoediting is comprised of three phases: Elimination, Equilibrium, and Escape. In the Elimination phase, developing tumors are recognized and destroyed by the cooperative actions of innate and adaptive immunity long before they become clinically apparent. If the immune system fails to eliminate the entire tumor, the surviving cells may enter the Equilibrium phase where their overall expansion is immunologically restrained but where net tumor cell destruction does not occur. It is in Equilibrium that immunological sculpting occurs and if the "edited" tumor cells are altered to such an extent that they can no longer be identified as foreign by the host immune system, they begin to grow progressively, establish an immunosuppressive tumor microenvironment and emerge as the clinically apparent disease we know as cancer.

These naturally occurring immune system–tumor interactions were not always accepted and, in fact, were the subject of much scientific debate for most of the 20th century. In 1909, Paul Ehrlich first suggested that the immune system repressed cancer development in long-lived mammals (Ehrlich, 1909). However, this hypothesis could not be stringently tested because so little was known about the composition and function of the immune system at the time and tractable experimental systems to objectively evaluate the cell-extrinsic processes that controlled cancer development had not yet been developed. Five decades later, after a deeper understanding of the immune system had been obtained and inbred strains of mice had been developed that permitted studies of the immune system's role in cancer development, F. MacFarlane Burnet and Lewis Thomas proposed the term "cancer immunosurveillance" to describe a process in which they envisaged that the immune system, and particularly T cells, could recognize and destroy transformed cells early in their development thereby protecting the host against cancer outgrowth (Burnet, 1957, 1970; Thomas, 1959). If the immune system was indeed capable of detecting and eliminating newly transformed tumor cells, then cancer would be expected to occur with higher frequencies in immunodeficient compared to immunocompetent individuals. However, when this hypothesis was put to the experimental test in the 1970s by Osias Stutman, no evidence was found to support its validity (Stutman, 1974, 1979). Specifically, nude mice on a CBA/N genetic background (the only immunodeficient mouse strain available at that time) did not display higher tumor rates of spontaneous cancers or cancers induced by the chemical carcinogen 3′-methylcholanthrene (MCA) than their wild-type counterparts. At the time, these experiments were considered so definitive that the concept of cancer immunosurveillance was summarily abandoned and the field developed arguments why the immune system could never see a developing tumor.

However, in the mid 1990s, it became clear that there were caveats to the Stutman conclusions that he could not have known at the time. Specifically, nude mice were subsequently found to possess some basal T cell function and thus were recognized as imperfect models of immunodeficiency (Hunig, 1983; Ikehara, Pahwa, Fernandes, Hansen, & Good, 1984; Maleckar & Sherman, 1987). The existence and antitumor functions of natural killer (NK) cells and other innate lymphocytes were also not known at the time (Herberman & Holden, 1978). The role of aryl hydroxylase isoforms in the bioconversion of MCA to its carcinogenic form was only appreciated two decades later together with the fact that CBA/N nude mice expressed the highest specific activity isoform of the enzyme (Heidelberger, 1975). The latter raised the possibility that carcinogenesis in the mice used by Stutman may have been too efficient for the immune system to control.

As a variety of better-characterized immunodeficient mouse strains became available, we and others subsequently showed that immunodeficient mice indeed develop more chemically induced and spontaneous tumors than their genetically matched immunocompetent counterparts. For example, *Rag2*$^{-/-}$ mice (which lack T, B, and natural killer T (NKT) cells), IFN-γ receptor-deficient mice (*IFNGR1*$^{-/-}$ mice), and mice lacking perforin (*pfp*$^{-/-}$ mice) treated with MCA develop tumors both more rapidly and with a higher frequency than wild-type mice (Kaplan et al., 1998; Shankaran et al., 2001; Smyth et al., 2000; Street, Cretney, & Smyth, 2001). The incidence of MCA sarcoma generation was lowest in wild-type mice, higher in *Rag2*$^{-/-}$ mice, and highest in *Rag2*$^{-/-}$ × $\gamma c^{-/-}$ mice (which lack all lymphocytes, including NK cells) implicating the innate immune system in the control of the outgrowth of developing tumors (O'Sullivan et al., 2012). In addition, when tumors derived from immunodeficient and immunocompetent mice were compared to one another, the former are more immunogenic and less tumorigenic than the latter (Shankaran et al., 2001). Thus the intact immune system not only protected against cancer development but also sculpted the immunogenicity of tumor cells that eventually formed, leading to cancers that were more fit to grow in an immunocompetent host. Tumors derived in immunodeficient mice were therefore highly immunogenic and were therefore called "unedited." In contrast, tumors derived from immunocompetent mice displayed reduced immunogenicity and were therefore called "edited" (Schreiber et al.,

2011). Consequently, we introduced the term "Cancer Immunoediting" to stress the fact that immunity manifests both host-protective and tumor-promoting effects on developing cancers. This conclusion thereby significantly broadened the concept of cancer immunosurveillance and better reflected the physiologic function of immunity in its interaction with cancer.

The concept of Cancer Immunoediting was solidified by clinical observations demonstrating that a similar process also occurred in humans. Based on historical data, it was long recognized that individuals with congenital immunodeficiencies displayed higher cancer rates, but many of these cancers were of infectious origins and therefore did not allow for unequivocal conclusions to be made (Penn, 1999). However, meta-analyses of clinical data revealed that organ transplant patients who were immunosuppressed as adults indeed displayed higher incidences of cancers with no known viral etiologies. For example, renal transplant patients from multiple institutions displayed higher incidences of colon, pancreas, lung, and endocrine cancers and melanoma compared to nontransplanted, non-immunosuppressed normal individuals (Birkeland et al., 1995) and reviewed in Dunn, Bruce, Ikeda, Old, and Schreiber (2002). In addition, cancer patients were often found to express T cells and antibodies specific for the tumors that they harbored (Dunn, Old, & Schreiber, 2004b). Some of the best-characterized cases were those involving paraneoplastic neurologic degenerations where patients presented with neurologic symptoms which were subsequently found to be the result of natural immune responses to cryptic neoplasia (Roberts, Perera, Lang, Vincent, & Newsom-Davis, 1985). Perhaps the best correlative evidence comes from the finding that cancer patients frequently show immune infiltrates into their tumors that are tumor specific and that the quantity, quality, and location of memory $CD8^+$ T cells in a patient's tumor can have prognostic value in determining the course of treatment for that patient (Galon et al., 2006). This approach has become known as the "Immunoscore" and, in the case of colorectal cancer, has been shown to have better predictive value than conventional tumor staging.

2.1 Elimination

Elimination, the first phase of Cancer Immunoediting, thus represents a modernized and expanded view of cancer immunosurveillance, where the molecules and cells of innate and adaptive immunity work together to recognize and destroy a developing tumor. The key components involved in the Elimination phase of Cancer Immunoediting include cells of both innate immunity [eg, NK, macrophages and dendritic cells (DCs)] and adaptive immunity (eg, NKT, $CD4^+$, and $CD8^+$ T cells; Smyth, Godfrey, & Trapani, 2001; Teng, Galon, Fridman, & Smyth, 2015). Similarly, host effector molecules such as tumor necrosis factor (TNF)-α, Fas/FasL, granzyme, perforin, TNF-related apoptosis-inducing ligand (TRAIL), as well as recognition molecules such as NKG2D in protective antitumor immunity have been shown to play critical roles in the Elimination Phase (Diefenbach, Jensen, Jamieson, & Raulet, 2001; Smyth, Cretney, et al., 2001) and reviewed in (Mittal, Gubin, Schreiber, & Smyth, 2014). Both type I interferons (IFN-α/β) and IFN-γ are required for the development of protective antitumor immune responses but play distinct roles in this phase of the process. Whereas IFN-γ targets both tumor and hematopoietic cells, IFN-α/β acts primarily on host cells (Diamond et al., 2011). Specifically, in the mouse, type I IFNs enhance cross-presentation activity of tumor antigens by $CD8\alpha^+/CD103^+$ DCs while IFN-γ promotes induction of $CD4^+$ T helper I (Th1) cells and $CD8^+$ cytotoxic T lymphocytes (CTL) and is the critical interferon for enhancing MHC I expression on tumor cells (Diamond et al., 2011; Fuertes et al., 2011). If all cancer cells are eliminated, then the Elimination phase represents the full extent of the Cancer Immunoediting process.

2.2 Equilibrium

However, if some cancer cells survive, then the process can progress to the second phase—Equilibrium—a period when immunity is able to control the net outgrowth of cancer cells and thereby keep them clinically unapparent without completely eliminating them. Anecdotal evidence for the Equilibrium phase came from observations of cancer transfer following organ transplantation. In a particularly well-documented case, two patients who received kidney transplants from the same cadaver donor both subsequently developed malignant melanoma (MacKie, Reid, & Junor, 2003). The origins of the cancer were traced back to the donor who had been diagnosed with melanoma that had been successfully treated 16 years before death and who had been presumed to be cancer free. However, by transfer of a kidney from this donor into "naïve" recipients who were then immunosuppressed to protect against graft rejection, it is presumed that tumor cells held in equilibrium by the donor's immune system were then released from their dormant state and began to grow in a progressive manner. This clinical scenario was recapitulated in a defined preclinical model in 2007 that provided the first experimental validation of the postulated Equilibrium phase (Koebel et al., 2007). In that study, 80% of mice treated with low doses of MCA remained free of clinically apparent cancers for greater than 200 d. However, if these mice were treated on day 200 with a cocktail of monoclonal antibodies that eliminated $CD4^+$ and $CD8^+$ T cells and blocked IFN-γ, they showed a rapid appearance of sarcomas at the original site of MCA injection. Subsequent studies showed that adaptive immunity was the driver of the Equilibrium phase since antibodies that inhibited adaptive immunity (specifically anti-CD4, or anti-CD8 or anti-IFN-γ or anti-IL-12) released the dormant tumor cells from their equilibrium state while mAb that inhibit innate immunity [such as those that deplete NK cells (anti-NK1.1), inhibit NK cell recognition (anti-NKG2D), or block NK cell effector function (anti-TRAIL)] did not. Interestingly, dormant cancer cells were found in lesions that contained actively proliferating lymphocytes. Tumor cells held in Equilibrium retained their highly immunogenic phenotype and thus remained unedited. In contrast, the rare dormant cancers that spontaneously progressed to actively growing tumors displayed reduced immunogenicity and thus had undergone editing. These results have been expanded to other tumor models including the use of mice lacking p53 (Teng et al., 2012) as well as a Tag-induced pancreatic cancer model where T cells arrested the growth of tumors via a mechanism dependent on IFN-γ and TNF (Braumuller et al., 2013). Equilibrium can represent an end stage of Cancer Immunoediting where cancer cells remain in a durable state of immunity-induced dormancy throughout the remaining lifespan of the host without progressing to clinically apparent cancer.

2.3 Escape

If editing results in a reduction of tumor immunogenicity such that the immune system can no longer control tumor cell outgrowth, an immunosuppressive tumor microenvironment develops resulting in the outgrowth of tumor cell variants that eventually become clinically apparent tumors (ie, Escape). Thus Escape from immune control (the third phase of Cancer Immunoediting) is now acknowledged to be one of the "Hallmarks of Cancer" (Hanahan & Weinberg, 2011).

Immune Escape can occur through many different mechanisms involving both changes in tumor cells and/or the microenvironment. Tumors may avoid immune recognition through loss of NKG2D ligands, downregulation of MHC I, beta 2 microglobulin and calreticulin, reduced expression of costimulatory molecules, and/or antigen loss (extensively reviewed in Dunn, Old, & Schreiber, 2004a; Vesely et al., 2011). Tumor cells also upregulate proteins that allow increased resistance to apoptosis and promotion of survival (such as STAT-3 or the antiapoptotic molecule Bcl2; Yu, Pardoll, & Jove, 2009). Development of an immunosuppressive tumor microenvironment through recruitment of suppressive cells such as myeloid-derived suppressor cells and regulatory T cells (Tregs), production of immunosuppressive cytokines such as IL-10 and transforming growth factor beta (TGFβ) or expression of immune checkpoints of the B7 family such as programmed death ligand 1 (PD-L1)/PD-1, cytotoxic T lymphocyte antigen-4 (CTLA-4), lymphocyte-activation gene 3 (LAG-3), T cell immunoglobulin and mucin domain 3 (TIM-3) by either tumor cells, immune cells, or both also promote immune escape (Mellman, Coukos, & Dranoff, 2011). Additionally a growing list of new moieties that contribute to tumor-induced immunosuppression such as T cell Immunoglobulin and ITIM Domain (TIGIT), CD73, V-domain Ig suppressor of T cell activation (VISTA), and B and T lymphocyte attenuator have been identified (Chauvin et al., 2015; Jin et al., 2010; Wang et al., 2011; Watanabe et al., 2003). Of these negative regulatory molecules, CTLA-4 was the first to be identified as a target to enhance T cell immunity in tumor-bearing mice (Leach et al., 1996) and was also the first to be targeted therapeutically in tumor-bearing patients (Hodi et al., 2003). CTLA-4 is a negative costimulatory receptor that is critical for maintaining immune homeostasis and preventing autoimmunity. Mice lacking CTLA-4 develop spontaneous lethal lymphoproliferative disease (Waterhouse et al., 1995) and humans treated with high-dose anti-CTLA-4 develop life-threatening immune complications (Gangadhar & Vonderheide, 2014). Importantly, landmark work by James Allison and colleagues revealed that CTLA-4 is responsible for the absence of reactivity of T cells against tumor antigens in tumor-bearing mice and patients and that T cell immunity to tumors can be enhanced following treatment with anti-CTLA-4 (Sharma & Allison, 2015; van Elsas, Hurwitz, & Allison, 1999; van Elsas et al., 2001). Allison's work revealed a clinical benefit to manipulating the complex balance between therapeutic enhancements of antitumor immunity while maintaining control over autoimmunity. CTLA-4 expression on $CD4^+$ and $CD8^+$ T cells is temporally delayed compared to expression of its activating counterpart CD28 (Pardoll, 2012). During normal T cell activation, CD28 interacts with CD80/86 (B7.1/B7.2) expressed on antigen presenting cells (APCs) and delivers a positive costimulatory signal to the responding T cell. However, this response is naturally regulated by expression of CTLA-4 that subsequently translocates to the T cell surface. CTLA-4 displays higher affinity to CD80/86 than CD28 and thus preferentially engages CD80/86 on target cells generating a negative costimulatory signal that shuts down T cell activation via mechanisms involving the protein phosphatases, SHP2 (PTPN11) and PP2A. Based on its mechanism of action, CTLA-4 is thought to primarily inhibit T cell priming. Thus in the context of a tumor-bearing individual, CTLA-4 expression on T cells blocks generation of new antitumor T cell specificities and thereby contributes significantly to the immunosuppressive nature of the microenvironment of edited, progressively growing tumors.

Subsequent work by others has revealed that a second inhibitory receptor, PD-1, is also involved in limiting the activity of activated T cells in tumor-bearing individuals (Dong et al., 2002; Dong, Zhu, Tamada, & Chen, 1999; Freeman et al., 2000). Rather than blocking T cell priming as affected by CTLA-4, it functions to dampen T cell effector functions. PD-1-dependent T cell inhibition results following engagement with its ligands, PD-L1 (B7-H1) or PD-L2 (B7-H2) that can be expressed on tumor cells as well as host cells in the tumor microenvironment (Latchman et al., 2001). PD-1 is upregulated upon antigen stimulation and becomes highly expressed upon continuous or chronic T cell receptor (TCR) signaling (Barber et al., 2006). In contrast, PD-L1 is constitutively expressed by a wide variety of immune and nonimmune cells (such as T cells, NK cells, monocytes, macrophages, DC, B cells, epithelial cells, murine hepatocytes, and vascular endothelial cells) and many other cells upregulate PD-L1 in the presence of strong inflammatory signals (such as IFN-γ), presumably to limit tissue damage induced by potent but potentially destructive T cell responses (Loke & Allison, 2003). Additionally, some human and mouse tumors constitutively express high levels of PD-L1 and this appears to be a mechanism by which tumors evade immune Elimination (Iwai et al., 2002). Thus like CTLA-4, PD-1 contributes significantly to the immunosuppressive nature of the tumor microenvironment and thus facilitates outgrowth of edited tumors despite the fact that they may still possess some degree of immunogenicity. It is this characteristic that has permits for the success of checkpoint blockade cancer immunotherapy.

3. ANTIGENIC TARGETS OF CANCER IMMUNOEDITING

A central tenet of Cancer Immunoediting is that recognition of tumor antigens by T cells drives the immunological sculpting of cancers. Tumor antigens can be divided into three broad categories: (a) tumor-associated antigens (TAA), (b) cancer-germline/cancer testis antigens (CTA), and (c) tumor-specific antigens (TSAs) (Coulie, Van den Eynde, van der Bruggen, & Boon, 2014; Heemskerk, Kvistborg, & Schumacher, 2013).

TAA are comprised of proteins encoded by genes encoded in the normal genome that may represent either normal differentiation antigens (such as rearranged Ig and TCR genes expressed in B and T lymphomas, respectively) or aberrantly expressed normal proteins [eg, melanosomal proteins such as tyrosinase, gp100, and melanoma antigen recognized by T cells 1 (MART-1)]. In the mid 1990s multiple groups identified a number of shared melanocyte differentiation antigens (Bakker et al., 1994; Kawakami et al., 1994; Wang, Robbins, Kawakami, Kang, & Rosenberg, 1995). A common feature of these melanoma antigens is their expression by normal melanocytes in the skin and eye as well as their overexpression in malignant melanoma cells. Overexpressed normal proteins that possess growth/survival-promoting functions [such as Wilms tumor 1 (WT1), a transcriptional regulator (Ohminami, Yasukawa, & Fujita, 2000); Survivin (an apoptosis inhibitor); Her2/neu (a growth factor receptor component) (Fisk, Blevins, Wharton, & Ioannides, 1995); or Telomerase (a senescence inhibitor)] represent TAA that directly participate in oncogenesis.

CTA is the second category of tumor antigens which are normally expressed in germ cells (testis and ovary) and trophoblast tissues as well as in cancer cells. Because of their relatively restricted tissue distribution, these antigens have represented attractive targets for immunotherapy. The first human CTA was identified using cDNA expression cloning in 1991 by Thierry Boon and colleagues. In this study, van der Bruggen et al. isolated a gene shared by a panel of melanoma cell lines that could be recognized by CTL in an HLA-A*01 restricted manner, and is now known as melanoma antigen family A1 (MAGE-A1; van der Bruggen et al., 1991) NY-ESO-1, subsequently identified by Lloyd Old, Ugur Sahin, and colleagues using serological analysis of recombinant cDNA expression libraries (SEREX), was cloned from an esophageal tumor and is one of the best-characterized human CTA with respect to its immunology (Chen et al., 1997). In addition to esophageal cancers, NY-ESO-1 is expressed in a wide range of tumors including hematopoietic cancers (eg, acute myeloid leukemia (AML), acute lymphoid leukemia (ALL), and myeloma) and solid tumors (eg, breast, lung, melanoma, ovarian, sarcoma, urinary bladder, and uterine cancers). Both natural and therapeutically induced humoral and cellular immune responses against NY-ESO-1 have been well documented in cancer patients. In humans, CTA are particularly diverse where over 100 family members have been identified (Simpson, Caballero, Jungbluth, Chen, & Old, 2005). In contrast, CTA in mice are much less polydisperse (De Backer et al., 1995).

The third antigen category includes genes that are uniquely expressed in tumor cells and may represent either oncogenic viral proteins or abnormal proteins that arise as a consequence of somatic mutations or posttranslational modifications. In the former case, gene products of oncogenic viruses can represent TSA such as EBNA1 and LMP1/2 found in Hodgkin's lymphoma and nasopharyngeal carcinoma; or Human papillomavirus (HPV) E6 and E7 expressed in cervical cancers. Spontaneously arising mutations, from exposure to carcinogens and/or from the genomic instability that is characteristic of neoplastic cells, can produce mutated proteins that function as TSA. These neoantigenic peptides can result from point mutations (missense mutations), alterations in the reading frame, extending the coding sequence beyond the normal stop codon (nonstop mutations), DNA insertions and deletions (Indels), or by chromosomal translocations (Heemskerk et al.). In contrast to TAA, TSA are almost exclusively unique to an individual. Support for this idea was originally documented in the first half of the 20th century by Foley, Gross, Prehn, Old, and colleagues using carcinogen-induced mouse tumors (Foley, 1953; Gross, 1943 Old, 1982; Prehn & Main, 1957). When mice were cured of their tumors by surgical resection and rechallenged with the same tumor cells, they were protected against rechallenge but not against challenge with independent tumors. Mice that were immunized with irradiated tumor cells were also protected against challenge with the same nonirradiated tumor, whereas mice pretreated with normal donor tissue were not protected. Other groups subsequently confirmed these results, leading to the widespread acceptance that mouse tumors and potentially human tumors could be specifically recognized by the immune system, at least under certain conditions. Additionally, it was found that tumor challenge and resection or tumor immunization was usually effective only when the immunizing tumor was the same as that used for the challenge, thus providing compelling evidence that the response was indeed tumor specific. Additional experiments performed in the 1970s by Thierry Boon and colleagues supported the notion that the immune system could recognize TSA and provided some of the first experimental evidence that the response in part could be directed at mutant antigens (De Plaen et al., 1988). Upon treatment of a mouse carcinoma cell line in vitro with a strong mutagen, some tumor cell line variants from the treated population could not form progressively growing tumors when injected into naïve syngeneic mice. Strikingly, when the same mice that had rejected the tumors were rechallenged with the parental carcinoma line, those mice were protected against tumor growth, even though the parental carcinoma line was seemingly nonimmunogenic. These results were confirmed and extended by Hans Schreiber and colleagues using preclinical models of ultraviolet (UV)-induced mouse tumors with paired normal tissue from the same mouse in which the tumor originated, unequivocally demonstrating that somatic mutations could form TSA (Dubey et al., 1997; Monach, Meredith, Siegel, & Schreiber, 1995). The first human TSAs were discovered in 1995 when Wölfel et al. identified an R24C mutation in CDK4 by screening a cDNA library isolated from cultured melanoma cells while Coulie *et al.* isolated a mutation at an intron/exon boundary, both of which formed immunogenic peptides that could be recognized by autologous CTL (Coulie et al., 1995; Wolfel et al., 1995). While most cancer mutations are private, a fraction of mutations are indeed shared between different cancers and different patients. In some cases, driver mutations can be immunogenic including those formed from mutant RAS (Linard et al., 2002) or BRAF (Somasundaram et al., 2006), as well as chromosomal fusions such as BCR-ABL or TEL-AML (Greco et al., 1996).

Additional neoantigens that may be shared between different cancers can result from aberrant phosphorylation. During transformation, protein kinase activity becomes dysregulated, leading to hyperphosphorylation of signaling proteins and changes in proliferation, differentiation, and cell growth. Phosphorylated residues can enhance the stability of individual peptides for both MHC I (Mohammed et al., 2008; Zarling et al., 2000; Zarling et al., 2006) and MHC II (Li et al., 2010). This observation suggests that phosphoproteins may be a particularly desirable shared target for cancer immunotherapy, as aberrantly phosphorylated residues may not have been subject to central tolerance. Cobbold et al. identified 10 phosphopeptides presented by HLA-A*02:01 and 85 presented by HLA-B*07:02 from a panel of hematologic malignancies including both leukemia and lymphoma specimens using a mass spectrometry approach (Cobbold et al., 2013). Many of these phosphopeptides were derived from signaling molecules with well-established roles as drivers of transformation. Interestingly, there were more than two-fold more phosphopeptides detected from aggressive malignancies (AML and ALL) compared with more indolent cancers [chronic lymphocytic leukemia

(CLL) and hairy cell leukemia (HCL)] or from healthy tissue. In addition, CTL lines could be derived from healthy individuals specific for only the phosphorylated form of LSP-1, a lymphoma marker, which were capable of recognizing HLA-A*02:01 positive cell lines derived from AML and CLL patients. Phosphopeptide-specific T cells predominantly had the phenotype of the central memory compartment, suggesting that most healthy individuals had mounted responses to tumor-associated phosphopeptides during their lifetimes, at frequencies similar to those responding to nonpersistent viruses. However, responses against a panel of phosphopeptides were reduced or absent in patients with active CLL, suggesting that patients with intact responses against phosphopeptides had improved survival. However, due to the small sample size available, this analysis did not reach statistical significance. Perhaps most significantly, a profound recovery in responses against phosphopeptides was observed in several patients with AML after undergoing an allogeneic stem cell transplant, and a phosphoprotein-specific CTL line from a patient isolate was able to kill an AML cell line in vitro. These findings suggest that the selection of transplant donors by the presence of phosphoprotein-specific responses in addition to standard criteria may be an attractive option to prevent disease relapse. Phosphopeptides expressed by solid tumors have also been identified, and TCRs specific for phosphoproteins may serve as additional epitope determinants for transgenic T cells (Zarling et al., 2014). Techniques allowing detection of potential phosphoneoantigens with mass spectrometry approaches using small amounts of starting material are available (Abelin et al., 2015).

In additional to aberrant phosphorylation, dysregulated glycosylation of proteins can lead to formation of tumor neoantigens. As normal tissues transform, changes in glycosylation in proteins integral to cell adhesion, motility, invasiveness, and signaling occur with the potential to form antigens for immune detection (Ono & Hakomori, 2004). Therapeutic antibodies can strongly bind to glycan epitopes that are uniquely expressed on the target tumor cell population as compared to normal tissues to disrupt molecules required for neoplastic cell growth, as well as to mediate antibody-dependent cellular cytotoxicity (Dingjan et al., 2015).

4. SETTING THE GROUNDWORK: GENOMIC APPROACHES TO CANCER ANTIGEN IDENTIFICATION

Advances in next generation sequencing allowed for whole genome sequencing of cancers and a better understanding of the mutational landscape present in many cancers (Koboldt, Steinberg, Larson, Wilson, & Mardis, 2013). In 2008, James Allison and Bert Vogelstein performed in silico analysis combining breast and colorectal cancer-sequencing data with epitope prediction algorithms and hypothesized that breast and colorectal cancers accumulate unique HLA epitopes (Segal et al., 2008). They proposed that as cancer is a process where transformed tissues accumulate genetic changes over time, all cancers would contain mutations with a potential to form epitopes recognizable by the immune system. Subsequent studies have further demonstrated that cancers over a broad spectrum contain a remarkable number of mutations that could form epitopes (Alexandrov et al., 2013). Lessons from preclinical models have helped shape our understanding of the mutational landscape that is surveyed by the immune system.

Two independent reports in 2012 used genomic sequencing and epitope prediction algorithms to identify mutant neoantigens responsible for rejection of a highly immunogenic unedited tumor (Matsushita et al., 2012) or an edited progressively growing tumor in a mouse prophylactically vaccinated with neoantigen-specific synthetic long peptides (SLP) (Castle et al., 2012). The highly immunogenic unedited MCA sarcoma line (d42m1) derived from immunodeficient *Rag2*$^{-/-}$ mice was subjected to cDNA capture sequencing to identify expressed missense mutations. These mutations were then computationally translated into corresponding proteins and pipelined into MHC I binding algorithms to predict strong MHC I binders. A mutation in the highly expressed protein spectrin-β2 was predicted and subsequently validated as a major rejection antigen responsible for the spontaneous rejection of the d42m1 tumor when transplanted into syngeneic wild-type immunocompetent mice. Importantly, when d42m1 was passaged through immunocompetent mice, it underwent cancer immunoediting, leading to outgrowth of preexisting tumor cells that lacked mutant spectrin-β2. A complementary study from Jacks and colleagues reached similar conclusions about Cancer Immunoediting using a genetically engineered mouse model of cancer (DuPage et al., 2012). Both of these studies demonstrated that selection for tumor variants that do not express strong antigens is one mechanism of Cancer Immunoediting. Importantly, although edited tumors from both mice and human may lack strong antigens required for *spontaneous* rejection, some still retain antigens that confer residual immunogenicity to the tumor that can be accessed by the proper type of cancer immunotherapy.

Using a weakly immunogenic melanoma tumor line derived from an immunocompetent mouse (B16-F10), Sahin and colleagues developed prophylactic personalized neoantigen-specific SLP vaccines based on predicted MHC I binding scores of mutant peptides identified by genomic sequencing. Analysis of 50 peptides revealed 16 to be immunogenic. Furthermore, two of the peptides provided prophylactic protection against tumor growth when incorporated into an SLP vaccine. These studies were the first to experimentally demonstrate that genomic sequencing and epitope prediction algorithms could identify mutant rejection antigens. These studies were soon to be followed by an explosion in genomic analyses to inform antitumor immune responses.

5. DEVELOPING CANCER IMMUNOTHERAPIES BASED ON GENOMIC IDENTIFICATION OF TUMOR-SPECIFIC NEOANTIGENS

Using neoantigens for therapeutic benefit has significant conceptual advantages over the use of TAA. The former are expressed exclusively by transformed cells and therefore are similar to foreign proteins in that they are not subject to central immunological tolerance. Perhaps equally important, neoantigens are tumor specific and therefore targeting them obviates concerns about cytotoxicity toward healthy tissue. Indeed, accumulating data suggest that neoantigens are important components of cancer immunotherapy. In 2005 a seminal study by Lennerz et al. used a cDNA library from a patient-derived melanoma line to screen autologous T cells and identified several neoantigens that induced T cell responses as assessed by IFN-γ ELISPOT (Lennerz et al., 2005). Reactivity against these neoantigens dominated the tumor-specific T cell response in the patient. That same year, Rosenberg and Robbins identified multiple neoantigens recognized by adoptively transferred tumor infiltrating lymphocytes (TIL) in a single patient with metastatic melanoma treated with adoptive T cell immunotherapy (Zhou et al., 2005). Importantly the neoantigen-specific T cells persisted in the patient. These data were also some of the first to suggest that the relevant T cell clones existed before ex vivo expansion and that the autologous T cell therapy was amplifying preexisting T cell responses.

The reverse immunology approaches for mutant tumor antigen identification used in the Matsushita et al. and Castle et al. studies made the possibility of designing patient-specific treatments exploiting the full repertoire of a patient's antigenome a reality. For this purpose, mutations are identified through deep sequencing, the region surrounding the mutation are virtually "translated," and then input into epitope prediction algorithms. Critical to this endeavor is the need to be able to correctly identify somatic mutations using next generation deep sequencing and accurately predict those mutations that form immunogenic neoantigens.

In order for an antigen to be immunogenic, it must be presented by MHC and recognized by T cells through their TCR. Whereas MHC I binds antigens of 8–11 amino acids in length and presents them to cytotoxic $CD8^+$ T cells, MHC II presents antigens of 11–20 amino acids to $CD4^+$ T cells (Babbitt et al., 1985; Bjorkman et al., 1987). The MHC alleles are remarkably diverse and the number of potential peptides processed from a given pathogen or tumor is also large, with a small minority actually binding to the MHC. This makes predicting which peptides will bind MHC challenging. Fortunately, multiple computational algorithms for prediction of antigen processing, presentation, and immunogenicity exist.

5.1 Epitope Prediction Algorithms

Multiple prediction tools for MHC I binding exist, with SYFPEITHI developed by Hans-Georg Rammensee being the first widely used and validated method (Rammensee et al., 1999). Subsequently, other prediction algorithms have been developed including those available from the Immune Epitope Database and Analysis Resource (IEDB; www.iedb.org; Vita et al., 2015). The IEDB is an online comprehensive database of T cell epitopes and tools for predicting MHC binding with the most commonly used prediction tools for MHC I tumor antigens being: (a) artificial neural networks (ANN)/NetMHC (Lundegaard et al., 2008; Nielsen et al., 2003), (b) NetMHCpan (Nielsen et al., 2007), and (c) SMMPMBEC and SMM (Kim, Sidney, Pinilla, Sette, & Peters, 2009; Peters & Sette, 2005). ANNs, like NetMHC, are algorithms modeled after the neural connections in the brain, learning from a set of input training data. NetMHC is one of the most commonly used and best validated epitope prediction programs available. While common allele predictions are often quite accurate, rare alleles are trained and validated on fewer data and thus are usually less accurate. To address this issue, pan-specific programs, such as NetMHCpan, were created to extrapolate from existing data to less common alleles. SMM and SMMPMBEC, as described by Sette and Peters, calculate matrices from affinity data of peptides binding to MHC. This allows for suppression of noise caused by the inevitable experimental error as well as limited data points present in the training data. Other prediction tools available from IEDB include ARB, Comblib_Sidney2008, Pickpocket, and Consensus.

For most studies, the primary selection criterion for predicting epitopes is the binding affinity of the peptide epitope for MHC I. However, whether the peptide is even available to bind MHC I, that is how efficiently the antigen is processed, also needs to be considered. Antigen processing involves the degradation of proteins within the cytoplasm by the proteasome (a specialized proteasomal complex induced by IFN-γ called the immunoproteasome is primarily responsible for the degradation of proteins into peptides that are optimal in size for MHC binding), followed by transportation of the peptides to the endoplasmic reticulum via the transporter associated with antigen processing (TAP) proteins (Blum, Wearsch, & Cresswell, 2013; Rock et al., 1994). Algorithms exist to predict both proteasomal cleavage and TAP transport. NetChop uses a neural network to predict proteasomal processing (Nielsen, Lundegaard, Lund, & Kesmir, 2005), whereas NetCTL and NetCTLpan also use a neural network to predict T cell epitopes but they combine predicted MHC binding, proteasomal cleavage, and TAP transport to generate a score (Peters, Bulik, Tampe, Van Endert, & Holzhutter, 2003). The MHC-NP algorithm assesses the probability that a peptide is naturally processed and binds to a given MHC based on data obtained from MHC elution experiments (Giguere et al., 2013).

In contrast to MHC I, predicting MHC II epitopes has remained a more difficult challenge. MHC II prediction methods have consistently underperformed MHC I, in part due to a paucity of data sets for MHC II training. Nevertheless, multiple MHC II binding algorithms are available with the most commonly used being NetMHCII (Nielsen & Lund, 2009) and TEPITOPEpan (Hammer et al., 1994).

There are also tools available that predict the relative ability of a peptide–MHC complex to elicit a T cell response, taking into account the amino acid properties as well as position in the peptide sequence. Peters and Sette have developed a model of peptide–MHC properties that enhance immunogenicity and this prediction tool is available from IEDB (Calis et al., 2013). Other T cell reactivity predictors include POPI, iMatrix (models the TCR-peptide and peptide–MHC interface), and CTLPred. However, these algorithms have limitations, as they have not been extensively validated. Combining both antigen processing and MHC binding should result in an increased accuracy of predicted epitopes. However, these algorithms are limited by the available data and are only as good as the data used to generate them. Since these models are currently based on restricted data sets from either in vitro studies or data sets based on previously identified T cell epitopes, the accuracy of the results should continue to improve as more antigens are identified.

5.2 Retrospective Bioinformatic Analyses of Previously Identified Cancer Neoantigens

The ability to identify mutant neoantigens permits a deeper understanding of the immune responses to cancer and raises the promise of therapeutic use of these antigens. Insights into the immunogenic mutant neoantigen landscape of many human cancers have come from recent studies from Wu and Hacohen (Fritsch et al., 2014). They determined that a large majority of known mutated neoantigens in multiple tumor types from patients experiencing long-term survival or tumor regression had strong or moderate predicted MHC I binding affinity for their respective alleles. In a separate report, they used bioinformatic and experimental approaches to explore the epitope landscape of 91 CLLs and predicted an average of 22 mutated HLA-binding peptides per CLL (Rajasagi et al., 2014). Further analysis of two patients that achieved long-term remission revealed CTL responses against predicted neoantigens could be detected. Application of their epitope prediction approach to sequencing data from many different cancer types revealed a range of predicted neoantigens per individual tumor, providing evidence that neoantigens are frequent in most human cancers.

A similar retrospective analysis was performed by Schumacher and colleagues, where data sets of known human cancer neoantigens were analyzed to determine whether they would have been identified using genomic approaches (van Buuren, Calis, & Schumacher, 2014). Specifically the following criteria were assessed: (1) sequencing coverage to allow confident calling of the mutant base, (2) NetChop cleavage probability of 0.5 or greater, (3) predicted binding affinity [predicted half-maximal inhibitory concentration (IC_{50})] of less than 500 nM using the NetMHCpan algorithm, and (4) low "similarity-to-self" of the mutant compared to wild-type epitope. The authors describe this "similarity-to-self" test as one that determines the likelihood that the mutant and wild-type epitopes can be distinguished by the T cells, either by altered levels of mutant antigen presentation or by an altered structure of the MHC-presented mutant antigen. Here again, algorithms were used to assess whether the potential mutant neoepitope would be presented at higher levels than the wild-type epitope. If the wild-type parental sequence was not predicted to be presented by MHC either because it was not processed or did not bind MHC, the peptide was considered different than self. Additionally, they explored whether the mutations altered the peptide–MHC/TCR interaction. In general, the TCR exposed surface lies in the core region of the epitope, which is the peptide sequence between the two anchor residues. If the core region of the mutant epitope is different than that of the parental sequence using a peptide:MHC binding energy covariance (PMBEC) value of ≤ 0.05, the mutant peptide is considered different than self. This analysis suggested that the current available methods for neoantigen prediction are relatively accurate and these methods would identify most of the previously known neoantigens.

5.3 Experimental Evidence from Preclinical Cancer Models That Neoantigens Form the Basis for Effective Personalized Cancer Immunotherapy

Experimental validation that mutant neoantigens identified by genomic and bioinformatics approaches can function in a therapeutic setting came from three studies that were published in 2014–2015. One study, stemming from work in our laboratory, employed a MCA sarcoma line (T3) that forms progressively growing tumors when transplanted into naïve syngeneic immunocompetent mice but is rejected in tumor-bearing mice following treatment with monoclonal antibodies blocking CTLA-4 or PD-1 (Gubin et al., 2014). Genomic sequencing analysis of T3, as illustrated in Fig. 2, followed by epitope prediction revealed two predominant H-2K^b epitopes [a G1254V mutation in Laminin α subunit 4 (mLama4) and a A506T mutation in Asparagine-linked glycosylation 8 (α-1,3-glucosyltransferase) (mAlg8)] as being the most likely targets of T cells activated by checkpoint blockade therapy. This prediction was validated by ex vivo screening of TIL isolated directly from tumors using either a panel of H-2K^b MHC I tetramers carrying one of the top 62 predicted H-2K^b epitopes or four top predicted H-2D^b epitopes as well as testing the eluted T cells for antigen-specific stimulation as detected by intracellular cytokine staining following coincubation with irradiated splenocyte feeder cells pulsed with the individual predicted peptides (Fig. 3). The validity of these findings were further confirmed by the following criteria: (a) the same epitopes were identified when tested on CTL lines generated from mice that had rejected T3 tumors following anti-PD-1 treatment, (b) the mutant epitopes were identified by mass spectrometry on IFN-γ-treated T3 tumors propagated in vitro, (c) tetramer positive staining T cells accumulated temporally in progressively growing tumors in vivo in mice treated with anti-PD-1 and reached maximal levels just prior to tumor rejection, and (d) prophylactic vaccination of mice with a combination of mLama4 and mAlg8 peptides protected the mice from subsequent challenge with T3 tumor cells. Perhaps most importantly, growing T3 tumors were rejected in mice treated with a therapeutic vaccine comprised of SLP encompassing

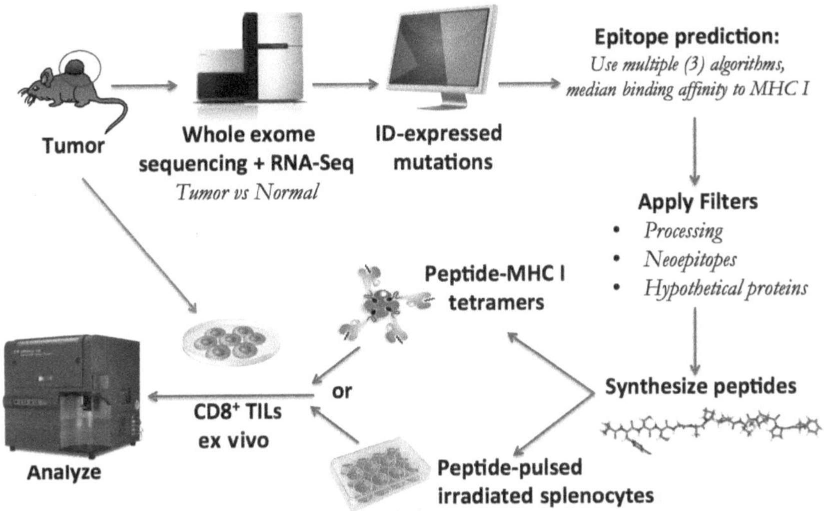

Figure 2 Genomics- and bioinformatics-based identification of mutant neoantigens. Tumor cells and normal tissue are subjected to whole exome and RNA-sequencing to identify expressed nonsynonymous somatic mutations. Corresponding mutant epitopes are then analyzed in silico for MHC class I binding. Filters are then applied for antigen processing, whether the mutant epitope has a stronger predicted binding affinity than the corresponding wild-type peptide, and deprioritization of hypothetical proteins. Peptides corresponding to predicted epitopes are then synthesized and used to identify mutant neoantigen-specific T cells in freshly explanted TIL using MHC I multimer-based screens or functional assays (eg, cytokine release, ELISPOT, or intracellular cytokine staining) by peptide stimulation. (See the color plate.)

the mLama4 and mAlg8 mutations together with the adjuvant Poly I:C. Rejection induced by the therapeutic vaccine was nearly as effective as treatment of tumor-bearing mice with checkpoint antibodies. Rejection was observed only rarely with Poly I:C alone or with an irrelevant SLP vaccine plus Poly I:C.

Similar results were obtained in a contemporary study by Yadav et al. who used mass spectrometry in combination with whole-exome and transcriptome sequencing to predict immunogenic TSA expressed by the carcinogen-induced colon adenocarcinoma MC-38 and the model prostate cancer TRAMP-C1 (Yadav et al., 2014). Of the 1290 and 67 expressed mutations found in MC-38 and TRAMP-C1, respectively, 7 were found to be presented on MHC I in MC-38 and none by TRAMP-C1 MHC I. All but one of the identified neoepitopes were predicted by the NetMHC algorithm to bind MHC I ($IC_{50} < 500$ nM). Of these identified mutant neoantigens, mutant forms of Reps1, Adpgk, and Dpagt1 protected mice

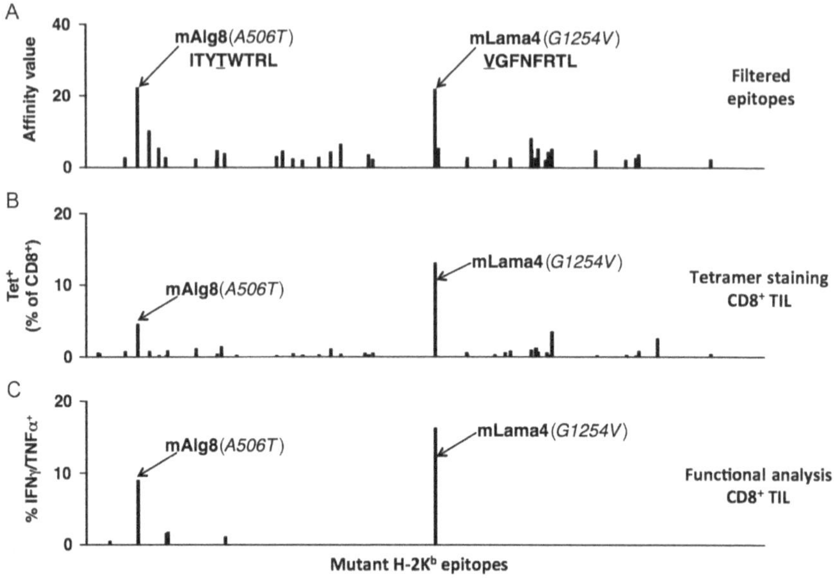

Figure 3 (A) Predicted MHC I binding affinity of filtered epitopes predicted by in silico analysis of missense mutations in the T3 tumor line. (B) Screening for specificities of $CD8^+$ TIL from anti-PD-1-treated, T3 tumor-bearing mice using MHC I tetramers loaded with top predicted peptides. (C). IFN-γ and TNF-α induction in $CD8^+$ TIL from anti-PD-1-treated, T3 tumor-bearing mice following culture with irradiated splenocytes pulsed with the top predicted peptides. *Figure adapted from Gubin, M. M., Zhang, X., Schuster, H., Caron, E., Ward, J. P., Noguchi, T., … Schreiber, R. D. (2014). Checkpoint blockade cancer immunotherapy targets tumour-specific mutant antigens.* Nature, 515*(7528), 577–58. http://dx.doi.org/110.1038/nature13988.*

from subsequent tumor challenge, achieving therapeutic tumor protection when administered together with agonist anti-CD40 antibody.

In the most recent studies, the Sahin group demonstrated using three separate preclinical cancer models that the majority of the predicted TSA in fact elicit $CD4^+$ T cells responses upon vaccination, even when vaccine epitopes are predicted based on MHC I prediction algorithms (Kreiter et al., 2015). Strikingly, elicitation of $CD4^+$ T cell responses using either peptide or RNA vaccination mediated protection from established tumors and was also shown to induce responses against additional MHC I epitopes through epitope spreading. In a separate study, Platten and colleagues used an SLP containing a MHC II epitope corresponding to mutated isocitrate dehydrogenase type 1 (IDH1), a mutation commonly found in a subgroup of gliomas, to demonstrate immune control of preestablished syngeneic IDH1 (R132H)-expressing tumor cells transplanted into mice devoid of mouse MHC and transgenic for human MHC I and MHC II (Schumacher et al., 2014). Together these studies demonstrate that neoantigen vaccines can be highly effective in therapeutically controlling established tumors and even inducing their immune elimination when the antigens included in the vaccine include both MHC I and MHC II epitopes. While it is clear that $CD8^+$ T cells can directly kill tumors expressing MHC I and produce antitumor effector cytokines, the role of MHC II antigens is less obvious. $CD4^+$ T cells may exert antitumor effects through the production of antitumor effector cytokines, licensing of DCs, or direct effects on tumors expressing MHC II. More work is needed to delineate the mechanism behind the antitumor effects of $CD4^+$ T cells. This work has encouraged renewed enthusiasm for development of tumor-specific vaccines as a method to treat cancer that may be more specific, safer, and potentially more effective than the methodologies that are available to us today.

6. NEOANTIGENS AS THERAPEUTIC TARGETS IN HUMAN CANCER

As our understanding of the dual functions of the immune system to both eliminate and sculpt the development of progressively growing tumors evolved, so too did the capacity to use the immune system as a therapeutic tool to control cancer. The recognition that tumor antigens were key to the immune system's capacity to discriminate between cancer cells and normal self formed the basis for many early clinical vaccine trials targeting TAA and subsequently CTA as antigens. While occasional successes were observed in these approaches, the overall response rates were disappointing and at best were very tumor-type specific (Rosenberg, Yang, & Restifo, 2004).

However, two more recent immunotherapeutic modalities (adoptive T cell therapy and checkpoint blockade) are displaying significantly higher response rates and display efficacy toward a much wider range of tumor types. It is of great significance that these more successful new therapies are directed, at least in part, against tumor-specific mutant neoantigens and based on this finding, clinical trials are now ongoing in many institutions that are exploring the use of personalized cancer immunotherapies based on targeting cancer-specific neoantigens (Table 2).

Support for this latter concept has come, in part, from correlative studies of the mutational load in various cancers and the response of a patient bearing these cancers to immunotherapy. Despite the potential for durable responses with the newer types of cancer immunotherapy, only a percentage of patients achieve objective responses to cancer immunotherapy (Hodi

Table 2 Ongoing or Planned Clinical Studies of Neoantigen Vaccines

Tumor Type	Phase	Vaccine Platform	Institution	Start Date	ClinicalTrial.gov Identifier
Melanoma	1	Neoantigen polyepitope coding RNA vaccine	Biontech AG	December 2013	NCT02035956
Melanoma	1	Synthetic long neoantigen peptides plus poly-ICLC	Dana-Farber Cancer Institute	January 2014	NCT01970358
Glioblastoma	1	Neoantigen peptide plus poly-ICLC + GM-CSF	Immatics Biotechnologies	October 2014	NCT02149225
MGMT-unmethylated Glioblastoma, Glioblastoma Multiforme	1	Synthetic long neoantigen peptides plus poly-ICLC	Dana-Farber Cancer Institute	November 2014	NCT02287428
Triple-negative breast cancer	1	Neoantigen polyepitope DNA vaccine	Washington University School of Medicine	June 2015	NCT02348320
Triple-negative breast cancer	1	Synthetic long neoantigen peptides plus poly-ICLC	Washington University School of Medicine	September 2015	NCT02427581
Triple-negative breast cancer	1	Neoantigen polyepitope coding RNA vaccine	Biontech AG	September 2015	NCT02316457
Glioblastoma multiforme astrocytoma, Grade IV	0	Synthetic long neoantigen peptides plus poly-ICLC	Washington University School of Medicine	November 2015	NCT02510950
Non-small cell lung cancer	0	Neoantigen dendritic cell vaccine	Washington University School of Medicine	January 2016	NCT02419170
Pancreatic, colorectal	1	Peptide vaccine plus IFA	MD Anderson Cancer Center	March 2016	NCT02600949

et al., 2010; Rosenberg et al., 2011). Because of the stochastic process by which mutations that form neoantigens are generated during cellular transformation, and because cancer immunotherapy relies on expression of antigens for both $CD4^+$ and $CD8^+$ T cells, genomics approaches are being investigated to develop a predictive biomarker of response to therapy. The genomic landscape of some tumors such as melanoma is characterized by a high mutational load (Alexandrov et al., 2013) as a consequence of exposure to UV light, which results in expression of a significant number of aberrant proteins products never before seen by the immune system capable of functioning as antigenic targets of a tumor-specific immune response. Using next generation sequencing, Snyder et al. demonstrated a correlation between clinical benefit from CTLA-4 blockade and the mutational load in metastatic melanoma (Snyder et al., 2014). This finding was subsequently validated by Van Allen et al. who used larger patient cohorts (Van Allen et al., 2015). This finding is not limited to melanoma, as a similar analysis in patients with non-small cell lung cancer also found that a correlation between a tumor's nonsynonymous mutation burden and objective patient response to PD-1 blockade exists (Rizvi et al., 2015). Many other histologies that result in a sizable fraction of human malignancies have mutation ranges that fall between 1 and 10 somatic mutations per megabase and thus are likely to express sufficient neoantigenicity to render them immunogenic. It remains an open question if immunotherapy approaches can be designed to induce therapeutic responses against tumors that express lower antigen burdens.

In a 515 patient study, RNA-sequencing analysis revealed increased numbers of mutational epitopes were associated with increased patient survival, higher intratumoral CTL content, and upregulation of genes encoding the immune checkpoints PD-1 and CTLA-4 (Brown et al., 2014). Little evidence of CTL infiltration was present in tumors with few mutational epitopes. This study provided the foundation for an extensive genomic analysis by Hacohen and colleagues using TCGA data sets of solid tumor biopsies. Rooney et al. derived a cytolytic index matrix based on expression of perforin and granzyme B (Rooney et al., 2015). When compared across 18 tumor types, this cytolytic score correlated with neoantigen load, as well as expression of viral transcripts. In addition, fewer neoantigens were present in colorectal tumors (CRC) than would be expected based on their mutation rate, implying that strong immune pressure had exerted a sculpting effect on the tumors as they developed. Interestingly, despite their restricted expression, CTA did not correlate with cytolytic function.

Mutations in genes with clearly established immune functions, such as beta 2 microglobulin (Restifo et al., 1996), MHC I heavy chains (Shukla et al., 2015), and caspase 8 were also enriched in tumor tissues, which would be expected to be selected for in tumors that escape immune control.

Attempts have also been made to correlate neoantigen load and the likelihood of response to immunotherapies in gastrointestinal malignancies. Using a cohort of 103 colorectal cancers with microsatellite instability, Maby et al. showed that $CD8^+$ TIL density correlates with the total number of frameshift mutations (Maby et al., 2015). Peripheral $CD8^+$ T cells derived from patients with microsatellite unstable colon cancer could lyse target cells pulsed with predicted neoepitopes derived from frameshift mutations after in vitro culture. Taken together, these results suggest that immunogenic neoantigens are more likely to arise in genetically unstable tumors and drive the T cell-dependent cytolytic activity that is critical to effect Cancer Immunoediting and immunotherapy. Interestingly, microsatellite instability $(MSI)^{hi}$ colorectal cancers represent the only CRC subset that is susceptible to checkpoint blockade immunotherapy (Le et al., 2015), a result that once again supports the hypothesis that cancer-specific mutant neoantigens are the favored targets of T cells that can be reactivated by this type of immunotherapy.

Finally, transcriptomic analysis on a subset of tumors from melanoma patients demonstrated that a cytolytic gene signature, along with elevated transcript expression of PD-L2, correlated with neoantigen load and response to ipilimumab (Van Allen et al., 2015). Interestingly the expression of CTLA-4 itself was an indicator of response. These findings may reflect the ongoing preexisting T cell responses possibly against mutant neoantigens, especially considering the presence of subsets of melanoma patients with inflamed tumor microenvironments that are a result of $CD8^+$ T cell reactivity as demonstrated by the Gajewski laboratory (Spranger et al., 2013) as well as findings suggesting that increased numbers of PD-L1 positive $CD8^+$ T cells correlates with response to PD-1 blockade (Tumeh et al., 2014). This concept is not limited to prediction of response in melanoma, as a correlation between antigen load and response to pembrolizumab also exists in NSCLC (Rizvi et al., 2015).

6.1 Neoantigens in Adoptive Cellular Therapy in Humans

As cancer immunoediting of developing tumors progresses from equilibrium to escape, the balance shifts toward cancer progression as adaptive immunity loses its ability to control tumor growth. By removing tumor-specific T cells from the inhibitory tumor microenvironment and allowing them to regain their cytotoxic function ex vivo prior to transfer back into the patient, adoptive cellular therapy (ACT) with TIL attempts to reverse this transition and achieve tumor elimination. In a description of their recent experience, the Steven Rosenberg group at the Surgery Branch of the National Cancer Institute treated 93 metastatic melanoma patients with infusion of autologous T cells in conjunction with IL-2 and different lymphodepleting regimens (Rosenberg et al., 2011). Response rates in this patient cohort varied between 49% and 72%. More impressively, 19 of the 20 patients who displayed a complete remission had responses that were durable beyond 3 years. Similar results have been reported in smaller series from other centers (Radvanyi et al., 2012).

Whether TAA can serve as the targets of the immune response during ACT has been intensively investigated. Initial studies on TILs from melanoma patients focused on the identification of T cell populations specific for shared TAA such as gp100, MART-1, and tyrosinase-related protein 1 (reviewed by Coulie et al., 2014). Despite their presence in normal tissues, CTL targeting these TAA rarely caused severe autoimmune toxicities, but their frequencies in TILs were usually quite low (Kvistborg et al., 2012). In two more recent studies, transfer of T cells highly selected for the melanocyte differentiation antigens gp100 and MART-1 led to clonal engraftment and autoimmune dermatitis in the majority of patients, but no objective responses (Chandran et al., 2015). Experiences such as these led to attempts to design transgenic TCRs with higher affinity for TAAs in the hope that this would increase the efficacy of tumor cell killing by effector cells. The Rosenberg group developed approaches to isolate high-affinity TCRs against gp100 and MART-1 using immunization of mice transgenic for HLA-A*02 or selection for high-affinity TCRs from T cell clones, respectively (Johnson et al., 2009). The genes encoding these TCRs were then transduced into autologous peripheral blood T cells that were subsequently used for adoptive transfer. This approach successfully led to objective responses in nearly a third of patients with metastatic melanoma. However, vitiligo and loss of vision and hearing were frequently seen in this study as a consequence of the destruction of normal melanocytes present in the skin, eye, and ear, respectively. In another example, despite extensive preclinical testing without an indication of off-target effects, an affinity-enhanced TCR specific for MAGE A3 caused fatal toxicity in two patients due to cross-recognition of the muscle protein Titin that was only detectable when a beating myocyte culture was tested as a target (Cameron et al., 2013;

Linette et al., 2013). These results indicate that whereas strong T cell-dependent responses against tumor cells can be therapeutically effective, the utility of this approach may be limited by the associated devastating off-target destruction of normal tissues.

However, recent work has revealed that the ACT approach may be made more specific if it employs TIL that display specificity for tumor-specific neoantigens. In 2013 the groups of Robbins and Rosenberg identified seven unique mutant MHC I epitopes that were presented by autologous tumor cells and recognized by in vitro expanded TIL from metastatic lesions of three patients who had shown objective responses to TIL therapy (Robbins et al., 2013). In a follow-up study they showed that therapeutically effective TIL not only consists of $CD8^+$ T cells specific for MHC I-restricted neoantigens but also $CD4^+$ T cells specific for MHC II-restricted tumor neoantigens from gastrointestinal malignancies (Tran et al., 2014; Tran et al., 2015). In this second study, TIL derived from a metastatic lung lesion from a cholangiocarcinoma patient were found to contain $CD4^+$ T cells specific for an epitope derived from a mutation in erbb2 interacting protein (ERBB2IP). When these tumor-specific $CD4^+$ T cells were expanded ex vivo, the expanded population was found to contain approximately 25% mutant ERBB2IP-specific $CD4^+$ Th1 cells. Adoptive transfer of the expanded cell population back into the patient led to a transient objective tumor response before subsequent disease progression. The same patient then received another infusion of TIL containing more than 95% ERBB2IP-specific $CD4^+$ T cells and again experienced a response with a reduction in tumor burden that was durable after 20 months of follow up. No adverse events were reported by the therapeutic use of neoantigen-specific T cells. Similar studies on different patients bearing different tumors are ongoing at NCI and elsewhere to explore whether tumor neoantigen-specific T cells represent a preferred approach to ACT that will lead to improved efficacy and reduced toxicity. Certainly, if this approach is successful, one could envisage engineering neoantigen-specific T cells that express intracellular domains containing signaling cassettes that promote T cell survival and effector functions.

Of course hope remains that a set of shared tumor neoantigens will eventually be found that can be used globally for cancer immunotherapy. However, at this point in time, shared mutant neoantigens are extremely rare and even when identified are restricted to a limited number of HLA alleles. Nevertheless, viral antigens can function as a specific type of shared tumor neoantigen and have found use as the targets for ACT. TIL products selected for high numbers of $CD8^+$ T cells specific for HPV can lead to tumor regression in patients with metastatic cervical cancer (Stevanović et al., 2015). Using TIL derived from an excised metastatic anal carcinoma, Draper et al. isolated an HLA-A*02:01-restricted TCR specific for HPV-16 E6, which upon transduction of the TCR genes into peripheral blood T cells enabled the recognition and killing of HPV-16 positive cervical and head and neck cell lines (Draper et al., 2015). This result suggests that adoptive transfer of T cells engineered to recognize HPV-associated malignancies may be an effective off-the-shelf tumor neoantigen-specific treatment.

6.2 Neoantigens as Targets of T cells Activated by Checkpoint Blockade in Humans

Seminal studies done in the Allison laboratory using preclinical models of murine cancers identified the potential of antibody blockade of the immune checkpoint molecule CTLA-4 as a potentially curative treatment strategy by "unleashing" endogenous tumor-specific T cells to destroy cancer cells (Hurwitz, Yu, Leach, & Allison, 1998; Leach et al., 1996; van Elsas et al., 1999). These findings were rapidly translated into the clinic with ipilimumab, a humanized monoclonal antibody against CTLA-4, which became the first immunotherapy agent to improve survival in metastatic melanoma, gaining FDA approval in 2011 (Hodi et al., 2010). The success of ipilimumab spurred the development of monoclonal antibodies against other immunoinhibitory molecules. Two monoclonal antibodies that block PD-1, nivolumab and pembrolizumab, have been FDA approved for use in metastatic melanoma and NSCLC, and nivolumab also has an indication for renal cell carcinoma (Borghaei et al., 2015; Garon et al., 2015). In addition, antibodies targeting both PD-1 and PD-L1 are currently being evaluated in clinical trials for a wide range of other malignancies. Early phase clinical trials also suggest that checkpoint blockade is efficacious in other solid and hematopoietic malignancies, including bladder, stomach, head and neck carcinoma, and Hodgkin lymphoma (Sharma & Allison, 2015).

The presence of $CD8^+$ T cells at the invasive margin prior to anti-PD-1 therapy is a predictor of response, raising the question of the nature of the tumor antigens recognized by this cell subset (Tumeh et al., 2014). The role of TAA as the targets of the therapeutic immune response induced by ipilimumab was investigated by Kvistborg et al. who screened a cohort of 40 melanoma patients treated with ipilimumab for the presence of $CD8^+$ T cells specific for a panel of 145 HLA-A*02 restricted shared antigens (Kvistborg et al., 2014). As a control, the frequency of $CD8^+$ T cells specific for cytomegalovirus (CMV), influenza, or Epstein-Barr virus (EBV) viral epitopes was also assessed. The latter remained stable before and after therapy. In contrast, the number of melanoma-specific responses increased in many patients, with as many as six new specificities recognized. However, the magnitude of preexisting melanoma-specific responses that was detectable before therapy did not change after ipilimumab therapy. Thus it was not possible to correlate changes in the magnitude of preexisting antitumor responses to that of clinical responses.

In contrast to these results, a positive correlation has been observed in T cell responses to tumor neoantigens in patients who respond to checkpoint blockade. In the first case report to describe the identification of neoantigen-specific CTL in a patient after ipilimumab therapy, van Rooij et al. used exome sequencing, RNA sequencing (RNASeq), and epitope prediction algorithms together with a screening approach utilizing MHC I tetramers to identify T cells specific for a mutant form of ataxia telangiectasia and Rad3-related gene (ATR) from a culture of TIL in a patient with melanoma (van Rooij et al., 2013). Although mutant antigen-specific CTL could be identified at low frequencies more than a year preceding ipilimumab therapy, their frequency increased more than fivefold within 5 wks after treatment, coinciding with reduction in tumor burden. Similar findings were seen in a patient with NSCLC who had a prolonged response to pembrolizumab, with neoantigen-specific $CD8^+$ T cells identifiable in the peripheral blood within 3 wks of treatment initiation and reaching their zenith within 6 wks, before decreasing (Rizvi et al., 2015). Again, the increase in antigen-specific effector cells correlated with a response in overall tumor burden as measured radiographically. These neoantigen-specific CTL were characterized as displaying a polyfunctional phenotype expressing IFN-γ, TNF-α, the degranulation marker CD107a, and the chemokine CCL4 after stimulation with mutant, but not the wild-type peptide. Both of these studies indicate that blockade of both CTLA-4 and PD-1 induces the proliferation of $CD8^+$ T cells specific for neoantigens.

6.3 The Use of Neoantigen Cancer Vaccines in Humans

The introduction of vaccines against common infectious diseases is the crowning achievement of the field of immunology. Unfortunately, thus far, this success has not translated into therapeutic benefits for patients with cancer. It has proven difficult to raise robust immune responses capable of overcoming the inhibitory environment present in tumors that have escaped immune control. For successful immune responses to develop in tumor-bearing individuals, a series of complex molecular and cellular events must take place in a highly coordinated fashion (Chen & Mellman, 2013). Tumor antigens must be taken up, processed, and presented by APC that also must express the appropriate costimulatory signals. Responding T cells must express TCRs specific for tumor antigen/MHC and must come into contact with a suitably activated APC in appropriate topographical locations within the host. T cells that receive the appropriate signals must then expand in sufficient numbers to destroy progressively growing tumor cells despite the myriad of inhibitory molecules and barriers that are present in a developing tumor. A separate point to note is that peripheral T cells specific for tumor-associated self-peptides have been subjected to negative thymic selection, and their TCRs often exhibit lower mean binding affinities than typical foreign antigens (Stone, Harris, & Kranz, 2015). Therefore the affinities of TCRs for self-peptide MHC tumor-associated epitopes may be too low for optimal $CD4^+$ and $CD8^+$ T cell effector function.

Numerous platforms geared at inducing immune responses to vaccination with TAA have been developed, including those that employ whole tumor cells, peptides together with adjuvants, DNA and RNA constructs, or cellular vaccines using APCs pulsed with tumor antigens among others. As a whole, clinical trials using TAAs as targets have been disappointing, and this topic has been reviewed extensively elsewhere (Rosenberg et al., 2004). This is not to say that vaccines targeting TAAs have not been thoroughly investigated in clinical trials. In one of the largest studies of this kind, although a vaccine targeting MAGE-A3 demonstrated a trend toward a clinical benefit preventing relapse after resection of early stage non-small cell lung cancer NSCLC in a phase II trial (Vansteenkiste et al., 2013), a subsequent phase III study with a goal enrollment of more than 2000 patients using this platform was recently terminated due to lack of efficacy during a midtrial preplanned data analysis. It is instructive that only one cellular vaccine has gained FDA approval, a preparation consisting of autologous peripheral blood mononuclear cells exposed ex vivo to a recombinant fusion protein of prostatic acid phosphatase and granulocyte-macrophage colony-stimulating factor (GM-CSF; sipuleucel-T). In a randomized phase III study, an overall survival benefit of 4.1 months was seen among patients with castration-resistant prostate cancer, despite no change in the time to disease progression (Kantoff et al., 2010). Due to this limited benefit, the need for leukapheresis, and high cost, adoption of sipuleucel-T therapy has been limited.

Given the limitations of TAA vaccines and clear preclinical and clinical data suggesting that endogenous antitumor responses can be targeted to TSA, considerable interest has been stoked in developing personalized, tumor-specific vaccination approaches. As previously mentioned, we demonstrated in our preclinical MCA sarcoma model that vaccination against TSA could be as therapeutically effective as checkpoint blockade immunotherapy. These results have been generalized to other mouse tumor models by other groups. But what about the use of neoantigen vaccines in human cancer patients?

Given the difficulty in the past with the identification of truly private tumor-specific mutant antigens, the first trial of a personalized neoantigen vaccine targeted shared mutations in oncogenic drivers. In a 1995 study, five patients with pancreatic adenocarcinoma were vaccinated with autologous DC pulsed with an RAS peptide containing their tumor-specific mutation (Gjertsen et al., 1995). Two of the five patients mounted a transient proliferative response against the vaccine, although in one patient a response against wild-type RAS was also detectable, but both patients eventually succumbed to their disease. Two later studies reported vaccine responses in 58% and 85% of patients, with the later study reporting 20% survival at 10 years in a cohort of patients vaccinated after complete resection of their pancreatic cancer, compared to no long-term survivals in a matched cohort treated without vaccine (Wedén et al., 2011). Given their incidence, approaches to target viral antigens associated with malignant transformation have reached clinical trials. SLP vaccines consisting of HPV E6 and E7 were effective at inducing an immune response that caused regression of vaginal intraepithelial neoplasia (Kenter et al., 2009) and similar vaccines are under study for more advanced cervical cancers.

The first study to use a genomics approach for neoantigen prediction and vaccination to generate neoantigen-specific $CD8^+$ T cell responses has been recently reported in three patients with previously resected melanoma. Carreno et al. used exome sequencing and in silico prediction to identify missense mutations that formed HLA-A*02:01 binding peptides which were confirmed biochemically (Carreno et al., 2015). Autologous DCs were pulsed with seven separate potential patient-specific neoantigens together with two known gp100 shared epitopes and infused into each patient. Each patient had evidence of a $CD8^+$ T cell response against one neoantigen, which could be identified in the peripheral blood prior to vaccination, and after vaccination each patient showed immune responses against two additional vaccine epitopes. At the time of this report, two of the patients had stable disease, with the remaining patient having no evidence of recurrence, without any evidence of autoimmune effects. This study suggests that vaccination with a panel of neoantigens can induce or enhance immune responses against tumor neoantigens. It is not possible at this point to determine whether the vaccine induced a de novo novel immune response against novel tumor antigens or boosted preexisting but undetectable immune responses to tumor antigens. It also remains unclear whether the antivaccine responses that appear after vaccination are capable of reacting with tumor cells since the analysis of T cell specificities in peripheral blood was performed using the vaccinating peptide and not tumor. Nevertheless, this study certainly indicates that a personalized cancer neoantigen vaccine approach is feasible, and multiple clinical trials have either begun enrollment or are planned at multiple institutions using different vaccine approaches to target neoantigens.

7. CONCLUDING REMARKS

Two parallel lines of investigation, one focused on the identification of endogenous immune responses to cancer, and the other on defining antigens that serve as therapeutically useful targets for immunotherapies, have both led to the same conclusion that tumor-specific neoantigens are ideal targets for immunotherapy. Where does the field go from here? The currently available bioinformatics approaches to identify neoantigens are clearly successful, but it is now apparent that the majority of in silico predictions do not induce tumor-reactive T cell responses which has necessitated the development of complex screening methods for their validation. Much work remains to be done in order to develop epitope prediction pipelines that are capable of predicting the TSAs that are recognized by endogenous and therapeutically induced immune responses to cancer with a high level of accuracy, which will be a necessary step forward before widespread translation of vaccine approaches into the clinic can occur.

Given the large numbers of identified TAA and evidence that immune responses against shared antigens can have therapeutic benefit, it would be unwise to dismiss this class of antigens entirely in the future design of cancer immunotherapies. More work needs to be done to directly compare situations where immune responses can be directed against TSA and TAA simultaneously in both preclinical models and clinically in order to define which type of antigen, if any, is more efficacious for a given situation. Another key point is that it is likely that combinations of vaccine, ACT, and/or checkpoint blockade approaches will be the most effective way to focus the immune system to eliminate cancer. Clinical trials attempting to induce new immune responses against neoantigens with vaccines while simultaneously modulating the tumor microenvironment with checkpoint blockade to foster the development of a robust neoantigen-specific immune response are already being planned at multiple cancer centers. In addition, combining neoantigen-based therapies with standard of care therapies (eg, chemotherapy or radiotherapy) may also find therapeutic usefulness especially in the case of tumors with low mutational loads where standard-of-care therapies could give rise to additional mutations that could be targeted by cancer-specific vaccines.

Decades of careful studies in preclinical model systems and clinical investigation has led the field of cancer immunology to the day where nearly all patients with melanoma and NSCLC will receive some form of immunotherapy during their course of treatment. The identification of neoantigens as the optimal targets of cancer immunotherapy promises to enter clinical practice to guide the diagnosis, prognosis, and treatment options for patients with a wide range of tumor histologies. We now sit on the brink of the introduction of standard of care immunotherapy approaches for nearly all patients with cancer.

ACKNOWLEDGMENTS

We are grateful to T. Noguchi for constructive criticism and comments. R.D.S. receives research support from the National Cancer Institute (RO1 CA043059, RO1 CA190700, U01 CA141541), the Cancer Research Institute, the WWWW Foundation, the Siteman Cancer Center/Barnes-Jewish Hospital (Cancer Frontier Fund), Bristol-Myers Squibb Inc., Janssen and Stand Up to Cancer. J.P.W. is supported by a T32 training grant in hematology (5T32HL007088-40) from the National Heart, Lung, and Blood Institute. M.M.G. is supported by a postdoctoral training grant (Irvington Postdoctoral Fellowship) from the Cancer Research Institute.

REFERENCES

Abelin, J. G., Trantham, P. D., Penny, S. A., Patterson, A. M., Ward, S. T., Hildebrand, W. H., … Hunt, D. F. (2015). Complementary IMAC enrichment methods for HLA-associated phosphopeptide identification by mass spectrometry. *Nature Protocols*, *10*(9), 1308–1318. http://dx.doi.org/10.1038/nprot.2015.086.

Alexandrov, L. B., Nik-Zainal, S., Wedge, D. C., Aparicio, S. A. J. R., Behjati, S., Biankin, A. V., … Stratton, M. R. (2013). Signatures of mutational processes in human cancer. *Nature*, *500*(7463), 415–421. http://dx.doi.org/10.1038/nature12477.

Altman, J. D., Moss, P. A., Goulder, P. J., Barouch, D. H., McHeyzer-Williams, M. G., Bell, J. I., … Davis, M. M. (1996). Phenotypic analysis of antigen-specific T lymphocytes. *Science*, *274*(5284), 94–96.

Babbitt, B. P., Allen, P. M., Matsueda, G., Haber, E., & Unanue, E. R. (1985). Binding of immunogenic peptides to Ia histocompatibility molecules. *Nature, 317*(6035), 359–361.

Bakker, A. B., Schreurs, M. W., de Boer, A. J., Kawakami, Y., Rosenberg, S. A., Adema, G. J., & Figdor, C. G. (1994). Melanocyte lineage-specific antigen gp100 is recognized by melanoma-derived tumor-infiltrating lymphocytes. *Journal of Experimental Medicine, 179*(3), 1005–1009.

Barber, D. L., Wherry, E. J., Masopust, D., Zhu, B., Allison, J. P., Sharpe, A. H., … Ahmed, R. (2006). Restoring function in exhausted CD8 T cells during chronic viral infection. *Nature, 439*(7077), 682–687. http://dx.doi.org/10.1038/nature04444.

Birkeland, S. A., Storm, H. H., Lamm, L. U., Barlow, L., Blohme, I., Forsberg, B., et al. (1995). Cancer risk after renal transplantation in the Nordic countries, 1964–1986. *International Journal of Cancer, 60*(2), 183–189.

Bjorkman, P. J., Saper, M. A., Samraoui, B., Bennett, W. S., Strominger, J. L., & Wiley, D. C. (1987). Structure of the human class I histocompatibility antigen, HLA-A2. *Nature, 329*(6139), 506–512. http://dx.doi.org/10.1038/329506a0.

Blum, J. S., Wearsch, P. A., & Cresswell, P. (2013). Pathways of antigen processing. *Annual Review of Immunology, 31*, 443–473. http://dx.doi.org/10.1146/annurev-immunol-032712-095910.

Borghaei, H., Paz-Ares, L., Horn, L., Spigel, D. R., Steins, M., Ready, N. E., … Brahmer, J. R. (2015). Nivolumab versus docetaxel in advanced nonsquamous non-small-cell lung cancer. *The New England Journal of Medicine, 373*(17), 1627–1639. http://dx.doi.org/10.1056/NEJMoa1507643.

Braumuller, H., Wieder, T., Brenner, E., Assmann, S., Hahn, M., Alkhaled, M., … Rocken, M. (2013). T-helper-1-cell cytokines drive cancer into senescence. *Nature, 494*(7437), 361–365. http://dx.doi.org/10.1038/nature11824.

Brown, S. D., Warren, R. L., Gibb, E. A., Martin, S. D., Spinelli, J. J., Nelson, B. H., & Holt, R. A. (2014). Neo-antigens predicted by tumor genome meta-analysis correlate with increased patient survival. *Genome Research, 24*(5), 743–750. http://dx.doi.org/10.1101/gr.165985.113.

Burnet, M. (1957). Cancer: A biological approach. III. Viruses associated with neoplastic conditions. IV. Practical applications. *British Medical Journal, 1*(5023), 841–847.

Burnet, F. M. (1970). The concept of immunological surveillance. *Progress in Experimental Tumor Research, 13*, 1–27.

Calis, J. J., Maybeno, M., Greenbaum, J. A., Weiskopf, D., De Silva, A. D., Sette, A., … Peters, B. (2013). Properties of MHC class I presented peptides that enhance immunogenicity. *PLoS Computational Biology, 9*(10). http://dx.doi.org/10.1371/journal.pcbi.1003266.

Cameron, B. J., Gerry, A. B., Dukes, J., Harper, J. V., Kannan, V., Bianchi, F. C., … Jakobsen, B. K. (2013). Identification of a Titin-derived HLA-A1-presented peptide as a cross-reactive target for engineered MAGE A3-directed T cells. *Science Translational Medicine, 5*(197), 197ra103. http://dx.doi.org/10.1126/scitranslmed.3006034.

Carreno, B. M., Magrini, V., Becker-Hapak, M., Kaabinejadian, S., Hundal, J., Petti, A. A., … Linette, G. P. (2015). A dendritic cell vaccine increases the breadth and diversity of melanoma neoantigen-specific T cells. *Science*. http://dx.doi.org/10.1126/science.aaa3828.

Castle, J. C., Kreiter, S., Diekmann, J., Lower, M., van de Roemer, N., de Graaf, J., … Sahin, U. (2012). Exploiting the mutanome for tumor vaccination. *Cancer Research, 72*(5), 1081–1091. http://dx.doi.org/10.1158/0008-5472.CAN-11-3722.

Chandran, S. S., Paria, B. C., Srivastava, A. K., Rothermel, L. D., Stephens, D. J., Dudley, M. E., … Kammula, U. S. (2015). Persistence of CTL clones targeting melanocyte differentiation antigens was insufficient to mediate significant melanoma regression

in humans. *Clinical Cancer Research, 21*(3), 534–543. http://dx.doi.org/10.1158/1078-0432.CCR-14-2208.
Chauvin, J.-M., Pagliano, O., Fourcade, J., Sun, Z., Wang, H., Sander, C., ... Zarour, H. M. (2015). TIGIT and PD-1 impair tumor antigen-specific CD8+ T cells in melanoma patients. *The Journal of Clinical Investigation, 125*(5), 2046–2058. http://dx.doi.org/10.1172/JCI80445.
Chen, D. S., & Mellman, I. (2013). Oncology meets immunology: The cancer-immunity cycle. *Immunity, 39*(1), 1–10. http://dx.doi.org/10.1016/j.immuni.2013.07.012.
Chen, Y. T., Scanlan, M. J., Sahin, U., Tureci, O., Gure, A. O., Tsang, S., ... Old, L. J. (1997). A testicular antigen aberrantly expressed in human cancers detected by autologous antibody screening. *Proceedings of the National Academy of Sciences of the United States of America, 94*(5), 1914–1918.
Cobbold, M., De La Pena, H., Norris, A., Polefrone, J. M., Qian, J., English, A. M., ... Engelhard, V. H. (2013). MHC class I-associated phosphopeptides are the targets of memory-like immunity in leukemia. *Science Translational Medicine, 5*(203), 203ra125. http://dx.doi.org/10.1126/scitranslmed.3006061.
Coulie, P. G., Lehmann, F., Lethe, B., Herman, J., Lurquin, C., Andrawiss, M., & Boon, T. (1995). A mutated intron sequence codes for an antigenic peptide recognized by cytolytic T lymphocytes on a human melanoma. *Proceedings of the National Academy of Sciences of the United States of America, 92*(17), 7976–7980.
Coulie, P. G., Van den Eynde, B. J., van der Bruggen, P., & Boon, T. (2014). Tumour antigens recognized by T lymphocytes: At the core of cancer immunotherapy. *Nature Reviews Cancer, 14*(2), 135–146. http://dx.doi.org/10.1038/nrc3670.
Cox, A. L., Skipper, J., Chen, Y., Henderson, R. A., Darrow, T. L., Shabanowitz, J., ... Slingluff, C. L., Jr. (1994). Identification of a peptide recognized by five melanoma-specific human cytotoxic T cell lines. *Science, 264*(5159), 716–719.
De Backer, O., Verheyden, A. M., Martin, B., Godelaine, D., De Plaen, E., Brasseur, R., ... Boon, T. (1995). Structure, chromosomal location, and expression pattern of three mouse genes homologous to the human MAGE genes. *Genomics, 28*(1), 74–83. http://dx.doi.org/10.1006/geno.1995.1108.
De Plaen, E., Lurquin, C., Van Pel, A., Mariame, B., Szikora, J. P., Wolfel, T., ... Boon, T. (1988). Immunogenic (tum-) variants of mouse tumor P815: Cloning of the gene of tum- antigen P91A and identification of the tum- mutation. *Proceedings of the National Academy of Sciences of the United States of America, 85*(7), 2274–2278.
Diamond, M. S., Kinder, M., Matsushita, H., Mashayekhi, M., Dunn, G. P., Archambault, J. M., ... Schreiber, R. D. (2011). Type I interferon is selectively required by dendritic cells for immune rejection of tumors. *The Journal of Experimental Medicine, 208*(10), 1989–2003. http://dx.doi.org/10.1084/jem.20101158.
Diefenbach, A., Jensen, E. R., Jamieson, A. M., & Raulet, D. H. (2001). Rae1 and H60 ligands of the NKG2D receptor stimulate tumour immunity. *Nature, 413*(6852), 165–171. http://dx.doi.org/10.1038/35093109.
Dingjan, T., Spendlove, I., Durrant, L. G., Scott, A. M., Yuriev, E., & Ramsland, P. A. (2015). Structural biology of antibody recognition of carbohydrate epitopes and potential uses for targeted cancer immunotherapies. *Molecular Immunology, 67*(2 Pt. A), 75–88. http://dx.doi.org/10.1016/j.molimm.2015.02.028.
Dong, H., Strome, S. E., Salomao, D. R., Tamura, H., Hirano, F., Flies, D. B., ... Chen, L. (2002). Tumor-associated B7-H1 promotes T-cell apoptosis: A potential mechanism of immune evasion. *Nature Medicine, 8*(8), 793–800. http://dx.doi.org/10.1038/nm730.
Dong, H., Zhu, G., Tamada, K., & Chen, L. (1999). B7-H1, a third member of the B7 family, co-stimulates T-cell proliferation and interleukin-10 secretion. *Nature Medicine, 5*(12), 1365–1369. http://dx.doi.org/10.1038/70932.
Draper, L. M., Kwong, M. L. M., Gros, A., Stevanović, S., Tran, E., Kerkar, S., ... Hinrichs, C. S. (2015). Targeting of HPV-16+ epithelial cancer cells by TCR gene

engineered T cells directed against E6. *Clinical Cancer Research, 21*(19), 4431–4439. http://dx.doi.org/10.1158/1078-0432.CCR-14-3341.

Dubey, P., Hendrickson, R. C., Meredith, S. C., Siegel, C. T., Shabanowitz, J., Skipper, J. C., … Schreiber, H. (1997). The immunodominant antigen of an ultraviolet-induced regressor tumor is generated by a somatic point mutation in the DEAD box helicase p68. *The Journal of Experimental Medicine, 185*(4), 695–705.

Dunn, G. P., Bruce, A. T., Ikeda, H., Old, L. J., & Schreiber, R. D. (2002). Cancer immunoediting: From immunosurveillance to tumor escape. *Nature Immunology, 3*(11), 991–998. http://dx.doi.org/10.1038/ni1102-991.

Dunn, G. P., Old, L. J., & Schreiber, R. D. (2004a). The immunobiology of cancer immunosurveillance and immunoediting. *Immunity, 21*(2), 137–148. http://dx.doi.org/10.1016/j.immuni.2004.07.017.

Dunn, G. P., Old, L. J., & Schreiber, R. D. (2004b). The three Es of cancer immunoediting. *Annual Review of Immunology, 22*, 329–360. http://dx.doi.org/10.1146/annurev.immunol.22.012703.104803.

DuPage, M., Mazumdar, C., Schmidt, L. M., Cheung, A. F., & Jacks, T. (2012). Expression of tumour-specific antigens underlies cancer immunoediting. *Nature, 482*(7385), 405–409. http://dx.doi.org/10.1038/nature10803.

Ehrlich, P. (1909). Ueber den jetzigen stand der Karzinomforschung. *Nederlands Tijdschrift voor Geneeskunde, 5*, 273–290.

Fisk, B., Blevins, T. L., Wharton, J. T., & Ioannides, C. G. (1995). Identification of an immunodominant peptide of HER-2/neu protooncogene recognized by ovarian tumor-specific cytotoxic T lymphocyte lines. *The Journal of Experimental Medicine, 181*(6), 2109–2117.

Foley, E. J. (1953). Antigenic properties of methylcholanthrene-induced tumors in mice of the strain of origin. *Cancer Research, 13*(12), 835–837.

Freeman, G. J., Long, A. J., Iwai, Y., Bourque, K., Chernova, T., Nishimura, H., … Honjo, T. (2000). Engagement of the PD-1 immunoinhibitory receptor by a novel B7 family member leads to negative regulation of lymphocyte activation. *The Journal of Experimental Medicine, 192*(7), 1027–1034.

Fritsch, E. F., Rajasagi, M., Ott, P. A., Brusic, V., Hacohen, N., & Wu, C. J. (2014). HLA-binding properties of tumor neoepitopes in humans. *Cancer Immunology Research, 2*(6), 522–529. http://dx.doi.org/10.1158/2326-6066.CIR-13-0227.

Fuertes, M. B., Kacha, A. K., Kline, J., Woo, S. R., Kranz, D. M., Murphy, K. M., & Gajewski, T. F. (2011). Host type I IFN signals are required for antitumor CD8+ T cell responses through CD8{alpha}+ dendritic cells. *The Journal of Experimental Medicine, 208*(10), 2005–2016. http://dx.doi.org/10.1084/jem.20101159.

Galon, J., Costes, A., Sanchez-Cabo, F., Kirilovsky, A., Mlecnik, B., Lagorce-Pages, C., … Pages, F. (2006). Type, density, and location of immune cells within human colorectal tumors predict clinical outcome. *Science, 313*(5795), 1960–1964. http://dx.doi.org/10.1126/science.1129139.

Gangadhar, T. C., & Vonderheide, R. H. (2014). Mitigating the toxic effects of anticancer immunotherapy. *Nature Reviews. Clinical Oncology, 11*(2), 91–99. http://dx.doi.org/10.1038/nrclinonc.2013.245.

Garon, E. B., Rizvi, N. A., Hui, R., Leighl, N., Balmanoukian, A. S., Eder, J. P., … Investigators, K.-. (2015). Pembrolizumab for the treatment of non-small-cell lung cancer. *The New England Journal of Medicine, 372*(21), 2018–2028. http://dx.doi.org/10.1056/NEJMoa1501824.

Giguere, S., Drouin, A., Lacoste, A., Marchand, M., Corbeil, J., & Laviolette, F. (2013). MHC-NP: Predicting peptides naturally processed by the MHC. *Journal of Immunological Methods, 400–401*, 30–36. http://dx.doi.org/10.1016/j.jim.2013.10.003.

Gillis, S., & Smith, K. A. (1977). Long term culture of tumour-specific cytotoxic T cells. *Nature, 268*(5616), 154–156.

Gjertsen, M. K., Bakka, A., Breivik, J., Saeterdal, I., Solheim, B. G., Søreide, O., … Gaudernack, G. (1995). Vaccination with mutant ras peptides and induction of T-cell responsiveness in pancreatic carcinoma patients carrying the corresponding RAS mutation. *Lancet*, *346*(8987), 1399–1400.

Greco, G., Fruci, D., Accapezzato, D., Barnaba, V., Nisini, R., Alimena, G., … Tosi, R. (1996). Two brc-abl junction peptides bind HLA-A3 molecules and allow specific induction of human cytotoxic T lymphocytes. *Leukemia*, *10*(4), 693–699.

Grivennikov, S. I., Greten, F. R., & Karin, M. (2010). Immunity, inflammation, and cancer. *Cell*, *140*(6), 883–899. http://dx.doi.org/10.1016/j.cell.2010.01.025.

Gross, L. (1943). Intradermal immunization of C3H mice against a sarcoma that originated in an animal of the same line. *Cancer Research*, *3*(5), 326–333.

Gubin, M. M., Artyomov, M. N., Mardis, E. R., & Schreiber, R. D. (2015). Tumor neoantigens: Building a framework for personalized cancer immunotherapy. *The Journal of Clinical Investigation*, *125*(9), 3413–3421. http://dx.doi.org/10.1172/JCI80008.

Gubin, M. M., Zhang, X., Schuster, H., Caron, E., Ward, J. P., Noguchi, T., … Schreiber, R. D. (2014). Checkpoint blockade cancer immunotherapy targets tumour-specific mutant antigens. *Nature*, *515*(7528), 577–581. http://dx.doi.org/10.1038/nature13988.

Hadrup, S. R., Bakker, A. H., Shu, C. J., Andersen, R. S., van Veluw, J., Hombrink, P., … Schumacher, T. N. (2009). Parallel detection of antigen-specific T-cell responses by multidimensional encoding of MHC multimers. *Nature Methods*, *6*(7), 520–526. http://dx.doi.org/10.1038/nmeth.1345.

Hammer, J., Bono, E., Gallazzi, F., Belunis, C., Nagy, Z., & Sinigaglia, F. (1994). Precise prediction of major histocompatibility complex class II-peptide interaction based on peptide side chain scanning. *The Journal of Experimental Medicine*, *180*(6), 2353–2358.

Hanahan, D., & Weinberg, R. A. (2011). Hallmarks of cancer: The next generation. *Cell*, *144*(5), 646–674. http://dx.doi.org/10.1016/j.cell.2011.02.013.

Heemskerk, B., Kvistborg, P., & Schumacher, T. N. (2013). The cancer antigenome. *The EMBO Journal*, *32*(2), 194–203. http://dx.doi.org/10.1038/emboj.2012.333.

Heidelberger, C. (1975). Chemical carcinogenesis. *Annual Review of Biochemistry*, *44*, 79–121. http://dx.doi.org/10.1146/annurev.bi.44.070175.000455.

Herberman, R. B., & Holden, H. T. (1978). Natural cell-mediated immunity. *Advances in Cancer Research*, *27*, 305–377.

Herin, M., Lemoine, C., Weynants, P., Vessiere, F., Van Pel, A., Knuth, A., … Boon, T. (1987). Production of stable cytolytic T-cell clones directed against autologous human melanoma. *International Journal of Cancer*, *39*(3), 390–396.

Hodi, F. S., Mihm, M. C., Soiffer, R. J., Haluska, F. G., Butler, M., Seiden, M. V., … Dranoff, G. (2003). Biologic activity of cytotoxic T lymphocyte-associated antigen 4 antibody blockade in previously vaccinated metastatic melanoma and ovarian carcinoma patients. *Proceedings of the National Academy of Sciences of the United States of America*, *100*(8), 4712–4717. http://dx.doi.org/10.1073/pnas.0830997100.

Hodi, F. S., O'Day, S. J., McDermott, D. F., Weber, R. W., Sosman, J. A., Haanen, J. B., … Urba, W. J. (2010). Improved survival with ipilimumab in patients with metastatic melanoma. *The New England Journal of Medicine*, *363*(8), 711–723. http://dx.doi.org/10.1056/NEJMoa1003466.

Hunig, T. (1983). T-cell function and specificity in athymic mice. *Immunology Today*, *4*(3), 84–87. http://dx.doi.org/10.1016/0167-5699(83)90125-1.

Hurwitz, A. A., Yu, T. F., Leach, D. R., & Allison, J. P. (1998). CTLA-4 blockade synergizes with tumor-derived granulocyte-macrophage colony-stimulating factor for treatment of an experimental mammary carcinoma. *Proceedings of the National Academy of Sciences of the United States of America*, *95*(17), 10067–10071.

Ikehara, S., Pahwa, R. N., Fernandes, G., Hansen, C. T., & Good, R. A. (1984). Functional T cells in athymic nude mice. *Proceedings of the National Academy of Sciences of the United States of America*, *81*(3), 886–888.

Iwai, Y., Ishida, M., Tanaka, Y., Okazaki, T., Honjo, T., & Minato, N. (2002). Involvement of PD-L1 on tumor cells in the escape from host immune system and tumor immunotherapy by PD-L1 blockade. *Proceedings of the National Academy of Sciences of the United States of America*, *99*(19), 12293–12297. http://dx.doi.org/10.1073/pnas.192461099.

Jin, D., Fan, J., Wang, L., Thompson, L. F., Liu, A., Daniel, B. J., … Zhang, B. (2010). CD73 on tumor cells impairs antitumor T-cell responses: A novel mechanism of tumor-induced immune suppression. *Cancer Research*, *70*(6), 2245–2255. http://dx.doi.org/10.1158/0008-5472.CAN-09-3109.

Johnson, L. A., Morgan, R. A., Dudley, M. E., Cassard, L., Yang, J. C., Hughes, M. S., … Rosenberg, S. A. (2009). Gene therapy with human and mouse T-cell receptors mediates cancer regression and targets normal tissues expressing cognate antigen. *Blood*, *114*(3), 535–546. http://dx.doi.org/10.1182/blood-2009-03-211714.

Kantoff, P. W., Higano, C. S., Shore, N. D., Berger, E. R., Small, E. J., Penson, D. F., … Investigators, I. S. (2010). Sipuleucel-T immunotherapy for castration-resistant prostate cancer. *The New England Journal of Medicine*, *363*(5), 411–422. http://dx.doi.org/10.1056/NEJMoa1001294.

Kaplan, D. H., Shankaran, V., Dighe, A. S., Stockert, E., Aguet, M., Old, L. J., & Schreiber, R. D. (1998). Demonstration of an interferon gamma-dependent tumor surveillance system in immunocompetent mice. *Proceedings of the National Academy of Sciences of the United States of America*, *95*(13), 7556–7561.

Kawakami, Y., Eliyahu, S., Delgado, C. H., Robbins, P. F., Rivoltini, L., Topalian, S. L., … Rosenberg, S. A. (1994). Cloning of the gene coding for a shared human melanoma antigen recognized by autologous T cells infiltrating into tumor. *Proceedings of the National Academy of Sciences of the United States of America*, *91*(9), 3515–3519.

Kenter, G. G., Welters, M. J. P., Valentijn, A. R. P. M., Lowik, M. J. G., Berends-van der Meer, D. M. A., Vloon, A. P. G., … Melief, C. J. M. (2009). Vaccination against HPV-16 oncoproteins for vulvar intraepithelial neoplasia. *The New England Journal of Medicine*, *361*(19), 1838–1847. http://dx.doi.org/10.1056/NEJMoa0810097.

Kim, Y., Sidney, J., Pinilla, C., Sette, A., & Peters, B. (2009). Derivation of an amino acid similarity matrix for peptide: MHC binding and its application as a Bayesian prior. *BMC Bioinformatics*, *10*, 394. http://dx.doi.org/10.1186/1471-2105-10-394.

Koboldt, D. C., Steinberg, K. M., Larson, D. E., Wilson, R. K., & Mardis, E. R. (2013). The next-generation sequencing revolution and its impact on genomics. *Cell*, *155*(1), 27–38. http://dx.doi.org/10.1016/j.cell.2013.09.006.

Koebel, C. M., Vermi, W., Swann, J. B., Zerafa, N., Rodig, S. J., Old, L. J., … Schreiber, R. D. (2007). Adaptive immunity maintains occult cancer in an equilibrium state. *Nature*, *450*(7171), 903–907. http://dx.doi.org/10.1038/nature06309.

Kreiter, S., Vormehr, M., van de Roemer, N., Diken, M., Lower, M., Diekmann, J., … Sahin, U. (2015). Mutant MHC class II epitopes drive therapeutic immune responses to cancer. *Nature*. http://dx.doi.org/10.1038/nature14426.

Kvistborg, P., Philips, D., Kelderman, S., Hageman, L., Ottensmeier, C., Joseph-Pietras, D., … Schumacher, T. N. (2014). Anti-CTLA-4 therapy broadens the melanoma-reactive CD8+ T cell response. *Science Translational Medicine*, *6*(254), 254ra128. http://dx.doi.org/10.1126/scitranslmed.3008918.

Kvistborg, P., Shu, C. J., Heemskerk, B., Fankhauser, M., Thrue, C. A., Toebes, M., … Schumacher, T. N. M. (2012). TIL therapy broadens the tumor-reactive CD8(+) T cell compartment in melanoma patients. *OncoImmunology*, *1*(4), 409–418.

Latchman, Y., Wood, C. R., Chernova, T., Chaudhary, D., Borde, M., Chernova, I., ... Freeman, G. J. (2001). PD-L2 is a second ligand for PD-1 and inhibits T cell activation. *Nature Immunology, 2*(3), 261–268. http://dx.doi.org/10.1038/85330.

Le, D. T., Uram, J. N., Wang, H., Bartlett, B. R., Kemberling, H., Eyring, A. D., ... Diaz, L. A., Jr. (2015). PD-1 blockade in tumors with mismatch-repair deficiency. *The New England Journal of Medicine, 372*(26), 2509–2520. http://dx.doi.org/10.1056/NEJMoa1500596.

Leach, D. R., Krummel, M. F., & Allison, J. P. (1996). Enhancement of antitumor immunity by CTLA-4 blockade. *Science, 271*(5256), 1734–1736.

Lennerz, V., Fatho, M., Gentilini, C., Frye, R. A., Lifke, A., Ferel, D., ... Wolfel, T. (2005). The response of autologous T cells to a human melanoma is dominated by mutated neoantigens. *Proceedings of the National Academy of Sciences of the United States of America, 102*(44), 16013–16018. http://dx.doi.org/10.1073/pnas.0500090102.

Ley, T. J., Mardis, E. R., Ding, L., Fulton, B., McLellan, M. D., Chen, K., ... Wilson, R. K. (2008). DNA sequencing of a cytogenetically normal acute myeloid leukaemia genome. *Nature, 456*(7218), 66–72. http://dx.doi.org/10.1038/nature07485.

Li, Y., Depontieu, F. R., Sidney, J., Salay, T. M., Engelhard, V. H., Hunt, D. F., ... Mariuzza, R. A. (2010). Structural basis for the presentation of tumor-associated MHC class II-restricted phosphopeptides to CD4+ T cells. *Journal of Molecular Biology, 399*(4), 596–603. http://dx.doi.org/10.1016/j.jmb.2010.04.037.

Linard, B., Bézieau, S., Benlalam, H., Labarrière, N., Guilloux, Y., Diez, E., & Jotereau, F. (2002). A ras-mutated peptide targeted by CTL infiltrating a human melanoma lesion. *Journal of Immunology (Baltimore, Md: 1950), 168*(9), 4802–4808.

Linette, G. P., Stadtmauer, E. A., Maus, M. V., Rapoport, A. P., Levine, B. L., Emery, L., ... June, C. H. (2013). Cardiovascular toxicity and titin cross-reactivity of affinity-enhanced T cells in myeloma and melanoma. *Blood, 122*(6), 863–871. http://dx.doi.org/10.1182/blood-2013-03-490565.

Linnemann, C., van Buuren, M. M., Bies, L., Verdegaal, E. M., Schotte, R., Calis, J. J., ... Schumacher, T. N. (2015). High-throughput epitope discovery reveals frequent recognition of neo-antigens by CD4+ T cells in human melanoma. *Nature Medicine, 21*(1), 81–85. http://dx.doi.org/10.1038/nm.3773.

Loke, P., & Allison, J. P. (2003). PD-L1 and PD-L2 are differentially regulated by Th1 and Th2 cells. *Proceedings of the National Academy of Sciences of the United States of America, 100*(9), 5336–5341. http://dx.doi.org/10.1073/pnas.0931259100.

Lundegaard, C., Lamberth, K., Harndahl, M., Buus, S., Lund, O., & Nielsen, M. (2008). NetMHC-3.0: Accurate web accessible predictions of human, mouse and monkey MHC class I affinities for peptides of length 8-11. *Nucleic Acids Research, 36*(Web Server issue), W509–W512. http://dx.doi.org/10.1093/nar/gkn202.

Maby, P., Tougeron, D., Hamieh, M., Mlecnik, B., Kora, H., Bindea, G., ... Latouche, J.-B. (2015). Correlation between density of CD8+ T-cell infiltrate in microsatellite unstable colorectal cancers and frameshift mutations: A rationale for personalized immunotherapy. *Cancer Research, 75*(17), 3446–3455. http://dx.doi.org/10.1158/0008-5472.CAN-14-3051.

MacKie, R. M., Reid, R., & Junor, B. (2003). Fatal melanoma transferred in a donated kidney 16 years after melanoma surgery. *The New England Journal of Medicine, 348*(6), 567–568. http://doi.org/10.1056/NEJM200302063480620.

Maleckar, J. R., & Sherman, L. A. (1987). The composition of the T cell receptor repertoire in nude mice. *The Journal of Immunology, 138*(11), 3873–3876.

Mantovani, A., Allavena, P., Sica, A., & Balkwill, F. (2008). Cancer-related inflammation. *Nature, 454*(7203), 436–444. http://dx.doi.org/10.1038/nature07205.

Matsushita, H., Vesely, M. D., Koboldt, D. C., Rickert, C. G., Uppaluri, R., Magrini, V. J., ... Schreiber, R. D. (2012). Cancer exome analysis reveals a T-cell-dependent

mechanism of cancer immunoediting. *Nature*, *482*(7385), 400–404. http://dx.doi.org/10.1038/nature10755.
Mellman, I., Coukos, G., & Dranoff, G. (2011). Cancer immunotherapy comes of age. *Nature*, *480*(7378), 480–489. http://dx.doi.org/10.1038/nature10673.
Mittal, D., Gubin, M. M., Schreiber, R. D., & Smyth, M. J. (2014). New insights into cancer immunoediting and its three component phases—Elimination, equilibrium and escape. *Current Opinion in Immunology*, *27*, 16–25. http://dx.doi.org/10.1016/j.coi.2014.01.004.
Mohammed, F., Cobbold, M., Zarling, A. L., Salim, M., Barrett-Wilt, G. A., Shabanowitz, J., … Willcox, B. E. (2008). Phosphorylation-dependent interaction between antigenic peptides and MHC class I: A molecular basis for the presentation of transformed self. *Nature Immunology*, *9*(11), 1236–1243. http://dx.doi.org/10.1038/ni.1660.
Monach, P. A., Meredith, S. C., Siegel, C. T., & Schreiber, H. (1995). A unique tumor antigen produced by a single amino acid substitution. *Immunity*, *2*(1), 45–59.
Nielsen, M., & Lund, O. (2009). NN-align. An artificial neural network-based alignment algorithm for MHC class II peptide binding prediction. *BMC Bioinformatics*, *10*, 296. http://dx.doi.org/10.1186/1471-2105-10-296.
Nielsen, M., Lundegaard, C., Blicher, T., Lamberth, K., Harndahl, M., Justesen, S., … Buus, S. (2007). NetMHCpan, a method for quantitative predictions of peptide binding to any HLA-A and -B locus protein of known sequence. *PLoS One*, *2*(8). http://dx.doi.org/10.1371/journal.pone.0000796.
Nielsen, M., Lundegaard, C., Lund, O., & Kesmir, C. (2005). The role of the proteasome in generating cytotoxic T-cell epitopes: Insights obtained from improved predictions of proteasomal cleavage. *Immunogenetics*, *57*(1–2), 33–41. http://dx.doi.org/10.1007/s00251-005-0781-7.
Nielsen, M., Lundegaard, C., Worning, P., Lauemoller, S. L., Lamberth, K., Buus, S., … Lund, O. (2003). Reliable prediction of T-cell epitopes using neural networks with novel sequence representations. *Protein Science*, *12*(5), 1007–1017. http://dx.doi.org/10.1110/ps.0239403.
Ohminami, H., Yasukawa, M., & Fujita, S. (2000). HLA class I-restricted lysis of leukemia cells by a CD8(+) cytotoxic T-lymphocyte clone specific for WT1 peptide. *Blood*, *95*(1), 286–293.
Old, L. J. (1982). Cancer immunology: The search for specificity. *National Cancer Institute Monograph*, *60*, 193–209.
Ono, M., & Hakomori, S. (2004). Glycosylation defining cancer cell motility and invasiveness. *Glycoconjugate Journal*, *20*(1), 71–78. http://dx.doi.org/10.1023/B:GLYC.0000018019.22070.7d.
O'Sullivan, T., Saddawi-Konefka, R., Vermi, W., Koebel, C. M., Arthur, C., White, J. M., … Bui, J. D. (2012). Cancer immunoediting by the innate immune system in the absence of adaptive immunity. *The Journal of Experimental Medicine*, *209*(10), 1869–1882. http://dx.doi.org/10.1084/jem.20112738.
Pardoll, D. M. (2012). The blockade of immune checkpoints in cancer immunotherapy. *Nature Reviews. Cancer*, *12*(4), 252–264. http://dx.doi.org/10.1038/nrc3239.
Penn, I. (1999). Posttransplant malignancies. *Transplantation Proceedings*, *31*(1–2), 1260–1262.
Peters, B., Bulik, S., Tampe, R., Van Endert, P. M., & Holzhutter, H. G. (2003). Identifying MHC class I epitopes by predicting the TAP transport efficiency of epitope precursors. *The Journal of Immunology*, *171*(4), 1741–1749.
Peters, B., & Sette, A. (2005). Generating quantitative models describing the sequence specificity of biological processes with the stabilized matrix method. *BMC Bioinformatics*, *6*(1), 132. http://dx.doi.org/10.1186/1471-2105-6-132.
Prehn, R. T., & Main, J. M. (1957). Immunity to methylcholanthrene-induced sarcomas. *Journal of the National Cancer Institute*, *18*(6), 769–778.

Radvanyi, L. G., Bernatchez, C., Zhang, M., Fox, P. S., Miller, P., Chacon, J., … Hwu, P. (2012). Specific lymphocyte subsets predict response to adoptive cell therapy using expanded autologous tumor-infiltrating lymphocytes in metastatic melanoma patients. *Clinical Cancer Research*, *18*(24), 6758–6770. http://dx.doi.org/10.1158/1078-0432.CCR-12-1177.

Rajasagi, M., Shukla, S. A., Fritsch, E. F., Keskin, D. B., DeLuca, D., Carmona, E., … Wu, C. J. (2014). Systematic identification of personal tumor-specific neoantigens in chronic lymphocytic leukemia. *Blood*, *124*(3), 453–462. http://dx.doi.org/10.1182/blood-2014-04-567933.

Rammensee, H., Bachmann, J., Emmerich, N. P., Bachor, O. A., & Stevanovic, S. (1999). SYFPEITHI: Database for MHC ligands and peptide motifs. *Immunogenetics*, *50*(3–4), 213–219.

Restifo, N. P., Marincola, F. M., Kawakami, Y., Taubenberger, J., Yannelli, J. R., & Rosenberg, S. A. (1996). Loss of functional beta 2-microglobulin in metastatic melanomas from five patients receiving immunotherapy. *Journal of the National Cancer Institute*, *88*(2), 100–108.

Rizvi, N. A., Hellmann, M. D., Snyder, A., Kvistborg, P., Makarov, V., Havel, J. J., … Chan, T. A. (2015). Mutational landscape determines sensitivity to PD-1 blockade in non-small cell lung cancer. *Science (New York, N.Y.)*, aaa1348. http://dx.doi.org/10.1126/science.aaa1348.

Robbins, P. F., Lu, Y.-C., El-Gamil, M., Li, Y. F., Gross, C., Gartner, J., … Rosenberg, S. A. (2013). Mining exomic sequencing data to identify mutated antigens recognized by adoptively transferred tumor-reactive T cells. *Nature Medicine*, *19*(6), 747–752. http://dx.doi.org/10.1038/nm.3161.

Roberts, A., Perera, S., Lang, B., Vincent, A., & Newsom-Davis, J. (1985). Paraneoplastic myasthenic syndrome IgG inhibits 45Ca2+ flux in a human small cell carcinoma line. *Nature*, *317*(6039), 737–739.

Rock, K. L., Gramm, C., Rothstein, L., Clark, K., Stein, R., Dick, L., … Goldberg, A. L. (1994). Inhibitors of the proteasome block the degradation of most cell proteins and the generation of peptides presented on MHC class I molecules. *Cell*, *78*(5), 761–771.

Rooney, M. S., Shukla, S. A., Wu, C. J., Getz, G., & Hacohen, N. (2015). Molecular and genetic properties of tumors associated with local immune cytolytic activity. *Cell*, *160*(1–2), 48–61. http://dx.doi.org/10.1016/j.cell.2014.12.033.

Rosenberg, S. A., Packard, B. S., Aebersold, P. M., Solomon, D., Topalian, S. L., Toy, S. T., et al. (1988). Use of tumor-infiltrating lymphocytes and interleukin-2 in the immunotherapy of patients with metastatic melanoma. A preliminary report. *The New England Journal of Medicine*, *319*(25), 1676–1680. http://dx.doi.org/10.1056/NEJM198812223192527.

Rosenberg, S. A., Yang, J. C., & Restifo, N. P. (2004). Cancer immunotherapy: Moving beyond current vaccines. *Nature Medicine*, *10*(9), 909–915. http://dx.doi.org/10.1038/nm1100.

Rosenberg, S. A., Yang, J. C., Sherry, R. M., Kammula, U. S., Hughes, M. S., Phan, G. Q., … Dudley, M. E. (2011). Durable complete responses in heavily pretreated patients with metastatic melanoma using T-cell transfer immunotherapy. *Clinical Cancer Research*, *17*(13), 4550–4557. http://dx.doi.org/10.1158/1078-0432.CCR-11-0116.

Schreiber, R. D., Old, L. J., & Smyth, M. J. (2011). Cancer immunoediting: Integrating immunity's roles in cancer suppression and promotion. *Science*, *331*(6024), 1565–1570. http://dx.doi.org/10.1126/science.1203486.

Schumacher, T., Bunse, L., Pusch, S., Sahm, F., Wiestler, B., Quandt, J., … Platten, M. (2014). A vaccine targeting mutant IDH1 induces antitumour immunity. *Nature*, *512*(7514), 324–327. http://dx.doi.org/10.1038/nature13387.

Schumacher, T. N., & Schreiber, R. D. (2015). Neoantigens in cancer immunotherapy. *Science (New York, N.Y.)*, *348*(6230), 69–74. http://dx.doi.org/10.1126/science.aaa4971.

Segal, N. H., Parsons, D. W., Peggs, K. S., Velculescu, V., Kinzler, K. W., Vogelstein, B., & Allison, J. P. (2008). Epitope landscape in breast and colorectal cancer. *Cancer Research*, *68*(3), 889–892. http://dx.doi.org/10.1158/0008-5472.CAN-07-3095.

Shankaran, V., Ikeda, H., Bruce, A. T., White, J. M., Swanson, P. E., Old, L. J., & Schreiber, R. D. (2001). IFNgamma and lymphocytes prevent primary tumour development and shape tumour immunogenicity. *Nature*, *410*(6832), 1107–1111. http://dx.doi.org/10.1038/35074122.

Sharma, P., & Allison, J. P. (2015). Immune checkpoint targeting in cancer therapy: Toward combination strategies with curative potential. *Cell*, *161*(2), 205–214. http://dx.doi.org/10.1016/j.cell.2015.03.030.

Shukla, S. A., Rooney, M. S., Rajasagi, M., Tiao, G., Dixon, P. M., Lawrence, M. S., … Getz, G. (2015). Comprehensive analysis of cancer-associated somatic mutations in class I HLA genes. *Nature Biotechnology*, *33*(11), 1152–1158. http://dx.doi.org/10.1038/nbt.3344.

Simpson, A. J., Caballero, O. L., Jungbluth, A., Chen, Y. T., & Old, L. J. (2005). Cancer/testis antigens, gametogenesis and cancer. *Nature Reviews. Cancer*, *5*(8), 615–625. http://dx.doi.org/10.1038/nrc1669.

Smyth, M. J., Cretney, E., Takeda, K., Wiltrout, R. H., Sedger, L. M., Kayagaki, N., … Okumura, K. (2001). Tumor necrosis factor-related apoptosis-inducing ligand (TRAIL) contributes to interferon gamma-dependent natural killer cell protection from tumor metastasis. *The Journal of Experimental Medicine*, *193*(6), 661–670.

Smyth, M. J., Godfrey, D. I., & Trapani, J. A. (2001). A fresh look at tumor immunosurveillance and immunotherapy. *Nature Immunology*, *2*(4), 293–299. http://dx.doi.org/10.1038/86297.

Smyth, M. J., Thia, K. Y., Street, S. E., Cretney, E., Trapani, J. A., Taniguchi, M., … Godfrey, D. I. (2000). Differential tumor surveillance by natural killer (NK) and NKT cells. *The Journal of Experimental Medicine*, *191*(4), 661–668.

Snyder, A., Makarov, V., Merghoub, T., Yuan, J., Zaretsky, J. M., Desrichard, A., … Chan, T. A. (2014). Genetic basis for clinical response to CTLA-4 blockade in melanoma. *The New England Journal of Medicine*, *371*(23), 2189–2199. http://dx.doi.org/10.1056/NEJMoa1406498.

Somasundaram, R., Swoboda, R., Caputo, L., Otvos, L., Weber, B., Volpe, P., … Herlyn, D. (2006). Human leukocyte antigen-A2-restricted CTL responses to mutated BRAF peptides in melanoma patients. *Cancer Research*, *66*(6), 3287–3293. http://dx.doi.org/10.1158/0008-5472.CAN-05-1932.

Spranger, S., Spaapen, R. M., Zha, Y., Williams, J., Meng, Y., Ha, T. T., & Gajewski, T. F. (2013). Up-regulation of PD-L1, IDO, and T(regs) in the melanoma tumor microenvironment is driven by CD8(+) T cells. *Science Translational Medicine*, *5*(200), 200ra116. http://dx.doi.org/10.1126/scitranslmed.3006504.

Stevanović, S., Draper, L. M., Langhan, M. M., Campbell, T. E., Kwong, M. L., Wunderlich, J. R., … Hinrichs, C. S. (2015). Complete regression of metastatic cervical cancer after treatment with human papillomavirus-targeted tumor-infiltrating T cells. *Journal of Clinical Oncology*, *33*(14), 1543–1550. http://dx.doi.org/10.1200/JCO.2014.58.9093.

Stone, J. D., Harris, D. T., & Kranz, D. M. (2015). TCR affinity for p/MHC formed by tumor antigens that are self-proteins: Impact on efficacy and toxicity. *Current Opinion in Immunology*, *33*, 16–22. http://dx.doi.org/10.1016/j.coi.2015.01.003.

Street, S. E., Cretney, E., & Smyth, M. J. (2001). Perforin and interferon-gamma activities independently control tumor initiation, growth, and metastasis. *Blood*, *97*(1), 192–197.

Stutman, O. (1974). Tumor development after 3-methylcholanthrene in immunologically deficient athymic-nude mice. *Science*, *183*(4124), 534–536.
Stutman, O. (1979). Chemical carcinogenesis in nude mice: Comparison between nude mice from homozygous matings and heterozygous matings and effect of age and carcinogen dose. *Journal of the National Cancer Institute*, *62*(2), 353–358.
Teng, M. W. L., Galon, J., Fridman, W. H., & Smyth, M. J. (2015). From mice to humans: Developments in cancer immunoediting. *The Journal of Clinical Investigation*, *125*(9), 3338–3346. http://dx.doi.org/10.1172/JCI80004.
Teng, M. W., Vesely, M. D., Duret, H., McLaughlin, N., Towne, J. E., Schreiber, R. D., & Smyth, M. J. (2012). Opposing roles for IL-23 and IL-12 in maintaining occult cancer in an equilibrium state. *Cancer Research*, *72*(16), 3987–3996. http://dx.doi.org/10.1158/0008-5472.CAN-12-1337.
Thomas, L. (1959). Delayed hypersensitivity in health and disease. In H. S. Lawrence (Ed.), *Cellular and humoral aspects of the hypersensitive states* (pp. 529–532). New York: Hoeber-Harper.
Tran, E., Ahmadzadeh, M., Lu, Y.-C., Gros, A., Turcotte, S., Robbins, P. F., … Rosenberg, S. A. (2015). Immunogenicity of somatic mutations in human gastrointestinal cancers. *Science (New York, N.Y.)*, *350*(6266), 1387–1390. http://dx.doi.org/10.1126/science.aad1253.
Tran, E., Turcotte, S., Gros, A., Robbins, P. F., Lu, Y.-C., Dudley, M. E., … Rosenberg, S. A. (2014). Cancer immunotherapy based on mutation-specific CD4+ T cells in a patient with epithelial cancer. *Science (New York, N.Y.)*, *344*(6184), 641–645. http://dx.doi.org/10.1126/science.1251102.
Tumeh, P. C., Harview, C. L., Yearley, J. H., Shintaku, I. P., Taylor, E. J. M., Robert, L., … Ribas, A. (2014). PD-1 blockade induces responses by inhibiting adaptive immune resistance. *Nature*, *515*(7528), 568–571. http://dx.doi.org/10.1038/nature13954.
Van Allen, E. M., Miao, D., Schilling, B., Shukla, S. A., Blank, C., Zimmer, L., … Garraway, L. A. (2015). Genomic correlates of response to CTLA-4 blockade in metastatic melanoma. *Science (New York, N.Y.)*, *350*(6257), 207–211. http://dx.doi.org/10.1126/science.aad0095.
van Buuren, M. M., Calis, J. J., & Schumacher, T. N. (2014). High sensitivity of cancer exome-based CD8 T cell neo-antigen identification. *OncoImmunology*, *3*, e28836. http://dx.doi.org/10.4161/onci.28836.
Van den Eynde, B., Hainaut, P., Herin, M., Knuth, A., Lemoine, C., Weynants, P., … Boon, T. (1989). Presence on a human melanoma of multiple antigens recognized by autologous CTL. *International Journal of Cancer*, *44*(4), 634–640.
van der Bruggen, P., Traversari, C., Chomez, P., Lurquin, C., De Plaen, E., Van den Eynde, B., … Boon, T. (1991). A gene encoding an antigen recognized by cytolytic T lymphocytes on a human melanoma. *Science*, *254*(5038), 1643–1647.
van Elsas, A., Hurwitz, A. A., & Allison, J. P. (1999). Combination immunotherapy of B16 melanoma using anti-cytotoxic T lymphocyte-associated antigen 4 (CTLA-4) and granulocyte/macrophage colony-stimulating factor (GM-CSF)-producing vaccines induces rejection of subcutaneous and metastatic tumors accompanied by autoimmune depigmentation. *The Journal of Experimental Medicine*, *190*(3), 355–366.
van Elsas, A., Sutmuller, R. P., Hurwitz, A. A., Ziskin, J., Villasenor, J., Medema, J. P., … Allison, J. P. (2001). Elucidating the autoimmune and antitumor effector mechanisms of a treatment based on cytotoxic T lymphocyte antigen-4 blockade in combination with a B16 melanoma vaccine: Comparison of prophylaxis and therapy. *The Journal of Experimental Medicine*, *194*(4), 481–489.

van Rooij, N., van Buuren, M. M., Philips, D., Velds, A., Toebes, M., Heemskerk, B., …Schumacher, T. N. (2013). Tumor exome analysis reveals neoantigen-specific T-cell reactivity in an ipilimumab-responsive melanoma. *Journal of Clinical Oncology, 31*(32), e439–e442. http://dx.doi.org/10.1200/JCO.2012.47.7521.

Vansteenkiste, J., Zielinski, M., Linder, A., Dahabreh, J., Gonzalez, E. E., Malinowski, W., …Brichard, V. G. (2013). Adjuvant MAGE-A3 immunotherapy in resected non-small-cell lung cancer: Phase II randomized study results. *Journal of Clinical Oncology, 31*(19), 2396–2403. http://dx.doi.org/10.1200/JCO.2012.43.7103.

Vesely, M. D., Kershaw, M. H., Schreiber, R. D., & Smyth, M. J. (2011). Natural innate and adaptive immunity to cancer. *Annual Review of Immunology, 29*, 235–271. http://dx.doi.org/10.1146/annurev-immunol-031210-101324.

Vita, R., Overton, J. A., Greenbaum, J. A., Ponomarenko, J., Clark, J. D., Cantrell, J. R., …Peters, B. (2015). The immune epitope database (IEDB) 3.0. *Nucleic Acids Research, 43*(Database issue), D405–D412. http://dx.doi.org/10.1093/nar/gku938.

Wang, R. F., Robbins, P. F., Kawakami, Y., Kang, X. Q., & Rosenberg, S. A. (1995). Identification of a gene encoding a melanoma tumor antigen recognized by HLA-A-31-restricted tumor-infiltrating lymphocytes. *Journal of Experimental Medicine, 181*(2), 799–804.

Wang, L., Rubinstein, R., Lines, J. L., Wasiuk, A., Ahonen, C., Guo, Y., …Noelle, R. J. (2011). VISTA, a novel mouse Ig superfamily ligand that negatively regulates T cell responses. *The Journal of Experimental Medicine, 208*(3), 577–592. http://dx.doi.org/10.1084/jem.20100619.

Watanabe, N., Gavrieli, M., Sedy, J. R., Yang, J., Fallarino, F., Loftin, S. K., …Murphy, K. M. (2003). BTLA is a lymphocyte inhibitory receptor with similarities to CTLA-4 and PD-1. *Nature Immunology, 4*(7), 670–679. http://dx.doi.org/10.1038/ni944.

Waterhouse, P., Penninger, J. M., Timms, E., Wakeham, A., Shahinian, A., Lee, K. P., …Mak, T. W. (1995). Lymphoproliferative disorders with early lethality in mice deficient in Ctla-4. *Science, 270*(5238), 985–988.

Wedén, S., Klemp, M., Gladhaug, I. P., Møller, M., Eriksen, J. A., Gaudernack, G., & Buanes, T. (2011). Long-term follow-up of patients with resected pancreatic cancer following vaccination against mutant K-ras. *International Journal of Cancer, 128*(5), 1120–1128. http://dx.doi.org/10.1002/ijc.25449.

Wolfel, T., Hauer, M., Schneider, J., Serrano, M., Wolfel, C., Klehmann-Hieb, E., …Beach, D. (1995). A p16INK4a-insensitive CDK4 mutant targeted by cytolytic T lymphocytes in a human melanoma. *Science, 269*(5228), 1281–1284.

Wood, L. D., Parsons, D. W., Jones, S., Lin, J., Sjoblom, T., Leary, R. J., …Vogelstein, B. (2007). The genomic landscapes of human breast and colorectal cancers. *Science, 318*(5853), 1108–1113. http://dx.doi.org/10.1126/science.1145720.

Yadav, M., Jhunjhunwala, S., Phung, Q. T., Lupardus, P., Tanguay, J., Bumbaca, S., …Delamarre, L. (2014). Predicting immunogenic tumour mutations by combining mass spectrometry and exome sequencing. *Nature, 515*(7528), 572–576. http://dx.doi.org/10.1038/nature14001.

Yu, H., Pardoll, D., & Jove, R. (2009). STATs in cancer inflammation and immunity: A leading role for STAT3. *Nature Reviews. Cancer, 9*(11), 798–809. http://dx.doi.org/10.1038/nrc2734.

Zarling, A. L., Ficarro, S. B., White, F. M., Shabanowitz, J., Hunt, D. F., & Engelhard, V. H. (2000). Phosphorylated peptides are naturally processed and presented by major histocompatibility complex class I molecules in vivo. *Journal of Experimental Medicine, 192*(12), 1755–1762.

Zarling, A. L., Obeng, R. C., Desch, A. N., Pinczewski, J., Cummings, K. L., Deacon, D. H., … Engelhard, V. H. (2014). MHC-restricted phosphopeptides from insulin receptor substrate-2 and CDC25b offer broad-based immunotherapeutic agents for cancer. *Cancer Research*, *74*(23), 6784–6795. http://dx.doi.org/10.1158/0008-5472.CAN-14-0043.

Zarling, A. L., Polefrone, J. M., Evans, A. M., Mikesh, L. M., Shabanowitz, J., Lewis, S. T., … Hunt, D. F. (2006). Identification of class I MHC-associated phosphopeptides as targets for cancer immunotherapy. *Proceedings of the National Academy of Sciences of the United States of America*, *103*(40), 14889–14894. http://dx.doi.org/10.1073/pnas.0604045103.

Zhou, J., Dudley, M. E., Rosenberg, S. A., & Robbins, P. F. (2005). Persistence of multiple tumor-specific T-cell clones is associated with complete tumor regression in a melanoma patient receiving adoptive cell transfer therapy. *Journal of Immunotherapy*, *28*(1), 53–62.

CHAPTER THREE

Tumor and Host Factors Controlling Antitumor Immunity and Efficacy of Cancer Immunotherapy

Stefani Spranger, Ayelet Sivan, Leticia Corrales, Thomas F. Gajewski[1]
The University of Chicago, Chicago, IL, United States
[1]Corresponding author: e-mail address: tgajewsk@medicine.bsd.uchicago.edu

Contents

Abstract

Despite recent clinical advances in immunotherapy, a fraction of cancer patients fails to respond to these interventions. Evidence from preclinical mouse models as well as clinical samples has provided evidence that the extent of activated T cell infiltration within the tumor microenvironment is associated with clinical response to immunotherapies including checkpoint blockade. Therefore, understanding the molecular mechanisms mediating the lack of T cell infiltration into the tumor microenvironment will be instrumental for the development of new therapeutic strategies to render those patients immunotherapy responsive. Recent data have suggested that major sources of intersubject heterogeneity include differences in somatic mutations in specific oncogene pathways between cancers of individual subjects and also environmental factors including commensal microbial composition. Successful identification of such causal factors

Advances in Immunology, Volume 130
ISSN 0065-2776
http://dx.doi.org/10.1016/bs.ai.2015.12.003

should lead to new therapeutic approaches that may facilitate T cell entry into non-inflamed tumors and expand the fraction of patients capable of responding to novel immunotherapies.

1. RATIONALE FOR STUDYING REGULATION OF THE T CELL-INFLAMED TUMOR MICROENVIRONMENT PHENOTYPE

Interrogation of the tumor microenvironment in human melanoma patients was initially pursued to identify mechanisms that might explain clinical response vs resistance to cancer vaccines in individual patients. To this end, baseline biopsies of melanoma metastases were analyzed by transcriptional profiling and confirmatory assays. This analysis revealed two major subsets of tumor microenvironment that were largely characterized by the presence or absence of a gene signature indicative of a T cell-inflamed tumor microenvironment. The T cell-inflamed subset of tumors showed presence of T cell transcripts, chemokines that likely mediate effector T cell recruitment, macrophage activation markers, and a type I IFN transcriptional profile (Harlin et al., 2009). Immunohistochemistry confirmed the presence of $CD8^+$ T cells, macrophages, and some B cells in these tumors. Interestingly, patients with clinical benefit to these vaccines exclusively showed the T cell-inflamed tumor microenvironment phenotype. Thus, the ability of cells within the tumor microenvironment to produce chemokines and recruit activated T cells appears to be instrumental for clinical benefit. Similar results have been observed in patients treated with high-dose IL-2 (Sullivan et al., 2009) and also with the anti-CTLA-4 mAb ipilimumab (Hamid et al., 2011).

The question arises as to why tumors infiltrated with T cells are not rejected spontaneously. More detailed analysis of the T cell-inflamed subset of tumors revealed the presence of transcripts encoding indoleamine-2, 3-dioxygenase (IDO), PD-L1, and FoxP3, all markers of immune-inhibitory pathways. IHC confirmed that PD-L1 and IDO protein expression, and also nuclear $FoxP3^+CD4^+$ cells, were found within T cell-inflamed tumors in the same region as $CD8^+$ T cells. Mouse mechanistic studies confirmed that $CD8^+$ T cells were required for the upregulation of all of these three factors within the tumor microenvironment. For PD-L1 and IDO induction, the requisite factor produced by the $CD8^+$ T cells was interferon (IFN)-γ. For $FoxP3^+$ Tregs, production of the chemokine CCL22 was identified, which mediated Treg recruitment into tumor sites (Spranger et al., 2013). Together, these data suggest that the involvement of these three immune-inhibitory mechanisms in T cell-inflamed tumors is driven by the activated $CD8^+$ T cells themselves, and likely explains failed spontaneous tumor elimination.

Importantly, clinical response with anti-PD-1 mAb, which is blocking PD-L1/PD-1 interactions directly within the tumor microenvironment, was found to occur almost exclusively in patients with preexisting T cell infiltrates in the region of PD-L1 upregulation (Topalian et al., 2012). Following anti-PD-1 administration, these $CD8^+$ T cells seemed to proliferate and expand to penetrate throughout the tumor, an event associated with tumor regression (Tumeh et al., 2014). These observations are consistent with preclinical data indicating that tumor regression upon checkpoint blockade was almost completely mediated by reactivation of $CD8^+$ T cells directly within the tumor site to be able to proliferate and produce IL-2 (Spranger et al., 2014).

Based on the observation that multiple immune regulatory pathways appear to be operational within the same tumor microenvironment, combination immunotherapies are being pursued. In preclinical models, concurrent doublets of anti-CTLA-4 ± anti-PD-L1 ± an IDO inhibitor were found to be synergistic in the B16 melanoma model in vivo (Spranger et al., 2014). Interestingly, each of these combination therapies involved reacquisition of IL-2 production and proliferation by $CD8^+$ T cells directly within the tumor microenvironment. In support of the importance of reactivation of T cells already present at the tumor site, therapeutic effects were preserved even in the presence of FTY720 blockade, which prevents exit of new T cells from lymph nodes. All three of these combination doublets are being tested clinically, and the combination of anti-CTLA-4 + anti-PD-1 was recently approved by the FDA for the treatment of patients with advanced melanoma (Larkin et al., 2015). With these ongoing successes in hand, increased attention is being invested into studying the non-T cell-inflamed tumor microenvironment, as a strategy to identify mechanisms of primary resistance to these successful immunotherapies.

2. MOLECULAR AND CELLULAR DRIVERS OF THE T CELL-INFLAMED TUMOR MICROENVIRONMENT

One step toward understanding what is failing to occur in the non-T cell-inflamed tumor microenvironment subset has been to investigate the mechanistic steps involved in successful generation of a T cell-inflamed phenotype when it does occur (Fig. 1). A critical process in spontaneous T cell priming against tumor-associated antigens involves the recruitment and activation of Batf3-lineage dendritic cells (DCs), expressing the markers CD8α or CD103 in the mouse (Edelson et al., 2010; Engelhardt et al., 2012; Fuertes et al., 2011; Hildner et al., 2008). By using knockout mice deficient for the transcription factor Batf3, it has been proven that lack of this particular subset of DCs is resulting in a significant reduction of systemic activation antitumor $CD8^+$ T cells (Fuertes et al., 2011). In addition, successful priming of antitumor immune responses requires this subset of DCs to receive signaling by type I IFNs (IFN-α/β), which are induced to be produced by host cells in the presence of tumor challenge (Fuertes et al., 2011). In concordance with these preclinical findings, analysis of human melanoma samples showed that the absence of type I interferons was also associated with the absence of T cells from the tumor microenvironment (Fuertes, Woo, Burnett, Fu, & Gajewski, 2013).

The critical involvement of host type I IFNs and Batf3-lineage DCs raised the next level question as to what tumor-derived factors might be responsible for inducing the requisite type I IFN production by host immune cells. Using a series of gene-targeted mice deficient in specific

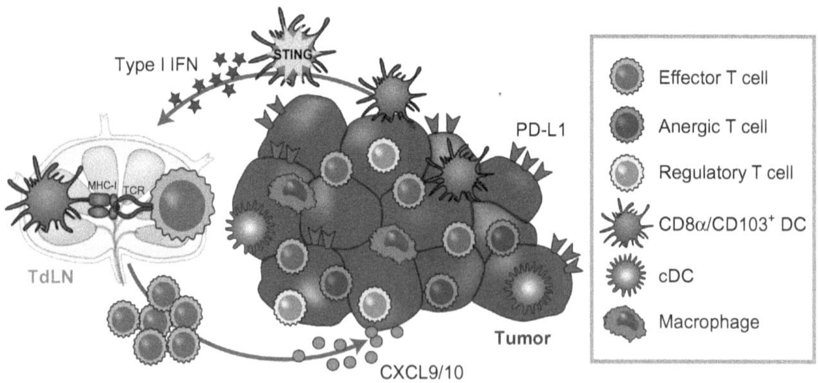

Figure 1 Processes involved in generation of the T cell-inflamed tumor microenvironment phenotype. Innate immune sensing of tumors appears to be dominantly mediated by the host STING pathway, which results in type I IFN production and activation of Batf3-lineage DCs. This DC subset cross-presents antigens to $CD8^+$ T cells, which once activated can be recruited into the inflamed tumor microenvironment and contribute to tumor control. The same STING pathway engagement likely leads to production of CXCL9 and CXCL10, which are the key chemokines that support effector T cell recruitment. (See the color plate.)

innate immune sensing pathways, recent evidence has implicated the DNA-sensing cGAS/STING pathway as critical for the spontaneous priming of antitumor T cells in vivo (Woo et al., 2014). Consistent with these results, intratumoral injection of STING agonists has resulted in profound immune-mediated tumor elimination in preclinical models in vivo (Corrales et al., 2015). Thus, one strategy with the potential to initiate de novo immune responses in non-T cell-inflamed tumors is through the use of STING agonists.

Once effector T cells are activated differentiated in peripheral lymphoid organs, the presence of appropriate chemokines within the tumor microenvironment is critical for those T cells to traffic into tumor sites. Gene expression profiling of human melanoma metastases identified a range of chemokines associated with $CD8^+$ T cell presence (Erdag et al., 2012; Harlin et al., 2009; Salerno, Olson, McSkimming, Shea, & Slingluff, 2014). Recent work has indicated that the CXCR3/CXCL9/10 chemokine axis is mandatory for T cell entry into the tumor microenvironment (Mikucki et al., 2015). Interestingly, one consequence of STING pathway activation in DCs is production of CXCL9 and CXCL10 (Woo et al., 2014), suggesting that properly activated DCs within the tumor microenvironment might also contribute to effector T cell recruitment. As mentioned earlier, once activated T cells successfully traffic into the tumor site, if antigen-expressing tumor cells are not immediately eradicated, immune-inhibitory mechanisms become upregulated that counteract chronic inflammation. Thus, immunotherapies aiming to block the function of those pathways are expected to be preferentially active in tumors displaying the T cell-inflamed phenotype. Uncovering molecular causes for absence of the T cell-inflamed tumor microenvironment should therefore enable new interventions to be developed that can expand the proportion of patients with immunotherapy-responsive tumors.

3. TUMOR-INTRINSIC ONCOGENE PATHWAYS MEDIATING THE NON-T CELL-INFLAMED TUMOR MICROENVIRONMENT

3.1 Tumor-Intrinsic WNT/β-Catenin Pathway Activation is Causal for T Cell Exclusion in Melanoma

The possibility that somatic differences at the level of the tumor cells themselves might explain the lack of a T cell-inflamed tumor microenvironment in a subset of cases was recently pursued through gene expression profiling of

266 individual melanoma metastases in concert with exome sequencing of the same tumors (Spranger, Bao, & Gajewski, 2015). Exome-sequencing data revealed that gain of function mutations in the β-catenin gene were found in seven tumors (14%) and loss-of-function mutations in negative regulators of the β-catenin pathway (APC, Axin1, TCF1) were identified in an additional ten of the non-T cell-inflamed tumors (23%). In addition, gene expression analysis of defined β-catenin target genes revealed that 48% of the non-T cell-inflamed tumors showed evidence for activation of the WNT/β-catenin pathway. Additional analysis of those β-catenin target genes showed a negative correlation with CD8α expression in the tumor, whereas PD-L1 expression showed an expected positive correlation with CD8α (Spranger et al., 2013). Through confirmatory immunohistochemistry, high β-catenin protein expression was associated with the absence of $CD8^+$ T cells within the tumor, indicating a significant correlation between activation of the WNT/β-catenin pathway and a non-T cell-inflamed tumor microenvironment (Spranger et al., 2015).

To interrogate the mechanistic relevance of tumor-intrinsic β-catenin signaling in controlling the host immune response to melanoma, genetically engineered mice were developed using a tamoxifen-regulated Cre-driven by the tyrosinase promoter in combination with active Braf ($Braf^{V600E}$) and conditional PTEN deletion ($PTEN^{-/-}$), with or without a conditional active β-catenin mutant (CAT-STA) (Bosenberg et al., 2006; Damsky et al., 2011; Dankort et al., 2009). Melanomas arising from $Braf^{V600E}$/$PTEN^{-/-}$ mice did have a modest T cell infiltrate at baseline as analyzed by flow cytometry and immunohistochemistry. However, melanomas comprising $Braf^{V600E}$ combined with active β-catenin ($Braf^{V600E}$/CAT-STA) completely lacked a T cell infiltrate. Moreover, addition of PTEN deletion ($Braf^{V600E}$/$PTEN^{-/-}$/CAT-STA) to accelerate tumor growth resulted in melanomas also lacking a T cell infiltrate. These results demonstrate that activation of the β-catenin pathway within melanoma tumor cells can dominantly exclude T cells from the tumor microenvironment.

The step at which tumor-intrinsic β-catenin activation might be antagonizing antitumor T cell responses was pursued by integrating Cre-inducible expression of the model antigen SIY (SIYRYYGL) (Cheung, Dupage, Dong, Chen, & Jacks, 2008). The use of this model antigen allowed the use of adoptive transfer of SIY-specific 2C TCR-transgenic T cells in order to evaluate endogenous T cell activation. Indeed, mice with SIY^+ tumors driven by mutated Braf, PTEN deletion, and active β-catenin ($Braf^{V600E}$/$PTEN^{-/-}$/CAT-STA) showed absence of spontaneous T cell activation, while tumors without active β-catenin showed normal T cell activation measured by CFSE dilution. In-depth analysis revealed a complete lack of Batf3-lineage DCs expressing the surface markers CD103 or CD8α in tumors expressing β-catenin. This was correlated with lack of expression of CCL4, a major chemokine responsible for $CD103^+$ DC recruitment. Mechanistically, CCL4 expression in β-catenin-expressing tumor cells was suppressed by the transcriptional repressor ATF3. Consistent with the lack of baseline T cell infiltration, combination immunotherapy with αCTLA-4 and αPD-L1 showed no therapeutic benefit in mice-bearing $Braf^{V600E}/PTEN^{-/-}$/CAT-STA tumors, while reduction in tumor outgrowth was observed for $Braf^{V600E}/PTEN^{-/-}$ tumors. Therefore, these collective data identify the WNT/β-catenin pathway as the first defined tumor-intrinsic oncogene pathway that can prevent the induction of antitumor T cell responses, excluding T cell from the tumor microenvironment and resulting in resistance to immunotherapy (Fig. 2) (Spranger et al., 2015).

Collectively, these data suggest a new model in which tumor-intrinsic β-catenin activation mediates immune evasion from an antitumor immune response through tumor immune avoidance. Although the data supporting

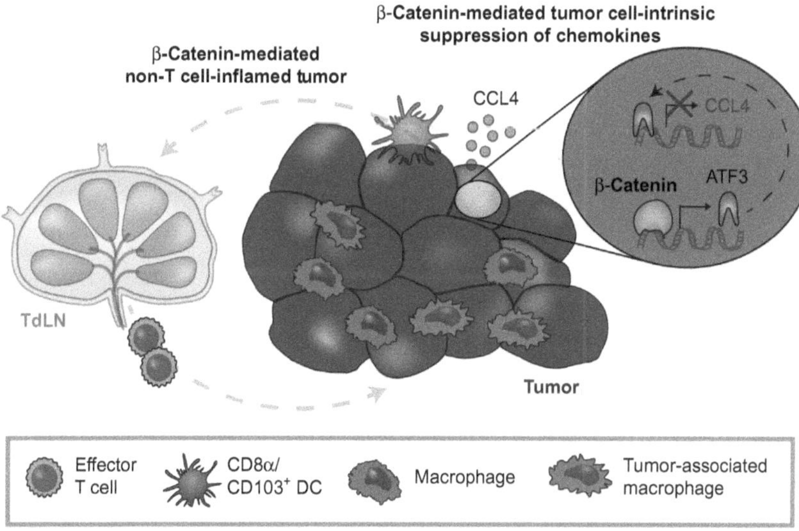

Figure 2 Model for how tumor-intrinsic β-catenin signaling mediates exclusion of T cells from the tumor microenvironment. Activation of the β-catenin pathway in tumor cells results in induction of ATF3, which serves as a transcriptional repressor for CCL4 gene expression. Lack of CCL4 accounts for failed recruitment of $CD103^+$ and $CD8\alpha^+$ DCs into the tumor microenvironment, thereby preventing cross-priming of host $CD8^+$ T cells. (See the color plate.)

this notion were obtained with β-catenin activation beginning early at the stage of tumor initiation, it is plausible that this process of immune avoidance could occur at any given time in tumor development if mutations activating the β-catenin pathway become acquired. In concordance with contemporary thinking around immune evasion (the three E hypothesis) (Dunn, Bruce, Ikeda, Old, & Schreiber, 2002), in which the tumor escapes from the immune system by elimination of immunogenic tumor cells with subsequent upregulation of immune-inhibitory mechanisms that suppress the function of residual T cells, we propose that tumor escape also can occur through selection of WNT/β-catenin-positive tumor cells. This new mechanism would result in T cell exclusion from the tumor microenvironment and could represent the second major tumor phenotype observed clinically (non-T cell-inflamed). An additional prediction of this model is that patients who encounter extreme immune selection (eg, following immunotherapies) might show acquisition of β-catenin pathway activation as a potential secondary escape mechanism. This possibility is testable clinically by analyzing tumors before successful treatment vs after secondary progression.

3.2 Other Candidate Oncogene Pathways That May Contribute to Immune Exclusion

It is very conceivable that additional molecular oncogenic pathways could result in similar effects to limit the accumulation of T cells within the tumor microenvironment. In particular since 48% of the non-T cell-inflamed melanomas showed evidence of WNT/β-catenin activation, the remaining 52% might have other molecular aberrations resulting in a similar functional outcome. One potential candidate pathway is mutant p53. In preclinical mouse models, restoration of intact p53 signaling has been found to increased recruitment and activation of innate immune cells (Xue et al., 2007). Additionally, restoration of p53 wildtype signaling was sufficient to induce tumor regression, accompanied with increased expression of proinflammatory cytokines, in a murine liver carcinoma model. In related work, activation and recruitment of natural killer (NK) cells were responsible for tumor regression in response to reexpression of wildtype p53 (Iannello, Thompson, Ardolino, Lowe, & Raulet, 2013). Mechanistically, NK cell recruitment was found to be driven by CCL2, which was expressed at higher levels in tumor cells after p53 wildtype restoration. In concordance with those findings in preclinical models, a recent study analyzing triple negative breast cancer indicated a correlation between mutant p53 and the absence of T cells in the tumor microenvironment (Quigley et al., 2015). Therefore, p53 signaling at steady state may enhance recruitment of innate immune cells as well as their activation, suggesting that mutations in the p53 pathway could contribute to the non-T cell-inflamed phenotype.

Another conceivable oncogenic pathway that could contribute to T cell exclusion is activation of STAT3 signaling. In transplantable tumor models, constitutively active STAT3 signaling results in decreased expression of proinflammatory mediators, while a dominant negative variant of STAT3 resulted in augmented expression of proinflammatory factors (Burdelya et al., 2005; Wang et al., 2004), including the chemokines CCL5 and CXCL10. Furthermore, recent studies using a carcinogen-induced lung cancer model as well as a genetically induced prostate cancer model have provided in vivo support for these notions (Ihara et al., 2012; Toso et al., 2014). Abolishing STAT3 signaling using a conditional knockout mouse model resulted in increased antitumor immune responses. This was associated with increased T cell accumulation and functionality within the tumor microenvironment, as well as increased expression of the chemokines CCL5 and CXCL10. Thus, activation of STAT3 signaling may represent another oncogenic pathway able to mediate reduction of immune cell recruitment into tumor sites.

The NFκB signaling pathway represents another important candidate that could impact on host immune responses. In general, cancer cell-intrinsic activation of this pathway has been associated with tumor progression (Baldwin, 2012; Basseres & Baldwin, 2006). Focusing on a more immunologically relevant aspect of this pathway, augmented NFκB signaling in hepatocellular carcinoma cells triggered by immune-derived TNF has been shown to promote tumor progression (Pikarsky et al., 2004). These data suggest that NFκB signaling in tumors could contribute to tumor growth despite immune system interactions. On the other hand, increased production of tumor-derived chemokines has been associated with activation of NFκB, an observation which could have a positive effect on T cell infiltration (Greten et al., 2004). This notion is also supported by studies on hyperactivation of NFκB within the tumor microenvironment, which has been reported to be associated with increased chemokine expression and recruitment of activated T cells (Muthuswamy et al., 2012). Therefore, the impact of tumor-intrinsic NFκB activation on the host immune response might depend on the cellular context of the tumor, and its role might be dependent on the type of immune infiltrate that develops. Additional cancer type-specific studies using GEM model systems will be required to fully determine the effects of NFκB activation on the immune response within the tumor microenvironment.

The PI3K/PTEN/AKT pathway is an additional interesting candidate to be considered as potentially impacting host immune responses. Studies of inflammation-induced cancer progression have investigated activation of the PI3K signaling, either through activating mutations in PIK3CA or loss-of-function mutations in PTEN. These studies have given indications that an increased accumulation of tumor-associated macrophages results, which in turn can contribute to an immune-suppressive microenvironment (Coussens, Zitvogel, & Palucka, 2013; Schmid et al., 2011). The accumulation of macrophages and differentiation into M2-like macrophages has been associated with increased tumor-derived production of TNF, IL-6, CSF-1 VEGF-A, and IL-8 (Bronte & Murray, 2015), which all can support tumor growth, either directly or indirectly. In contrast, recent reports analyzing tumor types which are not associated with inflammation-induced progression including triple negative breast cancer, have indicated that expression of PTEN was associated with the absence of T cells as well as low PD-L1 expression in the tumor microenvironment (Mittendorf et al., 2014), arguing that loss of PTEN expression (and constitutive PI3K activation) is associated with presence of T cells in the tumor microenvironment. Therefore, like NFκB signaling, the impact of tumor-intrinsic PI3K signaling on host T cell function will require further mechanistic studies in well-defined mouse models.

4. ENVIRONMENTAL FACTORS INFLUENCING THE TUMOR MICROENVIRONMENT: THE HOST MICROBIOTA

In addition to tumor-intrinsic oncogenic pathways that could impact on the nature of the host immune response against tumors, another source of intersubject heterogeneity that could theoretically regulate antitumor immunity is the commensal microbiota. A growing body of data has implicated the composition of intestinal commensal bacteria as a major environmental factor that varies between individuals, with the ability to impact systemic immunity. Commensal bacteria have been shown to direct differentiation of T cells leading to expansion of specific molecular subsets, thereby influencing systemic inflammatory processes that involve these T cell differentiation states (Hooper, Littman, & Macpherson, 2012; Ivanov & Honda, 2012). Other studies have highlighted a role for commensal bacteria in modulating the activation state of innate antigen-presenting cells (APCs), thereby impacting priming of systemic immune responses (Abt et al., 2012; Ganal et al., 2012).

While these studies establish a role for commensal bacteria in modulating systemic immune function in general, a potential impact of gut commensal microbes on systemic antitumor immune responses has only begun to be investigated. Two seminal studies showed that the therapeutic effect of chemotherapy was facilitated by the presence of commensal microbes. In one model, treatment with oral antibiotics reduced the therapeutic effect of oxaliplatin or the response to anti-IL-10R + CpG-ODN as an immunotherapy. Mechanistically, the effect was mapped to the level of inflammatory cytokine production by myeloid cells (Iida et al., 2013). In a second model, bacterial translocation from the gut into peripheral organs was implicated in the therapeutic effect of cyclophosphamide, as treatment with antibiotics reduced the generation of T helper 17 (Th17) that were involved with antitumor efficacy in this system (Viaud et al., 2013). In cancer models associated with inflammation, commensal bacteria also have been reported to promote tumor growth. In a recent study, TLR5-mediated sensing of gut commensals was shown to increase systemic levels of IL-6, leading to recruitment of MDSCs and γδ suppressor T cells into the tumor microenvironment, in a p53/K-ras-driven tumor model (Rutkowski et al., 2015). These data begin to highlight a role for commensal bacteria in shaping antitumor immune responses, with implications on cancer progression and treatment.

Whether commensal microbiota in fact influence spontaneous immune responses against tumors was initially investigated by comparing subcutaneous B16.SIY melanoma growth in genetically similar C57BL/6 mice derived from two different mouse facilities, Jackson Laboratory (JAX) and Taconic Farms (TAC), which have been shown to differ in their commensal microbes (Ivanov et al., 2009). These experiments revealed that C57BL/6 mice colonized by different microbial communities exhibited significant differences in B16.SIY melanoma growth rate and associated tumor antigen-specific T cell responses. Specifically, tumors in mice obtained from TAC grew more aggressively, accompanied by weaker antitumor T cell responses. Cohousing prior to tumor inoculation ablated the differences in tumor growth and tumor-specific immune responses between the two mouse populations, arguing for an environmental influence. Upon cohousing, TAC mice acquired the JAX phenotype, suggesting that JAX mice may harbor commensal microbes that dominantly facilitate improved antitumor immunity (Sivan et al., 2015).

In a more direct approach to assess the involvement of commensal microbes in modulating the antitumor immune response, the effects of fecal material transfer between the two mouse populations were investigated. This approach showed that prophylactic transfer of JAX fecal material into TAC recipients was sufficient to delay tumor growth and to enhance induction and infiltration of tumor-specific $CD8^+$ T cells, supporting a microbe- or microbial product-derived effect that was consistent with the effects observed upon cohousing. Administration of JAX fecal material to TAC mice-bearing established tumors showed a therapeutic effect that was comparable to treatment with systemic αPD-L1 monoclonal antibodies (mAbs), and combination treatment with both JAX feces and αPD-L1 mAbs was able to further enhance the tumor-specific T cell response, leading to near complete tumor elimination. Consistent with these results, αPD-L1 therapy alone was significantly more efficacious in JAX mice compared to TAC mice, which paralleled improved antitumor T cell responses. These data indicated that the commensal microbial composition can influence spontaneous antitumor immunity as well as response to immunotherapy with αPD-L1 mAb (Sivan et al., 2015).

Identification of the bacterium contained within JAX feces that elicited improved antitumor immune responses was pursued using 16S ribosomal RNA (rRNA) sequencing of fecal bacterial content collected over time from mice that were subjected to administration of fecal permutations. This approach revealed that administration of JAX fecal material to TAC mice induced a gradual, but consistent change in microbial community diversity, such that TAC mice became very similar to JAX mice in microbial composition. Comparative in-depth analyses of bacterial taxa abundance between the different fecal permutation groups pointed to *Bifidobacterium breve* and *Bifidobacterium longum* as being significantly associated with increased tumor antigen-specific T cell responses. Administration of a cocktail of *Bifidobacterium* species, which included *B. breve* and *B. longum*, to TAC mice-bearing established tumors led to an increase in *Bifidobacterium* levels in the fecal material of these mice, assessed by both 16S rRNA sequencing and quantitative PCR. This increase was accompanied by significantly improved tumor control and robust tumor-specific T cell responses in comparison to non-*Bifidobacterium* treated counterparts, either when *Bifidobacterium* was given alone or in combination with αPD-L1 mAbs, in a $CD8^+$ T cell-dependent manner (Sivan et al., 2015).

Mechanistically, $CD8^+$ SIY-specific 2C TCR Tg T cells exposed to tumors in JAX and *Bifidobacterium*-treated TAC mice produced markedly greater IFN-γ in both the tumor draining lymph node and the spleen, relative to TAC tumor-bearing mice, consistent with analyses of the endogenous T cell response, and pointed to an improvement in immune responses upstream of T cells, at the level of host DCs (Fig. 3). Indeed, DCs isolated from early tumors of JAX and *Bifidobacterium*-fed TAC showed increased maturation and expression of genes critical for antitumor responses including those involved in $CD8^+$ T cell activation and costimulation (Bak et al.,

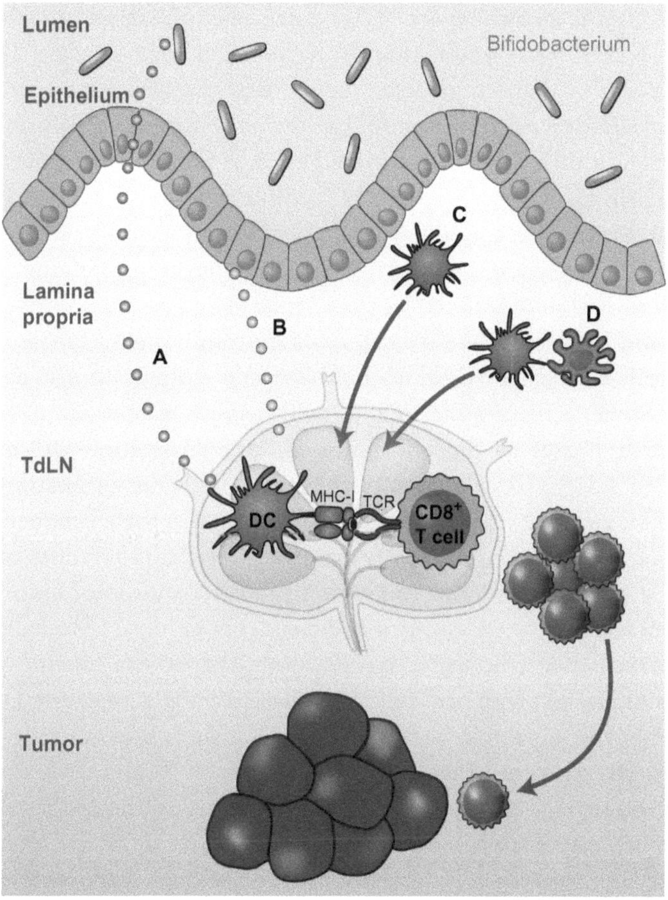

Figure 3 Potential mechanisms for Bifidobacterium-mediated improvement in antitumor immune responses. (A) and (B) A systemic factor might be released by Bifidobacterium (A) or by host cells, such as intestinal epithelial cells (B) that promotes global DC preactivation. (C) Altered DCs or DC precursors might be preactivated locally in the intestinal lamina propria and disseminate to distant sites. (D) A host cell, altered by crosstalk with *Bifidobacterium*, might interact with DCs to modulate DC function. (See the color plate.)

2012; Mackey et al., 1998; Scholer, Hugues, Boissonnas, Fetler, & Amigorena, 2008), DC maturation (Pan et al., 2004; Pettit et al., 1997), antigen processing and cross presentation (Compeer, Flinsenberg, van der Grein, & Boes, 2012; Jancic et al., 2007; Stober, Brode, White, Popoff, & Blackwell, 2007), chemokine-mediated recruitment of immune cells to the tumor microenvironment (Kabashima et al., 2007; Nukiwa et al., 2006; Zhang et al., 2003), and type I IFN signaling (Fuertes et al., 2011; Woo et al., 2014). *In vitro*, purified DCs isolated from lymphoid organs of JAX and *Bifidobacterium*-fed TAC mice elicited significantly elevated levels of T cell IFN-γ production in naïve $CD8^+$ SIY-specific 2C TCR Tg T cells. Taken together, these data suggest that commensal *Bifidobacterium*-derived signals modulate the activation of DCs in the steady state, which in turn supports improved effector function of tumor-specific $CD8^+$ T cells (Sivan et al., 2015).

These data support a model wherein commensal species belonging to the *Bifidobacterium* genus are involved in enhancing antitumor immunity in vivo, through calibration of DC activation, ultimately leading to improved T cell priming and increased $CD8^+$ T cell infiltration into distant tumor sites. An additive effect is observed when *Bifidobacterium* colonization is combined with antibodies targeting the αPD-1/αPD-L1 axis, likely due to targeting two distinct phases of the antitumor immune response—DC-mediated priming of $CD8^+$ T cells and reactivation of tumor-infiltrating $CD8^+$ T cells (Spranger et al., 2014), respectively. Given that these beneficial effects were observed in multiple tumor settings and that alteration of innate immune function was observed, this improved antitumor immunity is likely occurring in an antigen-independent fashion. Mechanistically, *Bifidobacterium* may colonize a specific compartment within the gut that enables it to interact with host cells that are critical for modulating DC function, or to release soluble factors that disseminate systemically leading to improved DC function (Fig. 3).

Based on this model, we propose that one source of intersubject heterogeneity with regard to spontaneous antitumor immunity and therapeutic effects of immune-based interventions may be the composition of gut microbes, which could be manipulated for therapeutic benefit. Importantly, commensal bacteria species other than *Bifidobacterium* may also have the capability to regulate antitumor immunity, either positively or negatively. Similar analyses can be performed in humans using 16S rRNA sequencing of stool samples from patients receiving checkpoint blockade or other immunotherapies, to identify commensals associated with clinical benefit.

It is therefore conceivable that clinical activity of these agents might similarly be improved through rational modulation of the commensal microbiota.

5. GERMLINE GENETIC DIFFERENCES AS AN ADDITIONAL SOURCE OF INTERPATIENT HETEROGENEITY

A third source of intersubject heterogeneity that could impact on the magnitude and quality of endogenous immune responses against tumors is germline polymorphisms in immune regulatory genes. Spontaneous antitumor immune responses have similarities to autoimmunity, and autoimmune diseases have been linked to a wide range of heritable genetic variants. The pursuit of single-nucleotide polymorphisms (SNPs) that might be associated with the degree of T cell infiltration into tumors or clinical response to immunotherapies is just in its infancy. The first germline polymorphism investigation described in this regard is a SNP in the gene encoding the chemokine receptor CCR5. Indeed, a CCR5 polymorphism was identified to be associated with clinical response to high-dose IL-2 (Bedognetti et al., 2013). More recently, a polymorphism in the IRF5 gene was identified that was associated with clinical benefit in a cohort of patients treated with T cell adoptive transfer (Uccellini et al., 2012), suggesting a link to the type I IFN pathway. A comprehensive analysis of germline SNPs in concert with the degree of T cell infiltration within the tumor microenvironment and/or clinical response to contemporary immunotherapies, such as anti-PD-1, is warranted.

6. CONCLUSIONS AND IMPLICATIONS

Novel immunotherapies, such as blocking Abs against PD-1, are showing clinical activity in a broad range of human cancers. Combination therapies targeting two immune regulatory checkpoints concurrently are expected to improve clinical benefit even further, particularly in patients showing the T cell-inflamed tumor microenvironment phenotype. Gaining additional clinical benefit with immunotherapies in patients with the non-T cell-inflamed tumor microenvironment will require a deep mechanistic understanding of tumor-intrinsic and host-derived processes that restrict generation of a spontaneous antitumor T cell response. Interrogation of these possibilities in patients has become technically feasible using modern genomics approaches. Analysis of tumor biopsies by gene expression profiling and exome sequencing, of germline SNPs from blood cells, and of commensal microbiota differences through 16S rRNA sequencing can be performed with respect to clinical response to defined immunotherapies. An integrated bioinformatics approach may also reveal functional interactions between these different axes. Identified resistance pathways that are validated mechanistically in preclinical models can be considered as targets for new therapeutic development. Recent data have identified tumor-intrinsic β-catenin pathway activation, as well as absence of host *Bifidobacterium* colonization, as two candidates ripe for forward translation back into the clinic. Similar analyses in patients bearing the full range of tumor types being treated with anti-PD-1 or other active immunotherapies should be performed in prospective studies, to identify the spectrum of regulatory mechanisms that might be manipulated therapeutically.

REFERENCES

Abt, M. C., et al. (2012). Commensal bacteria calibrate the activation threshold of innate antiviral immunity. *Immunity*, *37*, 158–170. http://dx.doi.org/10.1016/j.immuni.2012.04.011.

Bak, S. P., et al. (2012). Differential requirement for CD70 and CD80/CD86 in dendritic cell-mediated activation of tumor tolerized CD8 T cells. *Journal of Immunology*, *189*, 1708–1716. http://dx.doi.org/10.4049/jimmunol.1201271.

Baldwin, A. S. (2012). Regulation of cell death and autophagy by IKK and NF-kappaB: Critical mechanisms in immune function and cancer. *Immunological Reviews*, *246*, 327–345. http://dx.doi.org/10.1111/j.1600-065X.2012.01095.x.

Basseres, D. S., & Baldwin, A. S. (2006). Nuclear factor-kappaB and inhibitor of kappaB kinase pathways in oncogenic initiation and progression. *Oncogene*, *25*, 6817–6830. http://dx.doi.org/10.1038/sj.onc.1209942.

Bedognetti, D., et al. (2013). CXCR3/CCR5 pathways in metastatic melanoma patients treated with adoptive therapy and interleukin-2. *British Journal of Cancer*, *109*, 2412–2423. http://dx.doi.org/10.1038/bjc.2013.557.

Bosenberg, M., et al. (2006). Characterization of melanocyte-specific inducible Cre recombinase transgenic mice. *Genesis*, *44*, 262–267. http://dx.doi.org/10.1002/dvg.20205.

Bronte, V., & Murray, P. J. (2015). Understanding local macrophage phenotypes in disease: Modulating macrophage function to treat cancer. *Nature Medicine*, *21*, 117–119. http://dx.doi.org/10.1038/nm.3794.

Burdelya, L., et al. (2005). Stat3 activity in melanoma cells affects migration of immune effector cells and nitric oxide-mediated antitumor effects. *Journal of Immunology*, *174*, 3925–3931.

Cheung, A. F., Dupage, M. J., Dong, H. K., Chen, J., & Jacks, T. (2008). Regulated expression of a tumor-associated antigen reveals multiple levels of T-cell tolerance in a mouse model of lung cancer. *Cancer Research*, *68*, 9459–9468. http://dx.doi.org/10.1158/0008-5472.CAN-08-2634.

Compeer, E. B., Flinsenberg, T. W. H., van der Grein, S. G., & Boes, M. (2012). Antigen processing and remodeling of the endosomal pathway: Requirements for antigen cross-presentation. *Frontiers in Immunology*, *3*. http://dx.doi.org/10.3389/fimmu.2012.00037.

Corrales, L., et al. (2015). Direct activation of STING in the tumor microenvironment leads to potent and systemic tumor regression and immunity. *Cell Reports*, *11*, 1018–1030. http://dx.doi.org/10.1016/j.celrep.2015.04.031.
Coussens, L. M., Zitvogel, L., & Palucka, A. K. (2013). Neutralizing tumor-promoting chronic inflammation: A magic bullet? *Science*, *339*, 286–291. http://dx.doi.org/10.1126/science.1232227.
Damsky, W. E., et al. (2011). beta-Catenin signaling controls metastasis in Braf-activated Pten-deficient melanomas. *Cancer Cell*, *20*, 741–754. http://dx.doi.org/10.1016/j.ccr.2011.10.030.
Dankort, D., et al. (2009). Braf(V600E) cooperates with Pten loss to induce metastatic melanoma. *Nature Genetics*, *41*, 544–552. http://dx.doi.org/10.1038/ng.356.
Dunn, G. P., Bruce, A. T., Ikeda, H., Old, L. J., & Schreiber, R. D. (2002). Cancer immunoediting: From immunosurveillance to tumor escape. *Nature Immunology*, *3*, 991–998. http://dx.doi.org/10.1038/ni1102-991.
Edelson, B. T., et al. (2010). Peripheral CD103+ dendritic cells form a unified subset developmentally related to CD8alpha+conventional dendritic cells. *The Journal of Experimental Medicine*, *207*, 823–836. http://dx.doi.org/10.1084/jem.20091627.
Engelhardt, J. J., et al. (2012). Marginating dendritic cells of the tumor microenvironment cross-present tumor antigens and stably engage tumor-specific T cells. *Cancer Cell*, *21*, 402–417. http://dx.doi.org/10.1016/j.ccr.2012.01.008.
Erdag, G., et al. (2012). Immunotype and immunohistologic characteristics of tumor-infiltrating immune cells are associated with clinical outcome in metastatic melanoma. *Cancer Research*, *72*, 1070–1080. http://dx.doi.org/10.1158/0008-5472.CAN-11-3218.
Fuertes, M. B., Woo, S. R., Burnett, B., Fu, Y. X., & Gajewski, T. F. (2013). Type I interferon response and innate immune sensing of cancer. *Trends in Immunology*, *34*, 67–73. http://dx.doi.org/10.1016/j.it.2012.10.004.
Fuertes, M. B., et al. (2011). Host type I IFN signals are required for antitumor CD8+ T cell responses through CD8{alpha}+dendritic cells. *The Journal of Experimental Medicine*, *208*, 2005–2016. http://dx.doi.org/10.1084/jem.20101159.
Ganal, S. C., et al. (2012). Priming of natural killer cells by nonmucosal mononuclear phagocytes requires instructive signals from commensal microbiota. *Immunity*, *37*, 171–186. http://dx.doi.org/10.1016/j.immuni.2012.05.020.
Greten, F. R., et al. (2004). IKKbeta links inflammation and tumorigenesis in a mouse model of colitis-associated cancer. *Cell*, *118*, 285–296. http://dx.doi.org/10.1016/j.cell.2004.07.013.
Hamid, O., et al. (2011). A prospective phase II trial exploring the association between tumor microenvironment biomarkers and clinical activity of ipilimumab in advanced melanoma. *Journal of Translational Medicine*, *9*, 204. http://dx.doi.org/10.1186/1479-5876-9-204.
Harlin, H., et al. (2009). Chemokine expression in melanoma metastases associated with CD8+ T-cell recruitment. *Cancer Research*, *69*, 3077–3085. http://dx.doi.org/10.1158/0008-5472.CAN-08-2281.
Hildner, K., et al. (2008). Batf3 deficiency reveals a critical role for CD8alpha+dendritic cells in cytotoxic T cell immunity. *Science*, *322*, 1097–1100. http://dx.doi.org/10.1126/science.1164206.
Hooper, L. V., Littman, D. R., & Macpherson, A. J. (2012). Interactions between the microbiota and the immune system. *Science*, *336*, 1268–1273. http://dx.doi.org/10.1126/science.1223490.
Iannello, A., Thompson, T. W., Ardolino, M., Lowe, S. W., & Raulet, D. H. (2013). p53-Dependent chemokine production by senescent tumor cells supports NKG2D-dependent tumor elimination by natural killer cells. *The Journal of Experimental Medicine*, *210*, 2057–2069. http://dx.doi.org/10.1084/jem.20130783.

Ihara, S., et al. (2012). Inhibitory roles of signal transducer and activator of transcription 3 in antitumor immunity during carcinogen-induced lung tumorigenesis. *Cancer Research, 72*, 2990–2999. http://dx.doi.org/10.1158/0008-5472.CAN-11-4062.

Iida, N., et al. (2013). Commensal bacteria control cancer response to therapy by modulating the tumor microenvironment. *Science, 342*, 967–970. http://dx.doi.org/10.1126/science.1240527.

Ivanov, I. I., & Honda, K. (2012). Intestinal commensal microbes as immune modulators. *Cell Host & Microbe, 12*, 496–508. http://dx.doi.org/10.1016/j.chom.2012.09.009.

Ivanov, I. I., et al. (2009). Induction of intestinal Th17 cells by segmented filamentous bacteria. *Cell, 139*, 485–498. http://dx.doi.org/10.1016/j.cell.2009.09.033.

Jancic, C., et al. (2007). Rab27a regulates phagosomal pH and NADPH oxidase recruitment to dendritic cell phagosomes. *Nature Cell Biology, 9*, 367–378. http://dx.doi.org/10.1038/ncb1552.

Kabashima, K., et al. (2007). CXCL12-CXCR4 engagement is required for migration of cutaneous dendritic cells. *The American Journal of Pathology, 171*, 1249–1257. http://dx.doi.org/10.2353/ajpath.2007.070225.

Larkin, J., et al. (2015). Combined nivolumab and ipilimumab or monotherapy in untreated melanoma. *The New England Journal of Medicine, 373*, 23–34. http://dx.doi.org/10.1056/NEJMoa1504030.

Mackey, M. F., et al. (1998). Dendritic cells require maturation via CD40 to generate protective antitumor immunity. *Journal of Immunology, 161*, 2094–2098.

Mikucki, M. F. D., Matsuzaki, J., Skitzki, J., Gaulin, N., Muhitch, J., Ku, A., et al. (2015). Non-redundant requirement for CXCR3 signaling during tumoricidal T cell trafficking across tumor vascular checkpoints. *Nature Communications, 6*, 7458.

Mittendorf, E. A., et al. (2014). PD-L1 expression in triple-negative breast cancer. *Cancer Immunology Research, 2*, 361–370. http://dx.doi.org/10.1158/2326-6066.CIR-13-0127.

Muthuswamy, R., et al. (2012). NF-kappaB hyperactivation in tumor tissues allows tumor-selective reprogramming of the chemokine microenvironment to enhance the recruitment of cytolytic T effector cells. *Cancer Research, 72*, 3735–3743. http://dx.doi.org/10.1158/0008-5472.CAN-11-4136.

Nukiwa, M., et al. (2006). Dendritic cells modified to express fractalkine/CX3CL1 in the treatment of preexisting tumors. *European Journal of Immunology, 36*, 1019–1027. http://dx.doi.org/10.1002/eji.200535549.

Pan, J., et al. (2004). Interferon-gamma is an autocrine mediator for dendritic cell maturation. *Immunology Letters, 94*, 141–151. http://dx.doi.org/10.1016/j.imlet.2004.05.003.

Pettit, A. R., et al. (1997). Nuclear localization of RelB is associated with effective antigen-presenting cell function. *Journal of Immunology, 159*, 3681–3691.

Pikarsky, E., et al. (2004). NF-kappaB functions as a tumour promoter in inflammation-associated cancer. *Nature, 431*, 461–466. http://dx.doi.org/10.1038/nature02924.

Quigley, D., et al. (2015). Lymphocyte invasion in IC10/basal-like breast tumors is associated with wild-type TP53. *Molecular Cancer Research: MCR, 13*, 493–501. http://dx.doi.org/10.1158/1541-7786.MCR-14-0387.

Rutkowski, M. R., et al. (2015). Microbially driven TLR5-dependent signaling governs distal malignant progression through tumor-promoting inflammation. *Cancer Cell, 27*, 27–40. http://dx.doi.org/10.1016/j.ccell.2014.11.009.

Salerno, E. P., Olson, W. C., McSkimming, C., Shea, S., & Slingluff, C. L., Jr. (2014). T cells in the human metastatic melanoma microenvironment express site-specific homing receptors and retention integrins. *International Journal of Cancer. Journal international du cancer. 134*, 563–574. http://dx.doi.org/10.1002/ijc.28391.

Schmid, M. C., et al. (2011). Receptor tyrosine kinases and TLR/IL1Rs unexpectedly activate myeloid cell PI3kgamma, a single convergent point promoting tumor

inflammation and progression. *Cancer Cell, 19*, 715–727. http://dx.doi.org/10.1016/j.ccr.2011.04.016.
Scholer, A., Hugues, S., Boissonnas, A., Fetler, L., & Amigorena, S. (2008). Intercellular adhesion molecule-1-dependent stable interactions between T cells and dendritic cells determine CD8+ T cell memory. *Immunity, 28*, 258–270. http://dx.doi.org/10.1016/j.immuni.2007.12.016.
Sivan, A., et al. (2015). Commensal Bifidobacterium promotes antitumor immunity and facilitates anti-PD-L1 efficacy. *Science, 350*, 1084–1089. http://dx.doi.org/10.1126/science.aac4255.
Spranger, S., Bao, R., & Gajewski, T. F. (2015). Melanoma-intrinsic beta-catenin signalling prevents anti-tumour immunity. *Nature, 523*(7559), 231–235. http://dx.doi.org/10.1038/nature14404.
Spranger, S., et al. (2013). Up-regulation of PD-L1, IDO, and tregs in the melanoma tumor microenvironment is driven by CD8+ T cells. *Science Translational Medicine, 5*, 200ra116. http://dx.doi.org/10.1126/scitranslmed.3006504.
Spranger, S., et al. (2014). Mechanism of tumor rejection with doublets of CTLA-4, PD-1/PD-L1, or IDO blockade involves restored IL-2 production and proliferation of CD8(+) T cells directly within the tumor microenvironment. *Journal for Immunotherapy of Cancer, 2*, 3. http://dx.doi.org/10.1186/2051-1426-2-3.
Stober, C. B., Brode, S., White, J. K., Popoff, J., & Blackwell, J. M. (2007). Slc11a1, formerly Nramp1 is expressed in dendritic cells and influences major histocompatibility complex class II expression and antigen-presenting cell function. *Infection and immunity, 75*, 5059–5067. http://dx.doi.org/10.1128/iai.00153-07.
Sullivan, R. J., et al. (2009). A single center experience with high-dose (HD) IL-2 treatment for patients with advanced melanoma and pilot investigation of a novel gene expression signature as a predictor of response. *Journal of Clinical Oncology, 27*, 9003.
Topalian, S. L., et al. (2012). Safety, activity, and immune correlates of anti-PD-1 antibody in cancer. *The New England Journal of Medicine, 366*, 2443–2454. http://dx.doi.org/10.1056/NEJMoa1200690.
Toso, A., et al. (2014). Enhancing chemotherapy efficacy in Pten-deficient prostate tumors by activating the senescence-associated antitumor immunity. *Cell Reports, 9*, 75–89. http://dx.doi.org/10.1016/j.celrep.2014.08.044.
Tumeh, P. C., et al. (2014). PD-1 blockade induces responses by inhibiting adaptive immune resistance. *Nature, 515*, 568–571. http://dx.doi.org/10.1038/nature13954.
Uccellini, L., et al. (2012). IRF5 gene polymorphisms in melanoma. *Journal of Translational Medicine, 10*, 170. http://dx.doi.org/10.1186/1479-5876-10-170.
Viaud, S., et al. (2013). The intestinal microbiota modulates the anticancer immune effects of cyclophosphamide. *Science, 342*, 971–976. http://dx.doi.org/10.1126/science.1240537.
Wang, T., et al. (2004). Regulation of the innate and adaptive immune responses by Stat-3 signaling in tumor cells. *Nature Medicine, 10*, 48–54. http://dx.doi.org/10.1038/nm976.
Woo, S. R., et al. (2014). STING-dependent cytosolic DNA sensing mediates innate immune recognition of immunogenic tumors. *Immunity, 41*, 830–842. http://dx.doi.org/10.1016/j.immuni.2014.10.017.
Xue, W., et al. (2007). Senescence and tumour clearance is triggered by p53 restoration in murine liver carcinomas. *Nature, 445*, 656–660. http://dx.doi.org/10.1038/nature05529.
Zhang, L., et al. (2003). Intratumoral T cells, recurrence, and survival in epithelial ovarian cancer. *New England Journal of Medicine, 348*, 203–213. http://dx.doi.org/10.1056/NEJMoa020177.

CHAPTER FOUR

Immune Contexture, Immunoscore, and Malignant Cell Molecular Subgroups for Prognostic and Theranostic Classifications of Cancers

Etienne Becht[*,†,‡], **Nicolas A. Giraldo**[*,†,‡], **Claire Germain**[*,†,‡], **Aurélien de Reyniès**[§], **Pierre Laurent-Puig**[¶], **Jessica Zucman-Rossi**[||], **Marie-Caroline Dieu-Nosjean**[*,†,‡], **Catherine Sautès-Fridman**[*,†,‡], **Wolf H. Fridman**[*,†,‡,1]

[*]INSERM UMR_S 1138, Cancer, Immune Control and Escape, Cordeliers Research Centre, Paris, France
[†]Université Paris-Descartes, Sorbonne Paris Cité, UMR_S 1338, Cordeliers Research Centre, Paris, France
[‡]Sorbonne Universités, UPMC Univ Paris 06, UMR_S 1138, Cordeliers Research Centre, Paris, France
[§]Programme Cartes d'Identité des Tumeurs, Ligue Nationale Contre le Cancer, Paris, France
[¶]Université Paris Descartes, Sorbonne Paris Cite; INSERM, UMR-S1147, Paris, France
[||]INSERM, UMR-1162, Génomique fonctionnelle des tumeurs solides, Paris, France
[1]Corresponding author: e-mail address: herve.fridman@crc.jussieu.fr

Contents

Advances in Immunology, Volume 130
ISSN 0065-2776
http://dx.doi.org/10.1016/bs.ai.2015.12.002

Abstract

The outcome of tumors results from genetic and epigenetic modifications of the transformed cells and also from the interactions of the malignant cells with their tumor microenvironment (TME), which includes immune and inflammatory cells. For a given cancer type, the composition of the immunological TME is not homogeneous. Heterogeneity is found between different cancer types and also between tumors from patients with the same type of cancer. Some tumors exhibit a poor infiltration by immune cells, and others are highly infiltrated by lymphocytes. Among the latter, the architecture of the TME, with the localization of immune cells in the invasive front and the center of the tumor, the presence of tumor-adjacent organized lymphoid aggregates, and the type of inflammatory context, determines the prognostic impact of the infiltrating cells. The description and the understanding of the immune and inflammatory landscape in human tumors are of paramount importance at different levels of patient's care. It completes the mutational, transcriptional, and epigenetic patterns of the malignant cells and open paths to understand how tumor cells shape their immune microenvironment and are shaped by the immune reaction. It provides prognostic and theranostic markers, as well as novel targets for immunotherapies.

ABBREVIATIONS

ADCC antibody-dependent cell cytotoxcity
AID activation-induced cytidine deaminase
APC antigen-presenting cell
CAF cancer-associated fibroblast
ccRCC clear-cell renal cell carcinoma
CDC complement-dependent cascade
CIMP CpG-island methylator phenotype
CIN chromosomal instability
CRC colorectal cancer
CSR class switch recombination
CT center of the tumor
DC dendritic cell
DFS disease-free survival
FAP familial adenomatous polyposis
HEV high endothelial venule
IHC immunohistochemistry
IM invasive margin
MDSC myeloid-derived suppressor cell
MMR mismatch repair
MSI microsatellite instability

NK natural killer
non-TLS-DC dendritic cell nonassociated with a tertiary lymphoid structure
NSCLC non-small cell lung cancer
OS overall survival
PDGF platelet-derived growth factor
RCC renal cell carcinoma
ROS reactive oxygen species
TAM tumor-associated macrophage
TCGA The Cancer Genome Atlas
TCM central-memory T cell
TCR T-cell receptor
TEM effector-memory T cell
T_{fh} follicular-helper T cell
TIL tumor-infiltrating lymphocyte
TLS tertiary lymphoid structure
TLS-DC dendritic cell associated with a tertiary lymphoid structure
TME tumor microenvironment
T_{reg} regulatory T cell
$T_{\gamma\delta}$ gamma-delta T cell
VEGF vascular endothelial growth factor

1. INTRODUCTION

Tumors start with early dysplasic lesions, followed by malignant transformation into a locally invasive tumor and subsequent distant metastases. In this review, we will provide evidence for immune reactions within tumors and present the cellular composition and organization of the tumor microenvironment and the molecular mediators involved. We will summarize the genetic and epigenetic modifications of malignant cells and how tumor cells may influence their microenvironment. We will underline the novel prognostic markers and therapeutic targets provided by studies of the cross talk between tumors and their microenvironment. We will particularly focus on colorectal cancers (CRCs) and renal cell cancers (RCCs).

1.1 Control of Cancer Growth Through Adaptive Immunity

A fundamental aspect of cancer immunology is that the immune system, mainly through its adaptive arm, is able to identify and eliminate host cells undergoing malignant transformation and control tumor growth. Although the idea of a cancer-protecting function of the immune system was proposed more than a century ago (Ehrlich, 1909), it remained controversial during the 20th century (Stutman, 1974; Stutman, 1979). This idea is now widely accepted and considered a hallmark of cancer (Hanahan & Weinberg, 2011).

Schreiber and colleagues proposed the 3E theory (Dunn, Old, & Schreiber, 2004), standing for *E*limination, *E*quilibrium, and *E*scape. The idea is that most cells undergoing malignant transformation are recognized by immune cells and eliminated. Sometimes, some tumor cells proliferate fast enough to compensate for immune-mediated elimination, and the overall tumor size remains at equilibrium. In the third stage, tumor cells escape elimination from the immune system and grow uncontrolled. The challenge in studying how immune cells are able to eliminate cancer cells is that clinically detectable tumors have, by definition, managed to grow from a single mutated cell to a mass of derived malignant cells.

1.1.1 Immunosurveillance of Cancer

Epidemiological observations in immunodeficient patients indirectly hinted at a protective role of the immune system against cancer occurrence. In particular, it was observed in the 1980–90 decade that patients suffering from HIV/AIDS have higher chances of developing Kaposi's sarcoma (Haverkos & Drotman, 1985; Pape et al., 1983; Schreiber & Podack, 2009), a virus-associated cancer. Transplant recipients, who receive immunosuppressive therapies to avoid transplant rejection, have a consistently higher risk of developing this malignancy (Farge, 1993). Strikingly, transplant recipients are also at higher risk of developing other solid tumors, such as brain (Curtis et al., 1997) and other central nervous system malignancies (Curtis et al., 1997), thyroid (Curtis et al., 1997), bone (Curtis et al., 1997), colon (Curtis et al., 1997), lung (Kasiske, Snyder, Gilbertson, & Wang, 2004), prostate (Kasiske et al., 2004), stomach (Kasiske et al., 2004), esophagus (Kasiske et al., 2004), pancreas (Kasiske et al., 2004), ovary (Kasiske et al., 2004), breast (Kasiske et al., 2004), melanoma (Curtis et al., 1997; Kasiske et al., 2004), leukemia (Kasiske et al., 2004), hepatobiliary tumors (Curtis et al., 1997; Kasiske et al., 2004), cervical and vulvovaginal cancers (Kasiske et al., 2004), testicular (Kasiske et al., 2004), bladder (Kasiske et al., 2004), kidney (Kasiske et al., 2004), nonmelanoma skin cancers (Kasiske et al., 2004), oral cavity cancers (Curtis et al., 1997), and non-Hodgkin lymphoma (Kasiske et al., 2004).

Following these observations, several studies showed an increase in carcinogen-induced tumors numbers in immunodeficient mice compared to wild-type mice (Schreiber & Podack, 2009; Shankaran et al., 2001). Importantly, it was observed that a knockout of the Rag2 gene, which encodes the Recombination-activating gene 2, is sufficient to induce this effect. Rag2 is necessary for the V(D)J recombination and therefore for the survival of B and T cells in the periphery, supporting the idea that the adaptive arm of the immune system is responsible for the control of tumor occurrence. Direct evidence was later obtained using in vivo two-photon imaging: C57BL/6 mice were injected with the syngenic EL4 cell line expressing the exogenous OVA protein and latter injected with OT-I T cells specific for OVA antigens. In situ and in vivo imaging of the CD8 protein and the activity of CASP3, an apoptosis marker, showed that the cytotoxic lymphocytes $CD8^{+}$ T cells were actively killing tumor cells (Breart, Lemaître, Celli, & Bousso, 2008). The authors also noted that the elimination was a slow process, taking on average 6 h, and proposed that the amount of tumor-targeting $CD8^{+}$ T cells could be a limiting factor in the control of tumor growth.

1.1.2 Tumor Antigens

The adaptive immune system is able to discriminate between self and nonself. The distinction is clear in the case of invasive pathogens, but cancer cells are host's transformed cells and could therefore be thought as belonging to the self. The malignant transformation of normal cells involves mutations (Vogelstein et al., 2013) in pathways controlling cell proliferation, resistance to apoptosis, and other fundamental characteristics of cancers (Hanahan & Weinberg, 2011). These mutations encode epitopes that are not expressed by the host's normal cells and could therefore be considered as nonself, marking the malignant cell for elimination by the adaptive immune system (Fig. 1). Mutations encoding mutated peptides capable of eliciting an adaptive immune response are known as neoantigens. $CD8^{+}$ T cells clones specific for tumor antigens were found against the cyclin-dependent kinase 4 gene CDK4 (Wölfel et al., 1995) or mutated β-catenin (Robbins et al., 1996) in melanoma, and the tumor-expressed MHC Class II HLA-A2 gene in RCC (Brändle, Brasseur, Weynants, Boon, & Van den Eynde, 1996).

Interestingly, host's nonmutated proteins were also found to be targeted by the adaptive immune system. In particular, proteins that are expressed in MHC Class I-negative cells, such as sperm cells or trophoblasts, and aberrantly expressed by cancer cells, are able to elicit immune responses. The first example in human was an epitope of the MAGEA1 (van der Bruggen et al., 1991) testis-restricted protein aberrantly expressed in melanoma. Antigens were later found in proteins encoded by the MAGE family of antigens

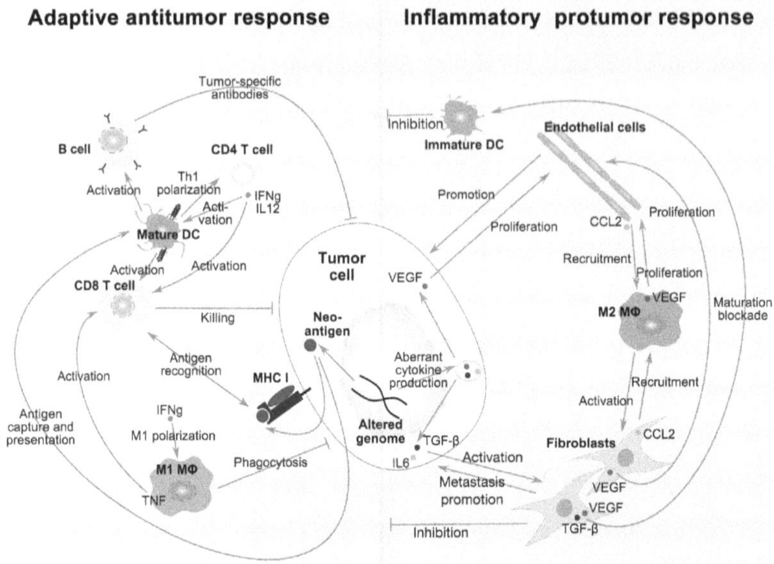

Figure 1 The two immune axes with their opposite effects on tumor evolution. Tumors' altered genomes can encode neoantigens that foster tumor-targeted adaptive immune responses. These are characterized by high densities of CD8$^+$ T cells, mature DC and B cells, as well as a Th1 polarization of the tumor microenvironment. On the other hand, genomic alterations lead to aberrant cytokine production in tumor cells, which can increase neoangiogenesis, activate fibroblasts and myeloid cells, result in the production of growth and survival factors, and inhibit the adaptive immune antitumor response. M1 MΦ: M1 macrophage. M2 MΦ, M2 macrophage; MHC I, Class I major histocompatibility complex; DC, dendritic cell. (See the color plate.)

and other germline-specific genes, such as BAGE1, GAGE1, XAGE1B, CTAG2, CTAG1, and SSX2, in melanoma, and also lung, colorectal, breast, and prostate carcinomas (Coulie et al., 1994). More surprisingly, some proteins constitutively expressed by nonmalignant cells were found to also elicit immune responses. These proteins are usually overexpressed in tumor cells, leading to a TCR-mediated activation of the corresponding specific lymphocytes (Coulie et al., 1994), while their expression is too low in normal cells to reach the threshold leading to T-cell activation. Examples of such antigens include the "prostate-specific antigen" protein in prostate cancer (Correale et al., 1997; Olson et al., 2010), the HER2/neu antigen (Fisk, Blevins, Wharton, & Ioannides, 1995) encoded by the amplified ERBB2 gene in breast and ovarian cancers, as well as the Melan-A protein in melanoma

(Coulie, Van den Eynde, van der Bruggen, & Boon, 2014; Kawakami, Eliyahu, Delgado, et al., 1994; Kawakami, Eliyahu, Sakaguchi, et al., 1994).

A more straightforward class of tumor-associated antigens are the peptides associated with carcinogenic viruses. These peptides remain expressed in the transformed cells and can elicit immune responses, as shown with Human Papilloma Virus infection in head and neck squamous cell carcinoma (van der Burg & Melief, 2011), Epstein–Barr virus in Hodgkin's lymphoma, nasopharyngeal carcinoma, NKT lymphoma, and Burkitt's lymphoma (Long, Parsonage, Fox, & Lee, 2010). Altogether, these examples show that although tumor cells are derived from normal cells, they express antigens that can be recognized as nonself.

1.1.3 Tumor Control by the Tumor Microenvironment (TME)

Tumor control resulting from genetic and epigenetic modifications of the transformed cells takes place in the TME which contains immune cells. Thus, in several situations in which a precancerous state can be individualized and studied, a shift from an immunological pattern with a T_{h1} orientation to a proinflammatory TME correlates with tumor invasiveness and aggressiveness. It is illustrated in cervical carcinoma in which high expression of genes encoding T_{h1} cytokines is evidenced in cervical in situ neoplasia, whereas IFN-γ expression is lost, and the expression of proinflammatory cytokines IL-6 is high in invasive and aggressive cervical carcinoma (Tartour et al., 1998). A change toward a T_{h2}-type cytokine pattern has also been reported in the evolution from intraepithelial neoplasia to invasive carcinoma (Bais et al., 2005). A similar shift has been described in pancreatic cancer, where there is a decrease in the density of $CD8^+$ T cells and mature dendritic cells (DCs) from low-grade premalignant lesions into invasive ductal adenocarcinoma (Hiraoka, Onozato, Kosuge, & Hirohashi, 2006; Hiraoka et al., 2011).

1.2 Inflammation and Cancer

It is well established that chronic inflammation in a tumor favors the outgrowth of malignant cells. Inflammation is associated with many carcinogenic events. For instance, mutation of the RET oncogene in thyrocytes is sufficient to induce papillary thyroid carcinoma and is accompanied with up-regulation of proinflammatory genes (Borrello et al., 2005). Mutation of RAS oncogenes in an ovarian cancer cell line xenograft in athymic mice is associated to the production of IL-8 (CXCL8) by the tumor cells, resulting in increased angiogenesis (Sparmann & Bar-Sagi, 2004). Activating mutation of the IL-6-receptor signal transducer gp130 (IL-6ST) triggers an inflammatory program in hepatocytes, favoring adenoma formation which can later develop into hepatocarcinoma (Rebouissou et al., 2009). Exogenous conditions can also trigger inflammatory signals. *Helicobacter pylori* is associated with gastric cancer (Forman et al., 1991; Nomura et al., 1991; Parsonnet et al., 1991) and mucosa-associated lymphoid tissue (MALT) lymphoma (Wotherspoon et al., 1993; Wotherspoon, Ortiz-Hidalgo, Falzon, & Isaacson, 1991), viral infection with Hepatitis B or C viruses is associated with the development of hepatocellular carcinoma (Grivennikov, Greten, & Karin, 2010), and tobacco smoke exposure triggers chronic lung inflammation which favors lung carcinoma development (Takahashi, Ogata, Nishigaki, Broide, & Karin, 2010). It has been proposed that 15% of all diagnosed cancers are caused by infection (Coussens & Werb, 2002). Inflammation is a very broad concept that refers to immune response-promoting conditions and can designate almost every immune cell population. In the context of tumor immunology, it usually refers to immune cell populations or cytokines that promote cancer growth. It is paradoxical as inflammation is associated with the promotion of immune responses, including adaptive immune responses. Some authors described adaptive immune responses as "good inflammation" and the protumor signals as "bad inflammation" (Mantovani, Allavena, Sica, & Balkwill, 2008) (Fig. 1). In this chapter, we use "inflammation" to refer to the latter. In the following section, we detail some of tumor-promoting effects associated with inflammation through direct effect on cancer cells.

1.2.1 Inflammation Promotes Carcinogenesis

The role of inflammation in cancer is particularly illustrated in inflamed tissues secondary to persisting infectious agents, whatever their types. Thus, as stated earlier, bacteria such as *H. pylori* support the emergence of gastric cancer (Salama, Hartung, & Müller, 2013) and MALT lymphoma (Parsonnet et al., 1994), without directly transforming the malignant cell. Viruses, such as Hepatitis B and C, create liver inflammation that may result in some cases in malignant hepatocellular carcinoma (Arzumanyan, Reis, & Feitelson, 2013). A similar phenomenon is observed in Kaposi's sarcoma, induced by Herpes Virus type 8, often in the context of immunodepressed HIV-positive patients (Haverkos & Drotman, 1985; Pape et al., 1983). The case of papilloma viruses and Epstein–Barr virus is more complex since they are oncogenic for the infected cells and also create an inflammatory TME (Castle et al., 2001). The association of parasites, such as *Schistosoma haematobium* and *Opisthorchis viverrini/Clonorchis sinensis* with higher incidence of bladder cancer and cholangiocarcinoma, respectively, reveals similar inflammation-favored carcinogenesis (Sripa et al., 2012; Vennervald & Polman, 2009). In addition to chronic infection, cancer-favoring inflammation can also be the consequence of external stimuli, such as smoke, asbestos, silica, alcohol, aflatoxin, or chronic inflammatory diseases, such as gastritis, pancreatitis, inflammatory bowel disease, thyroiditis, and osteomyelitis, which favor cancer emergence in the corresponding organ (reviewed in Fridman et al., 2014).

Preexisting inflammation can directly promote carcinogenesis through mutational effects. In particular, reactive oxygen species (ROS) have been shown to directly modify DNA sequences (Colotta, Allavena, Sica, Garlanda, & Mantovani, 2009). Inflammatory reactions can also lead to epigenetic modifications of the DNA, albeit no causal effect on carcinogenesis was observed (Hahn et al., 2008).

1.2.2 Inflammatory Mediators Promote Cell Survival and Cell Proliferation

Inflammatory factors, such as IL-6, IL-1β, or IL-22, induce inflammatory responses leading to the activation of transcription factors NF-κB and STAT3 in cancer cells. These pathways can lead to the expression of anti-apoptotic molecules (Al Zaid Siddiquee & Turkson, 2008; Fan, Yang, & Engelhardt, 2002), such as BCL2 (Mantovani et al., 2008) and BCL-X (BCL2L1) (Elinav et al., 2013) which promotes survival of malignant B cells in follicular lymphoma (Tsujimoto, Finger, Yunis, Nowell, & Croce, 1984). Mutation of STAT3 in nontumorigenic immortalized fibroblasts is able to induce successful xenograft in nude mice (Bromberg et al., 1999), and activation of the IL-6-STAT3 pathway has been shown to have procarcinogenesis activities in many malignancies (Elinav et al., 2013), including pancreatic ductal adenocarcinoma (Fukuda et al., 2011) and intraepithelial carcinoma (Lesina et al., 2011), and lung (Gao et al., 2007) and gastric (Bronte-Tinkew et al., 2009) adenocarcinomas. NF-κB activation promotes cancer cells' survival, for instance by inhibiting TNF-mediated apoptosis (Micheau & Tschopp, 2003).

Inflammation also triggers proliferation of malignant cells (Adachi et al., 2006), notably by increasing the expression of the cyclins B, D1, and D2 (Bollrath et al., 2009; Elinav et al., 2013; Grivennikov et al., 2009).

1.2.3 Inflammation Triggers Angiogenesis

When the tumor reaches a certain size, oxygen supply becomes too limited for it to diffuse in all the areas of the tumor and hypoxic conditions arise. Inflammation is one of the mechanisms subverted by tumors to sustain neoangiogenesis and increase blood supply. Inflammatory mediators released by either malignant, hematopoietic or other stromal cells can increase local angiogenesis. For instance, IL-1β released by cancer cells simultaneously triggers angiogenesis and the recruitment of inflammatory cells to the tumor bed in an MYC-dependent pancreatic β-cell cancer mouse model (Shchors et al., 2006). Mast cells in turn can promote angiogenesis (Mantovani et al., 2008; Soucek et al., 2007). While tumor cells have been known for a long time to induce local angiogenesis, the contribution of stromal cells, and in particular innate immune cells, to neoangiogenesis is now established (Bingle, Lewis, Corke, Reed, & Brown, 2006; Du et al., 2008; Giraudo, Inoue, & Hanahan, 2004; Lin et al., 2006; Rivera & Bergers, 2015; Shojaei et al., 2007), notably through the release of vascular endothelial growth factors (VEGFs), epidermal growth factor, fibroblast growth factor 2, TNF-α, TGF-β, platelet-derived growth factors (PDGFs), placental growth factor, neuropilin-1, and IL-8 (CXCL8) (Rivera & Bergers, 2015). Among innate immune cells, macrophages are the most abundant in tumors (Bindea et al., 2013) and were shown to control angiogenesis in a mouse model of breast cancer (Lin et al., 2006) through their ability to secrete VEGF-A (Lin et al., 2007) (Fig. 1). Inhibition of the colony-stimulating factor 1-receptor (CSF1R), which is required for macrophage differentiation and survival, was shown to inhibit neovascularization in a glioma model (Pyonteck et al., 2013).

1.2.4 Role of Antiinflammatory Agents

Modulating inflammation is a potential treatment modality for the treatment of some cancers. First, prophylactic use of antiinflammatory agents such as aspirin has been shown to be associated with a reduction in CRC incidence (Algra & Rothwell, 2012; Chan, Ogino, & Fuchs, 2007; Flossmann & Rothwell, 2007), a finding that has then been extended to breast (Gierach et al., 2008), esophageal, gastric, prostate, and lung cancers (Thorat & Cuzick, 2013). Second, suppressing inflammatory signals in highly inflammatory clinically detectable cancers can lead to a halt of tumor growth (Balkwill, 2009), potentially synergizing with cytotoxic agents (Green, Ferguson, Zitvogel, & Kroemer, 2009).

1.3 Tumor Escape of the Adaptive Immune Control

DNA instability in tumors, associated to selective pressure from immune-mediated elimination of tumor cells, leads to the emergence of escape mechanisms in tumor cells. These can either be directly induced by the tumor cells or through other cell populations of the TME.

1.3.1 Reduction of the Immunogenicity of the Malignant Cells

$CD8^+$ T cells are the main effector of antitumor immune responses, but their activity requires the presentation of peptides on target cells by Class I MHC molecules. These molecules are heterodimers consisting of two subunits, one of them being encoded by a single gene, B2M. Some tumor cells harbor inactivating mutations in the B2M gene, abrogating expression of any functional Class I MHC molecule and inhibiting the activity of $CD8^+$ T cells (Bernal, Ruiz-Cabello, Concha, Paschen, & Garrido, 2012). NK cells can sense the loss of Class I MHC expression and exert contact-dependent cytotoxicity. However, tumor-infiltrating NK cells have been reported to display inhibited phenotypes compared to NK cells populating nonmalignant tissue distant from the tumor. Indeed, NK cells infiltrating non-small cell lung cancer (NSCLC) tumors were shown to downregulate the expression of the activating receptors NKp30, NKp80, CD16, NKG2D, and DNAM-1 and consistently to have lower degranulation and cytotoxic capacities ex vivo (Platonova et al., 2011), possibly owing to TGF-β signaling (Donatelli et al., 2014). Similar results have been observed in melanoma (Pietra et al., 2012) and breast cancer (Mamessier et al., 2011).

Loss of Class I MHC expression is a striking illustration of tumor adaptation to immune pressure (Angell, Lechner, Jang, LoPresti, & Epstein, 2014). However, in most cases, immune escape occurs in a more subtle and slow way, through the selection of peptides with low immunogenicity. Mouse models have successfully illustrated this phenomenon, as 3-methylcholanthrene-induced tumors from immunocompetent mice have higher xenograft success rates in syngenic fully immunocompetent mice compared to those grown in $Rag^{-/-}$ mice lacking T and B lymphocytes (Shankaran et al., 2001). In human melanoma, vaccination based on the tumor-expressed gp100 peptide induced a reduction in tumor gp100 expression compared to prevaccination samples (Riker et al., 1999).

1.3.2 Expression of Inhibitory Molecules

As other immune cells, $CD8^+$ T cells express, either constitutively or after activation, inhibitory receptors (also known as immune checkpoints) that regulate their activity. Tumors are able to subvert this mechanism and avoid $CD8^+$ T cell-mediated elimination. A notable example includes the expression of PD-1 (PDCD1) ligands by tumor cells. PD-1 is an inhibitory receptor expressed by a variety of immune cells, including T cells. It can bind to either PD-L1 (CD274) or PD-L2 (PDCD1LG2), which results in reduced cytotoxic capacities (Blank et al., 2004), proliferative capacities (Blank et al., 2004; Carter et al., 2002), and response to TCR stimulation (Freeman et al., 2000). In physiological conditions, IFN-γ the major cytokine of the T_{h1} axis, produced by activated T_{h1} and $CD8^+$ cells, has been shown to induce PD-L1 and PD-L2 expression by surrounding cells (Lee et al., 2005; Mazanet & Hughes, 2002). Immune checkpoints include other molecules such as LAG-3, CTLA-4, and TIM-3. Contact-dependent mechanisms can even mediate T-cell elimination, such as Fas-ligand expressed by tumor cells which can bind to Fas expressed by surrounding lymphocytes, inducing their apoptosis (Chappell & Restifo, 1998; O'Connell, O'Sullivan, Collins, & Shanahan, 1996), but the importance of this effect is still debated (Igney, Behrens, & Krammer, 2000; Igney & Krammer, 2005).

Tumor cells can also release soluble factors that result in suppression of T-cell responses in the microenvironment. TGF-β and VEGF-A orientate the functionality of surrounding hematopoietic cells toward a suppressive phenotype. Other factors, such as the antiinflammatory interleukin IL-10, Galectin-1 (LGALS1) (Gabrilovich, Nadaf, Corak, Berzofsky, & Carbone, 1996), gangliosides (Rabinovich, Gabrilovich, & Sotomayor, 2007), and prostaglandin E2 (PGE2) (Chahlavi et al., 2005; Hahne et al., 1996), are implicated in the direct inhibition or elimination of infiltrating T cells.

1.3.3 Microenvironment-mediated Immunosuppression

Many suppressive pathways involve multiple cell populations from the TME. A critical step in the TCR-mediated activation of T lymphocytes is its interaction with an antigen-presenting cell (APC), mostly DCs and macrophages, but also some B-cell subsets. In addition to presenting Class I and Class II MHC-bound peptides to $CD4^+$ and $CD8^+$ T cells, respectively (primary activation signal), APCs deliver co-stimulatory signals. These signals depend on which ligand/receptor couples are engaged between the T cell and the APC. The type of signal depends mostly on the activation status of the APC. Notably after sensing of danger signals, DCs mature and consequently convey co-stimulatory signals to the T cell. On the other hand, immature DCs presenting a specific antigen to a T cell will transduce tolerogenic signals, by repressing the T cell's ability to respond to future TCR stimulation, apoptosis, or differentiation to a T_{reg} phenotype.

Several studies reported a defective presentation by DCs infiltrating tumors. First, the maturation of monocytes to DCs is dampened in favor of a macrophage differentiation through the action of IL-6 and macrophage colony-stimulating factor 1 (CSF1) (Chomarat, Banchereau, Davoust, & Palucka, 2000; Menetrier-Caux et al., 1998). Second, the maturation process of DCs is inhibited (Alcalay & Kripke, 1991; Chaux, Moutet, Faivre, Martin, & Martin, 1996; Engelhardt et al., 2012; Gabrilovich et al., 1996; Giraldo et al., 2015; Tas, Simons, Balm, & Drexhage, 1993) by several mechanisms. Molecular mediators of the response to hypoxia pathway such as VEGF-A have been implicated in impairment of DC maturation (Fig. 1) through the inhibition of the inflammatory transcription factor NF-κB (Oyama et al., 1998). The theory of immunogenic cell death proposes that activation of DC mostly depends on the type of cell death tumor cells underwent before their uptake by phagocytes (Green et al., 2009). Markers of immunogenic cell death notably include the translocation of the chaperone calreticulin from the cytosol to the plasma membrane and the release of adenosine triphosphate and of high-mobility group box 1 (HMGB1) protein in the extracellular milieu. Mice models have shown that after cancer cell line injection in an immunocompetent host and clearance due to cytotoxic chemotherapy, subsequent rechallenge using the same cell line will lead to rejection only in the case of immunogenic cell death (Casares et al., 2005). The fact that myeloid cells in the TME exert tolerogenic roles led to the functional definition of myeloid-derived suppressor cells (MDSCs) (Gabrilovich & Nagaraj, 2009). Lack of consensual markers in humans hamper comprehensive analyses of the MDSC populations in human tumors, and a wide variety of factors have been implicated in their expansion and polarization, including VEGF, GM-CSF, G-CSF, M-CSF, gangliosides, prostaglandins, IFN-γ, complement C5a, TGF-β, interleukins IL-1b, IL-6, IL-10, IL-12, and IL-13, and chemokines CCL2, CXCL5, and CXCL12 (Gabrilovich & Nagaraj, 2009). However, the mechanisms by which they exert their suppressive functions have been well studied, notably the depletion of L-arginine by the enzyme Arginase 1 which leads to impaired T-cell proliferation (Ochoa, Zea, Hernandez, & Rodriguez, 2007; Rodriguez et al., 2005).

Lymphoid cells and notably regulatory T cells (T_{reg}) are involved in antigen-specific suppressive function. Upon TCR activation, T_{reg} release the immunosuppressive cytokine IL-10 which downregulates Th1 cytokines and co-stimulatory molecules on APCs. T_{reg} are differentiated under the influence of TGF-β and IL-2 stimulation through the IL-2 high-affinity receptor CD122-CD25 heterodimer. Other cytokines, notably VEGF-A, have been shown to induce regulatory polarization of $CD4^+$ T cells (Terme et al., 2013).

Hypoxia is in general linked to increased immunosuppression. It directly inhibits T-cell responses, as hypoxic conditions inhibit IL-2 and IFN-γ release after TCR-mediated T-cell activation (Becker, Andersen, Schrama, & Thor Straten, 2013). In ovarian cancer, tumor cells response to hypoxia was shown to induce the expression of the chemokine CCL28 which attracts T_{reg} (Facciabene et al., 2011). In mice, hypoxic area has been shown to favor the M2 polarization of macrophages (Movahedi et al., 2010). In mice, VEGF-A signaling was also shown to directly induce T-cell expression of PD-1 and other immune checkpoints, notably Tim-3 and CTLA-4 (Voron et al., 2015).

Fibroblasts can also modify T-cell responses, through several mechanisms. As major producers of the extracellular matrix, they control the trafficking of T cells from the invasive margin (IM) to the tumor stroma (Salmon et al., 2012). Fibroblasts can directly inhibit TNF-α and IFN-γ-mediated antitumor immunity (Kraman et al., 2010), hamper DC maturation (Séguier et al., 2013), and inhibit T-cell proliferation (Bocelli-Tyndall et al., 2006; Haniffa et al., 2007; Jones, Horwood, Cope, & Dazzi, 2007). They have been shown to constitutively express the immune checkpoint ligand PD-L1 (Pinchuk et al., 2008), and this expression is upregulated upon IFN-γ stimulation (Pinchuk et al., 2008).

2. THE IMMUNE MICROENVIRONMENT OF DIFFERENT CANCER SUBTYPES

Once malignant cell transformation has occurred and tumors grow, invade locally, and spread in distant organs, they still interact with their TME, and particularly with immune and inflammatory cells in the organ where the primary tumors develop, but also in the organs invaded by metastatic cells.

The composition of the immunological TME is not homogeneous. Heterogeneity is found between different cancer types, and also between tumors from patients with the same type of cancers. Some tumors exhibit a low infiltration by immune cells; others are highly infiltrated by lymphocytes. Among the latter, the architecture of the TME, with the localization of immune cells in the invasive front and the center of the tumor, the presence of tumor-adjacent organized lymphoid aggregates, and the type of inflammatory context, determines the prognostic impact of the infiltrating cells. Therefore, the description and the understanding of the immune and inflammatory landscape in human tumors are of paramount importance at the different levels of patient's care. It completes the mutational, transcriptional, and epigenetic patterns of the malignant cells and opens paths to understand how tumor cells shape their immune microenvironment and are shaped by the immune reaction. It provides prognostic markers, such as the immunoscore, that are highly predictive of patient's survival in primary and metastatic tumor sites, and are independent of the classically used prognostic classifications such as those based on TNM, on blood lactate dehydrogenase levels, MSKCC nomogram and Fuhrman grade in RCC, Gleason score in prostate cancer. It helps to select patients who are likely to benefit from the exploding field of immunotherapies, by identifying tumors containing infiltrating immune cells which express inhibitory immune checkpoint molecules such as CTLA-4 and/or PD-1, or expressing novel targets for immunomodulation. In addition, poorly infiltrated tumors may identify patients that would benefit from adoptive T-cell therapies (Giraldo et al., 2014). Finally, increasing knowledge in the immune and inflammatory TME should reveal which molecules to target for innovative immunotherapeutic approaches.

2.1 Cellular Composition of the Tumor Microenvironment and Clinical Outcome of Cancer Patients

Tumors are composed of transformed malignant cells interacting with constituents of the tissue where they grow. Fibroblasts, blood, and lymphatic vessels, found in normal tissues, are present and dysfunctional in tumors. In addition, tumor cells are surrounded by infiltrating cells of hematopoietic origin, including lymphocytes, macrophages, DCs, mast cells, and polymorphonuclear cells. These different cellular populations are not scattered within tumors, but are located in selective areas of the tumors and can be found in the tumor core, or "center of the tumor" (CT), in the invasive margin (IM), and in organized lymphoid aggregates adjacent to the tumor nests. During the last decade, thanks to the development of robust antibodies adapted for immunohistochemistry (IHC), biobanks of large cohorts of annotated tumors, automated quantitative imaging technologies for IHC, and bio-informatics for high-throughput analyses, a precise view of the immune and inflammatory landscapes of human cancers has emerged.

2.1.1 Role of T Cells

Numerous studies have quantified the immune infiltrates in human cancers and investigated their correlation with prognosis. In the vast majority of cases, high densities of memory T lymphocytes with a T_{h1} orientation and a strong $CD8^+$ compartment correlate with favorable prognosis.

In 2012, our group published a review revising the studies assessing their relationships and concluded that high densities of $CD3^+$ T cells, $CD8^+$ cytotoxic T cells, and $CD45RO^+$ memory T cells were clearly associated with a longer disease-free survival (DFS) and/or overall survival (OS) in many different tumors (including melanoma, head and neck, breast, bladder, urothelial, ovarian, colorectal, renal, and prostatic and lung cancer), with the only exception being RCC (Fridman, Pagès, Sautès-Fridman, & Galon, 2012). Since 2012, our understanding of the clinical impact of the $CD8^+$ T-cell infiltration into the tumors has been enriched, with more than 60 new studies published on this topic (Table 1 and Fig. 2). The majority support a positive impact on prognosis associated with extensive memory T-cell infiltration in the same cancers we discussed earlier, eg, CRC (Anitei et al., 2014; Boissière-Michot et al., 2014; Correale et al., 2012; Kim, Bae, et al., 2015; Ling et al., 2014; Ong et al., 2015; Shinto et al., 2014), bladder (Ingels et al., 2014), breast (Chen et al., 2014; de Kruijf et al., 2013; García-Martínez et al., 2014; Kim et al., 2013; Mohammed et al., 2013; Park et al., 2012; Rathore et al., 2014; Seo et al., 2013; West et al., 2013), Merkel cell carcinoma (Paulson et al., 2014), glioma (Han et al., 2014), pancreatic (Ino et al., 2013; Tang et al., 2014), head and neck (Balermpas et al., 2014; Czystowska et al., 2013; Millrud et al., 2012; Wallis et al., 2015), cervical (Origoni et al., 2013), NSCLC (Goc et al., 2014; Hald et al., 2013), gastric (Chen et al., 2013; Feichtenbeiner et al., 2014; Kim et al., 2014), ovarian (Bachmayr-Heyda et al., 2013; Nielsen et al., 2012; Preston et al., 2013), and HCC (Cariani et al., 2012; Chen et al., 2012; Guo et al., 2007; Huang et al., 2012) (Table 1 and Fig. 2). In addition, the correlation between increased $CD8^+$ T-cell infiltration and good clinical outcome was recently described in the previous unstudied neoplasias, including GIST (Rusakiewicz et al., 2013), oropharyngeal (Jung et al., 2013; Näsman et al., 2012; Wansom et al., 2012), thyroid (Cunha et al., 2012), and biliary tract cancer (Goeppert et al., 2013).

Table 1 The Association Between Immune Cell Infiltrates and Patients' Prognosis in Cancer

	$CD8^+$ T Cells	Th1	Macrophages (All)	M1 Macrophages	M2 Macrophages	Dendritic Cells	NK Cells	B Cells
Melanoma	—		Poor (Jensen et al., 2009)		Poor (Jensen et al., 2009)	Good (Gerber et al., 2014; Ladányi et al., 2007; Neagu, Constantin, & Zurac, 2013; Simonetti et al., 2007)		Good (Ladányi et al., 2011)
Head and neck cancers	Good (Czystowska et al., 2013; Balermpas, Rödel, Weiss, Rödel, & Fokas, 2014; Millrud et al., 2012; Wallis, Stafford, & Greenman, 2015)					Poor (Jensen et al., 2012; Treilleux et al., 2004)		Good (van Herpen et al., 2008)
Breast cancer	Good (Chen et al., 2014; de Kruijf et al.,	Good (Gu-Trantien	Poor (Campbell et al., 2011;		Poor (Medrek, Pontén, Jirström, &	Good (Iwamoto et al., 2003; Lespagnard et al.,		Poor (DeNardo and Coussens, 2007; Good

Continued

Table 1 The Association Between Immune Cell Infiltrates and Patients' Prognosis in Cancer—cont'd

	$CD8^+$ T Cells	Th1	Macrophages (All)	M1 Macrophages	M2 Macrophages	Dendritic Cells	NK Cells	B Cells
	2013; García-Martínez et al., 2014; Kim et al., 2013; Mohammed, Going, Edwards, Elsberger, & McMillan, 2013; Park, Lee, & Yoon, 2012; Rathore et al., 2014; Seo et al., 2013; West et al., 2013)	et al., 2013; Kristensen et al., 2012)	DeNardo et al., 2011; García-Martínez et al., 2014; Leek et al., 1996; Mahmoud et al., 2012a, 2012b; Zhang et al., 2013)		Leandersson, 2012)	1999; Park et al., 2012; La Rocca et al., 2008)		(Kotlan et al., 2005; Mahmoud et al., 2012a, 2012b; Schmidt et al., 2008; Simsa, Teillaud, Stott, Tóth, & Kotlan, 2005)
Bladder cancer	Good (Ingels et al., 2014)		Poor (Hanada et al., 2000; Sjödahl et al., 2014)			Good (Inoue, Furihata, Ohtsuki, & Fujita, 1993)		
	None (Sjödahl et al., 2014)							

Ovarian cancer	Good (Bachmayr-Heyda et al., 2013; Nielsen et al., 2012; Preston et al., 2013)		Poor (Salvesen & Akslen, 1999)	Good (Zhang, He, et al., 2014)	Poor (Lan et al., 2013)	Good (Eisenthal et al., 2001; Zhang et al., 2015)		Good (Nielsen et al., 2012; Sato et al., 2005; Zhang et al., 2003)
Oesophageal cancer	—		Poor (Guo et al., 2007)		Poor (Shigeoka et al., 2013)	Good (Furihata et al., 1992; Ishigami et al., 2003a, 2003b; Lu et al., 2013)	Good (Hsia et al., 2005)	
Colorectal cancer	Good (Anitei et al., 2014; Boissière-Michot et al., 2014; Correale et al., 2012; Kim, Bae, Li, Cho, & Kang, 2015; Ling, Edin, Wikberg, Öberg, & Palmqvist, 2014; Ong et al., 2015; Shinto et al., 2014)	Good (Boissière-Michot et al., 2014; Chen et al., 2014)	Good (Algars et al., 2012; Chaput et al., 2013; Forssell et al., 2007; Khorana, Ryan, Cox, Eberly, & Sahasrabudhe, 2003; Zhou et al., 2010)	Good (Edin et al., 2012)	Poor (Herrera et al., 2013)	Good (Gulubova et al., 2012; Väyrynen et al., 2014; McMullen, Lai, Dabbagh, Wallace, & de Gara, 2010; Nakayama et al., 2003)	Good (Bindea et al., 2013; Menon et al., 2004)	Poor (Barbera-Guillem et al., 2000)

Continued

Table 1 The Association Between Immune Cell Infiltrates and Patients' Prognosis in Cancer—cont'd

	CD8$^+$ T Cells	Th1	Macrophages (All)	M1 Macrophages	M2 Macrophages	Dendritic Cells	NK Cells	B Cells
	None (Lim et al., 2014; Väyrynen et al., 2013)					Poor (Sandel et al., 2005)		
Renal cell carcinoma	Poor (Giraldo et al., 2015)				Poor (Dannenmann et al., 2013; Komohara et al., 2011; Xu et al., 2014)		Good (Donskov, 2007; Eckl et al., 2012)	
Prostatic adenocarcinoma	Poor (Ness et al., 2014)		Good (Shimura et al., 2000)					
Lung carcinoma	Good (Goc et al., 2014; Hald et al., 2013)		Good (Kawai et al., 2008; Ma et al., 2010; Ohri, Shikotra, Green, Waller, & Bradding, 2009; Welsh et al., 2005)	Good (Ma et al., 2010; Ohri et al., 2009)	Poor (Hirayama et al., 2012; Ohtaki et al., 2010; Zeni et al., 2007)	Good (Al-Shibli et al., 2009; Dai et al., 2010; Hald et al., 2013; Inoshima et al., 2002; Johnson et al., 2000)	Good (Al-Shibli et al., 2009; Johnson et al., 2000; Takanami, Takeuchi, & Giga, 2001; Villegas et al., 2002)	Good (Al-Shibli et al., 2008; Germain et al., 2014; Hernández-Prieto et al., 2015; Lohr et al., 2013; Pelletier, Edwardes, Michel, Halwani, & Morin, 2001)
	Poor (Tian et al., 2015)		Poor (Zhang et al., 2011)				None (Platonova et al., 2011)	

Pancreatic cancer	Good (Ino et al., 2013; Tang et al., 2014)		Poor (Hou, Chao, Tung, Wang, & Shan, 2014; Ino et al., 2013; Kurahara et al., 2011; Sugimoto et al., 2014; Sugimura et al., 2015; Zeng et al., 2014)		
	None (Tewari et al., 2013)				
Cervical cancer	Good (Origoni et al., 2013)	Good (Heller et al., 2003)			
Anal squamous cell carcinoma	—				
Oral squamous cell carcinoma					Good (Wirsing, Rikardsen, Steigen, Uhlin-Hansen, & Hadler-Olsen, 2014)

Continued

Table 1 The Association Between Immune Cell Infiltrates and Patients' Prognosis in Cancer—cont'd

	$CD8^+$ T Cells	Th1	Macrophages (All)	M1 Macrophages	M2 Macrophages	Dendritic Cells	NK Cells	B Cells
Glioma	Good (Han et al., 2014)							
	None (Yue et al., 2014)							
Hepatocellular carcinoma	Good (Cariani et al., 2012; Chen et al., 2012; Guo et al., 2007; Huang et al., 2012)	Good (Chew et al., 2012)	Good (Li et al., 2009; Zhu et al., 2008)		Poor (Kong et al., 2013)	Good (Cai et al., 2006; Yin, Lu, Lai, Liang, & Huang, 2003)	Good (Taketomi et al., 1998)	
	Poor (Che et al., 2014)		Poor (Ding et al., 2009; Fan et al., 2014)					
Gastric cancer	Good (Chen et al., 2013; Feichtenbeiner et al., 2014; Kim et al., 2014)	Good (Chang, Du, Zhao, Ma, & Cao, 2014)	Good (Ohno et al., 2003)	Good (Zhang et al., 2014)	Poor (Kawahara et al., 2010; Pantano et al., 2013)	Good (Amoueian, Attaranzadeh, & Montazer, 2015; Ananiev, Gulubova, & Manolova, 2011;		

				Kashimura et al., 2012; Tsukayama, Omura, Yoshida, Tanaka, & Watanabe, 2005)	
	None (Okita et al., 2014)	Poor (Haas et al., 2009) (Kawahara et al., 2010; Osinsky et al., 2011)		Poor (Ishigami et al., 2010)	Good (Ishigami et al., 2000; Takeuchi et al., 2001)
Medulloblastoma	—				
Merkel cell carcinoma	Good (Paulson et al., 2014)				
Urothelial cell carcinoma	—	Poor (Chai et al., 2008)	Poor (Ichimura et al., 2014)		
Follicular lymphoma and Hodgkin's lymphoma	Poor		Poor (Klein et al., 2014)		

Continued

Table 1 The Association Between Immune Cell Infiltrates and Patients' Prognosis in Cancer—cont'd

	$CD8^+$ T Cells	Th1	Macrophages (All)	M1 Macrophages	M2 Macrophages	Dendritic Cells	NK Cells	B Cells
GIST	Good (Rusakiewicz et al., 2013)						Good (Rusakiewicz et al., 2013)	
Biliary tract cancer	Good cancer (Goeppert et al., 2013)					Good (Furihata et al., 2005; Nakakubo et al., 2003)		Good (Goeppert et al., 2013)
Oropharingeal Carcinoma	Good (Jung et al., 2013; Wansom et al., 2012)		Poor (Liu et al., 2008; Wolf et al., 2015)		Poor (He et al., 2014)	Good (Reichert, Scheuer, Day, Wagner, & Whiteside, 2001)		
Thyroid	Good (Cunha et al., 2012)		Poor (Ryder, Ghossein, Ricarte-Filho, Knauf, & Fagin, 2008)					
Endometrial	None (Čermáková et al., 2014)		Poor (Čermáková et al., 2014; Ohno et al., 2004)		Poor (Kübler et al., 2014)	Good (Lijun et al., 2012; Honig, Schaller, Dietl, Backe, & Kammerer, 2005)		

Melanoma brain metastasis	Good (Hamilton et al., 2013; Kluger et al., 2015)							
Liver colorectal metastasis	Good (Brunner et al., 2014; Katz et al., 2013; Liu et al., 2012; Yoon et al., 2012)							Good (Meshcheryakova et al., 2014, colorectal cancer liver metastases)
Lung colorectal metastasis	Good (Remark et al., 2013)						Good (Remark et al., 2013)	
Lung renal cell carcinoma metastasis	Poor (Remark et al., 2013)						Good (Remark et al., 2013)	
	51 Good	6 Good	14 Good	5 Good	Good 0	35 Good	15 Good	17 Good
	7 None	0 Poor	24 Poor	0 Poor	24 Poor	4 Poor	1 None	2 Poor
	5 Poor							

The references for the prognosis associated with $CD8^+$ and Th1-lymphocyte infiltrates in human cancer before 2012 are included in Fridman et al. (2012).

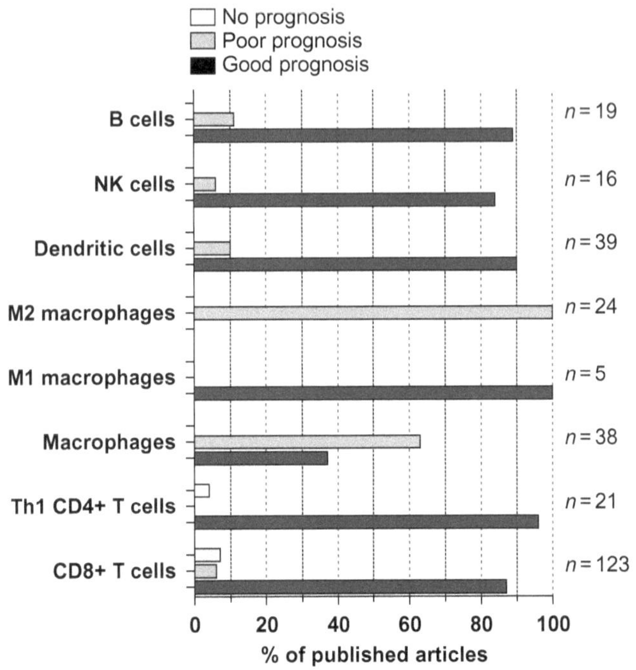

Figure 2 Prognosis impact associated with tumor infiltration by various hematopoietic populations. The references for the prognosis associated with $CD8^+$ and Th1-lymphocyte infiltrates in human cancer before 2012 are included in Fridman et al. (2012).

Interestingly, some studies have assessed this correlation also in the metastatic sites and have consistently found that increased $CD8^+$ T-cell densities correlated with prolonged OS in CRC liver (Brunner et al., 2014; Katz et al., 2013; Liu et al., 2012; Yoon et al., 2012), lung metastases (Remark et al., 2013), and CNS melanoma metastases (Hamilton et al., 2013; Kluger et al., 2015) (Table 1 and Fig. 2). Since 2012, few studies have been published on the role of infiltrating T_{h1}-oriented T cells in tumor control or progression. Nevertheless, they have consistently found that their increased infiltration is associated with favorable clinical outcome in CRC (Boissière-Michot et al., 2014; Chen & Chen, 2014), gastric (Chang et al., 2014), breast (Gu-Trantien et al., 2013; Kristensen et al., 2012), and HCC (Chew et al., 2012) (Table 1 and Fig. 2). Finally, an association between $CD8^+$ T-cell infiltration and poor clinical outcome has been observed in NSCLC (Tian et al., 2015), HCC (Che et al., 2014), prostate cancer (Ness et al., 2014), and RCC (Giraldo et al., 2015) (Table 1 and Fig. 2). Overall, most studies assessing the densities of immune cell populations agree in that tumors poorly infiltrated with $CD8^+$ T cells often display higher histologic grades, characterized by dedifferentiation, prominent vascularization, and inflammation.

For other T-cell subsets, the picture is less straightforward. There seems to be an inverse correlation between the clinical impact of T_{h2}- and T_{h17}-positive cells. For instance, in ovarian, pancreatic or gastric cancers, high densities of T_{h2} cells correlate with poor prognosis, whereas high densities of T_{h17} cells correlate with better prognosis in ovarian, gastric, or esophageal cancers. Conversely, high densities of T_{h2} cells are associated with good prognosis in CRC, NSCLC, or breast cancer, in which T_{h17} cell densities correlate with poor prognosis (reviewed in Teng, Galon, Fridman, & Smyth, 2015). Although these studies have been performed in different laboratories, on different cohorts of patients, and on tumors from different primary sites, they suggest that different tumors, in different organs, may contribute to shape different TMEs in which the impact of a given T-cell population depends on the overall organization of the immune landscape and on the balance between acute and chronic inflammation (Teng et al., 2015). The case of T_{reg} is more complex. Following the initial study of Curiel et al. (2004) demonstrating a deleterious impact of T_{reg} in ovarian cancer, reinforced by studies in melanoma, breast, lung, pancreatic, liver, or colorectal cancers, other studies conversely reported longer survival associated with high T_{reg} cell densities in colorectal, bladder, head and neck, or ovarian cancers (reviewed in Fridman et al., 2014). These discrepancies may be due to the relative lack of specificity of FOXP3 as a marker of T_{reg} and to the fact that T_{reg} cells may have different clinical impacts depending on the subtype of cancer investigated.

2.1.2 Role of B Cells

The deleterious or beneficial role of B cells in antitumor immunity has long been a matter of debate. A few studies in humans suggested that B cells could exert a protumoral function. Indeed, the detection of specific antibodies in the serum or in the tumor of breast cancer patients was associated with poor prognosis (DeNardo, Andreu, & Coussens, 2010). Similarly, in advanced CRC, a reduction of the tumor size following B-cell depletion could be observed in half of the patients treated with the antihuman CD20 antibody Rituximab (Barbera-Guillem et al., 2000). Many studies in mice proposed several mechanisms to explain this protumoral effect, including the maintenance of a chronic inflammation by B cells (de Visser, Korets, & Coussens, 2005) or the formation of immune complexes able to activate the complement system, which in turn triggers inflammation (Tan & Coussens, 2007), or to promote extracellular matrix degradation and angiogenesis in a granulocyte- and macrophage-dependent manner (Barbera-Guillem, May, Nyhus, & Nelson, 1999). B cells were also shown to inhibit cytotoxic T-cell responses in several tumor models (DeNardo et al., 2010), or to directly favor tumor growth and metastasis through the production of lymphotoxin and activation of LT receptor-expressing castration-resistant prostate cancer cells (Ammirante, Luo, Grivennikov, Nedospasov, & Karin, 2010). However, some of these studies have to be interpreted with caution, as they were performed in μMT B-cell-deficient mice, known to display severe immune system abnormalities (João, Ogle, Gay-Rabinstein, Platt, & Cascalho, 2004; Moulin et al., 2000). In addition, while some of these studies were based on total mouse B-cell depletion, several subpopulations of B cells have now been identified, including the immunomodulatory "Bregs" (Balkwill, Montfort, & Capasso, 2013; Rosser & Mauri, 2015). Through the secretion of TGF-β or IL-10, such B cells may favor the differentiation and the recruitment of T_{reg}, further amplifying the immunosuppressive environment (Horikawa, Minard-Colin, Matsushita, & Tedder, 2011; Olkhanud et al., 2011). In a very recent study, Shalapour et al. demonstrated that a subpopulation of B cells, defined as IL-10-producing PD-L1-expressing IgA plasmocytes, was responsible for oxaliplatin resistance of mouse prostate tumors, and suppression of $CD8^+$ T cells mediated antitumor immunity (Shalapour et al., 2015).

On the opposite, recent studies in mice and in humans have demonstrated that B cells play a positive role in the antitumor immune response. In a mouse model of B16 melanoma, DiLillo et al. demonstrated that B-cell depletion was associated with faster tumor growth, increased lung metastasis, and decreased numbers of both $CD4^+$ and $CD8^+$ effector T cells (DiLillo, Yanaba, & Tedder, 2010). These observations were confirmed by Kobayashi et al. in B-cell linker protein (BLNK)-deficient mice, another model of B-cell deficiency considered more relevant than μMT mice (Hayashi et al., 2000; Jumaa et al., 1999; Kobayashi et al., 2014; Minegishi et al., 1999; Pappu et al., 1999). In parallel, numerous clinical studies have demonstrated that a high density of B cells within the TME is associated with increased patient survival, for instance in CRC (Bindea et al., 2013). In primary cutaneous melanoma, Ladanyi et al. showed that high numbers of intratumoral and/or peritumoral B cells were associated with less visceral metastases and prolonged survival (Ladányi et al., 2011). Similarly, high densities of B cells have been associated with better prognosis in breast cancer (Kotlan et al., 2005; Mahmoud et al., 2012a; Schmidt et al., 2008; Simsa et al., 2005), NSCLC (Al-Shibli et al., 2008; Hernández-Prieto et al., 2015; Pelletier et al., 2001), head and neck cancer (van Herpen et al., 2008), ovarian cancer (Nielsen et al., 2012; Sato et al., 2005; Zhang et al., 2003), metastatic colorectal cancer (Meshcheryakova et al., 2014), and biliary tract cancer (Goeppert et al., 2013) (Table 1 and Fig. 2). In a very recent study, Hernandez-Prieto et al. identified, in untreated stage I and II NSCLC patients, a cluster of 50 genes that was associated with a low-risk recurrence. The 50 predictor genes were for the vast majority related to B-cell and plasma cell immune responses, with genes coding for unique B lineage-specific markers, and B-cell functional elements, or having a strong influence on B-cell homeostasis (Hernández-Prieto et al., 2015).

Interestingly, an increasing number of studies have now reported that not only the density but also the organization of tumor-infiltrating B cells may be positively associated with patient's outcome. We reported the presence of tertiary lymphoid structures (TLSs) in the TME of NSCLC, characterized by DC-Lamp$^+$ mature DC forming clusters with T cells, surrounded by CD20$^+$ B cells organized in a B-cell follicle (Dieu-Nosjean et al., 2008). Such TLSs (whose function is developed in the following section) were observed in many situations of chronic inflammation, including cancer (Aloisi & Pujol-Borrell, 2006; Dieu-Nosjean, Goc, Giraldo, Sautès-Fridman, & Fridman, 2014), and were shown to display a general organization very similar to that observed in secondary lymphoid organs (SLOs) (Germain, Gnjatic, & Dieu-Nosjean, 2015). Like in conventional SLOs, B-cell follicles within TLSs are composed of a mantle of IgD$^+$ naive B cells, surrounding a germinal center (GC) identified by highly proliferating Ki67$^+$ B cells, a network of CD21$^+$ follicular DCs, the presence of CD3$^+$ CD4$^+$ CXCL13$^+$ follicular T helper (T_{fh}) cells, and expression of activation-induced cytidine deaminase (AID) and of the transcription repressor BCL6, both of which associated with active B-cell proliferation, class switch recombination (CSR), and somatic hyper-mutation (Cipponi et al., 2012; Figenschau, Fismen, Fenton, Fenton, & Mortensen, 2015; Germain et al., 2014; Gottlin et al., 2011; Gu-Trantien et al., 2013; Shimabukuro-Vornhagen et al., 2014). Since then, several studies in cancer have described a positive association between the presence of B-cell follicles and patient outcome. For instance, we showed in NSCLC that a high density of TLS-CD20$^+$ B cells was associated with increased patient's OS, independently of the preoperative treatment received (Germain et al., 2014). Similar observations were subsequently made in oral squamous cell carcinoma

(Wirsing et al., 2014) and in metastatic colorectal cancer (Meshcheryakova et al., 2014). Indirect markers of B-cell follicles, namely T_{fh} cells or the B-cell chemoattractant CXCL13, were also associated with prolonged patient survival in breast cancer (Gu-Trantien et al., 2013).

Several mechanisms can explain the positive impact of intratumoral B cells on antitumor immunity. Consistent with Ki67, AID, and BCL6 expression within B-cell follicles, several studies demonstrated the existence of a specific antitumor B-cell response. Indeed, we, and others, described the presence of all stages of B-cell differentiation within NSCLC and colorectal tumors, including terminally differentiated memory B cells and plasmablasts/plasma cells (Germain et al., 2014; Shimabukuro-Vornhagen et al., 2014). Cipponi et al. in metastatic melanoma (Cipponi et al., 2012) and Nzula et al. in ductal breast carcinoma (Nzula, Going, & Stott, 2003) both showed that TLS-B cells are antigen experienced, as demonstrated by clonal amplification and high mutation rates. While serum reactivities against different tumor antigens have been observed in many cancers and potentially associated with good clinical outcome (Ayyoub et al., 2004; Gnjatic et al., 2010; Hamaï et al., 2011; Nielsen et al., 2012; Ohue et al., 2012; Reuschenbach, von Knebel Doeberitz, & Wentzensen, 2009), we recently demonstrated that ex vivo-cultured NSCLC-infiltrating B cells were able to produce high levels of specific antitumor antibodies (Germain et al., 2014). Besides their capacity to trigger complement-dependent and antibody-dependent cell cytotoxicity (CDC and ADCC), antibodies when bound to their antigen can also form immune complexes able to activate DCs and to favor the generation of antigen-specific $CD8^+$ T-cell immune responses (Carmi et al., 2015; Noguchi et al., 2012). Of note, Lohr et al. showed in NSLCL the existence of a positive correlation between the presence of tumor-infiltrating plasma cells and longer survival (Lohr et al., 2013). In parallel with their capacity to produce antigen-specific antibodies, B cells can also act as powerful APC (Yuseff, Pierobon, Reversat, & Lennon-Duménil, 2013). Thanks to several ligand/receptor pairs involved in B–T-cell interactions (CD80–CD86/CD28, CD40/CD40L, OX40L/OX40, CD27/CD70), B cells were shown to play a major role during initial priming and expansion of $CD4^+$ T cells (Gatto & Brink, 2010; Harp, Lovett-Racke, Racke, Frohman, & Monson, 2008; Parker Harp et al., 2015; Serreze et al., 1998), to be able to cross-present antigens to $CD8^+$ T cells (Ahmadi, Flies, Efebera, & Sherr, 2008; Gnjatic et al., 2003; de Wit et al., 2010), and to promote cytotoxic T lymphocyte survival and proliferation (Deola et al., 2008). In accordance with these last functions,

Nielsen et al. observed in ovarian cancer that tumor-infiltrating $CD20^+$ B cells were often found in close proximity to $CD8^+$ T cells, expressed MHC Class I, MHC Class II, CD80, and CD86 molecules, and that the presence of combined $CD20^+$ and $CD8^+$ lymphocytes was associated with prolonged patient's survival (Nielsen et al., 2012). The expression of surface markers' characteristic of APC by tumor-infiltrating B cells was also described in primary lung cancer (Yasuda et al., 2002) and in hepatocellular carcinoma (Shi et al., 2013). Finally, few studies suggested that some tumor-infiltrating B cells could act as true killer cells, through the expression of granzyme B and TRAIL (Shi et al., 2013), or in a Fas Ligand-dependent manner (Tao et al., 2015).

2.1.3 Role of NK Cells

Natural killer (NK) cells have been described as another major cellular population mediating antitumor immune responses (Vivier, Ugolini, Blaise, Chabannon, & Brossay, 2012). These cells express an array of receptors (including activating, inhibitory, adhesion and cytokine receptors) that enable them to identify tumor cells and spare normal cells. Overall, the integration of these signals determines whether or not NK cells become activated and eliminate its target. The two most important mechanisms of cancer cell recognition by NK cells are the downregulation of MHC I molecules and the expression of stress-induced ligands to NK activation receptors (eg, MICA or MICB, which bind to NKG2D expressed on the NK cell) by tumor cells.

The prognostic impact of NK infiltration has been studied in some tumors, and their increased densities have been consistently associated with favorable clinical outcome. This association has been demonstrated in CRC (Bindea et al., 2013; Coca et al., 1997; Menon et al., 2004), gastric (Ishigami et al., 2000; Takeuchi et al., 2001), vulvar squamous cell (Sznurkowski, Zawrocki, & Biernat, 2014), esophageal (Hsia et al., 2005), renal (Donskov, 2007; Eckl et al., 2012), HCC (Taketomi et al., 1998), NSCLC (Al-Shibli et al., 2009; Johnson et al., 2000; Takanami et al., 2001; Villegas et al., 2002), and GIST (Rusakiewicz et al., 2013), in addition to RCC lung metastases (Remark et al., 2013). Nevertheless, other studies in NSCLC and breast cancer have not found any association between NK cell infiltration and prognosis (Platonova et al., 2011), where they display an inhibited phenotype and decreased functional capacities (Carrega et al., 2008; Platonova et al., 2011) (Table 1 and Fig. 2). More studies assessing the prognostic impact associated with NK cells infiltration are needed.

2.1.4 Role of Tumor-Associated Macrophages

Due to their physiological role in amplifying the inflammatory cascade, it is not unreasonable to suspect that the infiltration by tumor-associated macrophages (TAMs) could be associated with poor prognosis in human cancers.

In fact, several studies have found that the clinical outcome associated with increased numbers of TAMs is mainly determined by the tumor type (Zhang et al., 2012). While the augmented densities of this population are associated with favorable clinical outcome in colorectal (Algars et al., 2012; Chaput et al., 2013; Forssell et al., 2007; Khorana et al., 2003; Zhou et al., 2010), gastric (Ohno et al., 2003), NSCLC (Kawai et al., 2008; Ma et al., 2010; Ohri et al., 2009; Welsh et al., 2005), HCC (Li, Qiu, et al., 2009; Zhu et al., 2008), prostate (Shimura et al., 2000), and cervical cancer (Heller et al., 2003), it has exactly the opposite association in endometrial (Čermáková et al., 2014; Ohno et al., 2004), esophageal (Guo et al., 2007), gastric (Ishigami et al., 2003b; Kawahara et al., 2010; Osinsky et al., 2011), urothelial (Chai et al., 2008), oral (Liu et al., 2008; Wolf et al., 2015), HCC (Ding et al., 2009; Fan et al., 2014), melanoma (Jensen et al., 2009), breast (Campbell et al., 2011; DeNardo et al., 2011; García-Martínez et al., 2014; Leek et al., 1996; Mahmoud et al., 2012b; Zhang et al., 2013), ovarian (Salvesen & Akslen, 1999), bladder (Hanada et al., 2000; Sjödahl et al., 2014), NSCLC (Zhang et al., 2011), thyroid (Ryder et al., 2008), and primary CRC tumors (Bindea et al., 2013) (Table 1 and Fig. 2).

The heterogeneous association between TAM densities and the clinical outcome among different tumors might reflect the plasticity of this population. A protective role in tumorigenesis has been proposed for M1 macrophages (through mechanisms including the activation of the T_{h1} response and by antagonizing the suppressive activities of regulatory immune cells), while M2 have been shown to promote tumor growth, invasion, metastasis, stroma remodeling, and angiogenesis (Biswas & Mantovani, 2010).

The absence of M1/M2 specific and consensual markers has been the major obstacle in the assessment of the clinical impact of each subtype. Up to date, this task has been accomplished using CD11c or NOS2 for M1 TAMs, and CD163, CD204, or CD206 for M2 TAMs. Indeed, increased M1 TAM densities seem to be associated with a favorable clinical outcome in NSCLC (Ma et al., 2010; Ohri et al., 2009), ovarian (Zhang, He et al., 2014), colorectal (Edin et al., 2012), and gastric cancer (Zhang, Wang, et al., 2014a), while those of M2 are linked to poor prognosis in several tumors, including NSCLC (Hirayama et al., 2012; Ohtaki et al., 2010;

Zeni et al., 2007), mesothelioma (Cornelissen et al., 2014), esophageal (Shigeoka et al., 2013), gastric cancer (Kawahara et al., 2010; Pantano et al., 2013), pancreatic (Hou et al., 2014; Ino et al., 2013; Kurahara et al., 2011; Sugimoto et al., 2014; Sugimura et al., 2015; Zeng et al., 2014), CRC (Herrera et al., 2013), HCC (Kong et al., 2013), Hodgkin lymphoma (Klein et al., 2014), renal (Dannenmann et al., 2013; Komohara et al., 2011; Xu et al., 2014), urothelial (Ichimura et al., 2014), breast (Medrek et al., 2012), endometrial (Kübler et al., 2014), ovarian (Lan et al., 2013), melanoma (Jensen et al., 2009), and squamous oral carcinoma (He et al., 2014). Additionally, some studies have demonstrated that, when associated with poor clinical outcome, $CD68^+$ cells are often correlated with the tumor microvessel density, in addition to HIF, VEGF (Chai et al., 2008), and matrix metalloproteinase expression (Bolat et al., 2006; Hanada et al., 2000; Leek et al., 1996; Osinsky et al., 2011; Valković et al., 2002), suggesting that they might have an M2 phenotype (Table 1 and Fig. 2).

Although not yet conclusive, studies linking TAM infiltration and patient's clinical outcome suggest that high densities of M2-oriented cells are associated with higher tumor stages and poor prognosis. In addition, it also indicates that the biological value of measuring the sole densities of $CD68^+$ cells should be revisited, as it gives no information of the cell function or polarization.

2.1.5 Myeloid-Derived Suppressor Cells

The role of MDSC in tumors has been reviewed extensively by others (Wesolowski, Markowitz, & Carson, 2013). Due to the lack of specific markers to quantitate human MDSC in tumors, we do not review them here.

2.1.6 Role of DCs

Upon encounter with an antigen and in the presence of danger signals, immature DCs go through a process called maturation, which allows them to migrate into the lymph node, where they can prime the naïve $CD4^+$ and $CD8^+$ T cells. The phenotype of the mature DC seems to be quite important for the orientation and strength of the subsequent immune response, and it can be modulated by the cytokine microenvironment, as well as the type of antigen (Palucka & Banchereau, 2012). The TME takes advantage of the DC plasticity, can induce a proinflammatory and/or a tolerogenic DC phenotype, or can block their maturation at different stages.

Due to the plasticity and heterogeneity of DC phenotypes, the quantification of infiltrating DC has been a difficult task. The more relevant markers that have been used for their quantification are S-100, CD83, DC-LAMP, or CD1a, in addition to CD207 for Langerhans cells. Their link with clinical prognosis has been reported in a variety of human solid tumors. The majority of these studies report that an extensive infiltration with DC is associated with increased OS and cover many tumors types, including melanoma (Gerber et al., 2014; Ladányi et al., 2007; Neagu et al., 2013; Simonetti et al., 2007), HCC (Cai et al., 2006; Yin et al., 2003), gallbladder (Furihata et al., 2005; Nakakubo et al., 2003), oral (Reichert et al., 2001), esophageal (Furihata et al., 1992; Ishigami et al., 2003a; Lu et al., 2013), gastric (Amoueian et al., 2015; Ananiev et al., 2011; Kashimura et al., 2012; Tsukayama et al., 2005), NSCLC (Al-Shibli et al., 2009; Dai et al., 2010; Hald et al., 2013; Inoshima et al., 2002; Johnson et al., 2000), colorectal (Gulubova et al., 2012; McMullen et al., 2010; Nakayama et al., 2003; Väyrynen et al., 2014), bladder (Inoue et al., 1993), breast (Iwamoto et al., 2003; La Rocca et al., 2008; Lespagnard et al., 1999; Park et al., 2012), endometrial (Honig et al., 2005; Lijun et al., 2012), and ovarian cancer (Eisenthal et al., 2001; Zhang et al., 2015). Nevertheless, the infiltration with $CD123^+$ plasmacytoid DC has been associated with shorter OS in breast cancer (Jensen et al., 2012; Treilleux et al., 2004), as well as the presence of $CD208^+$ and $CD1a^+$ DCs in colorectal (Sandel et al., 2005) and gastric cancer (Ishigami et al., 2010) (Table 1 and Fig. 2).

2.2 Tertiary Lymphoid Structures: Antitumor Immunity Generating Site

Lymphoid aggregates can be detected in the IM of most tumors. Some of them form TLSs. They exhibit properties of active immune sites, resembling TLSs arising from other tissues upon infection, or secondary to autoimmune or chronic inflammatory diseases (Dieu-Nosjean et al., 2014; Goc et al., 2014). Characteristically, TLSs exhibit a T-cell zone (composed mainly of $CD4^+CD62L^+CD45RO^+$ CM or $CD8^+CD62L^+CD45RA^+$ naïve T lymphocytes) with embedded mature DCs, GCs with proliferating B cells, and follicular DCs. They are surrounded by high endothelial venules (HEVs) (de Chaisemartin et al., 2011; Martinet et al., 2011). In addition, laser microdissection of these structures revealed that they are enriched in genes associated with T-cell chemoattractant molecules, such as CCL19, CCL21, CXCL13, IL-16, CCL22, and CCL17 (de Chaisemartin et al., 2011).

In view of their similarities with GCs in lymph nodes, it has been hypothesized that TLSs represent a site of where in situ antigen presentation and lymphocyte activation can occur in an environment protected from the immunosuppressive factors found in the rest of the TME (Dieu-Nosjean et al., 2014; Goc et al., 2014). Indeed, studies on primary melanoma and NSCLC have correlated the densities of mature DC (DC-Lamp^{+}) within TLSs with a strong infiltration by activated T cells and a T_{h1}-oriented response (Dieu-Nosjean et al., 2008; Goc et al., 2014; Ladányi et al., 2007). Moreover, the higher densities of these structures correlated with favorable clinical outcome in NSCLC (Dieu-Nosjean et al., 2014; Germain et al., 2014; Goc et al., 2014), colorectal (Di Caro et al., 2014; Väyrynen et al., 2014), melanoma (Cipponi et al., 2012), renal (Giraldo et al., 2015), and breast cancer (Martinet et al., 2011). The mechanisms underlying the neogenesis of these structures are still unclear in human tumors.

The main reason for the association of TLSs with longer patient's survival resides in the fact that they may be sites of induction and maintenance of immunity toward tumors (Dieu-Nosjean et al., 2014). The circulating naïve CCR7^{+} T cells enter TLSs directly through HEVs, thus escaping the immunosuppressive milieu of the TME. The most likely hypothesis is that CD4^{+} and CD8^{+} T cells are educated by mature DC, which may present tumor-associated antigens, and become mature T cells with specialized effector functions, the effector-memory T cells (TEM) migrating to the tumor nests to control the primary tumors whereas the long-lived central-memory T cells (TCMs) migrate through lymphatics to lymph nodes and enter the periphery. The latter may remain for long periods of time in the body, helping to control potential metastatic cells (Dieu-Nosjean et al., 2014). The CXCR5-positive B cells also enter through the HEVs and form a GC where, after a contact with antigen-presenting follicular DCs, they proliferate and may produce antitumor antibodies, as reported in NSCLC (Germain et al., 2014).

TLS density is heterogeneous among patients. Recent studies undertook the in-depth analysis of their impact on the good prognostic value of CD8^{+} T cells in NSCLC and clear-cell renal cell carcinoma (ccRCC). Although high TLS densities are associated with higher T-cell infiltration and with a Th1 response as well as a cytotoxic orientation in the IM and the tumor core, there are tumors with high infiltration of CD8^{+} T-cell infiltration and low TLS density. In NSCLC, where high CD8^{+} T-cell density globally correlates with longer patient's survival, recent studies on large series of patients at all stages of the disease (*n*=341) showed that the group of patients with tumors exhibiting low TLS densities identifies a cohort of poor prognosis within the patients with high $CD8^+$ T-cell densities, suggesting a functional heterogeneity among the $CD8^+$ T cells (Goc et al., 2014). One could hypothesize that the $CD8^+$ T cells that correlate with poor prognosis in NSCLC have not been educated in the immunity-generating TLS sites, but could have entered directly into the TME through blood vessels unprotected from the local immunosuppressive signals. In contrast with NSCLC where most if not all the DC-$Lamp^+$ DCs are fully mature and found within TLSs, the majority of the DC-$Lamp^+$ cells (80%) in ccRCC tumors do not express the DC maturation marker CD83 and high levels of MHC-Class II antigens and are found outside the TLS, close to the $CD31^+$ blood vessels (Giraldo et al., 2015). Conversely, the minor proportion of DC-$Lamp^+CD83^+$MHC-Class II^{Hi} fully activated/mature DCs (20%) are found within TLSs and close to the $PNAd^+$ HEVs in these tumors. The density of the non-TLS-DC-$Lamp^+$ cells is associated with higher densities of PD-1^+CD8^+ T cells, higher PD-L1, and PD-L2 expression on the tumor cells and poor DFS and OS, as shown by IHC in a series 80 ccRCC patients. The smaller group of ccRCC tumors with high TLS-DC densities identifies patients in which strong $CD8^+$ T-cell infiltration correlates with longer survival (Giraldo et al., 2015). These observations in NSCLC and in ccRCC support the hypothesis that T cells which enter into TLSs through HEVs are locally educated, and exhibit antitumor activity that persists for a long period of time. Thus, TLSs may govern the generation of TCMs that circulate and confer long-term protection after tumor removal, in addition to the tissue-resident effector-memory T cells. In contrast, T cells migrating into the tumor through blood vessels, without passing through TLSs, may be immunosuppressed, become exhausted, accelerate tumor growth, and are unable to control metastasis.

TLSs govern the functional orientation of the local immune reaction in primary, but also in metastatic tumor sites. The analysis of lung metastases from various primary tumors revealed that some metastatic tumors exhibited high TLS densities, such as lung metastases from colorectal, thyroid, and prostate cancers, while other exhibited intermediate numbers, such as breast, renal cell, and liver cancers, or were very low in TLS density such as leiomyosarcoma and osteosarcoma (Dieu-Nosjean et al., 2014). In colorectal and RCC, in which pairs between primary and metastatic sites have been analyzed, the TLS density in the metastasis mirrors that of the primary site (Remark et al., 2013), suggesting that the malignant cells rather than the organ shape major components of the immune contexture. Moreover, the clinical impacts of the immune contexture correlate with the primary and metastatic sites. In CRC, high memory $CD8^+$ T cell, TLS-DC infiltration, and a Th1 orientation correlate with longer survival from the times of surgery for the primary and the metastatic tumors (Remark et al., 2013). In contrast, high memory $CD8^+$ T cells and non-TLS-DC infiltration correlated with shorter survival (Remark et al., 2013). In these cases, in addition to low TLS numbers and the presence ofDC-Lamp positive not fully mature DC outside the TLS, there was a strong expression of genes encoding proinflammatory (IL-6, IL-8, IL-1A) and immunosuppressive (TGF-ß, IL-10) cytokines, myeloid cell-attracting chemokines (CCL2, CCL3, CCL5), and proangiogenic factors (VEGF). The immune contexture therefore reflects a complex landscape in which the ultimate favorable situation, the presence of high quantities of functional memory cytotoxic T cells into the tumor, as reflected by the immunoscore, is modulated by the other components of the TME, including the secretion of the chemokines necessary to form TLSs, the state of tumor vascularization outside of TLSs, the presence of immunosuppressed lymphocytes and myeloid cells, and that of tumor-associated stromal and fibroblastic cells. The malignant cells highly participate in shaping these immune contextures by the neoantigens they potentially express and present, as well as proinflammatory, proangiogenic, immune-activating, or immunosuppressive cytokines they produce and release into their microenvironments (Becht et al., 2014). The fine analysis of molecularly defined subgroups of human cancers may therefore help understanding the shaping of the various immune contextures with different clinical impacts.

2.3 The CRC Tumor Microenvironment

The most comprehensive analysis of the immune landscape has been performed in CRC. CRC may indeed represent a paradigm of many human cancers. It may be preceded by chronic inflammatory states, such as inflammatory bowel disease or influenced by the gut microbiota (Grivennikov et al., 2012). In patients presenting with Lynch syndrome, as well as in about 15% of sporadic tumors (Fig. 3A), microsatellite instability (MSI) defines a group of tumors in which defects in DNA-repair enzymes result in a high mutational load, creating new tumor-derived peptides that bind the Class I MHC molecules and elicit a tumor-specific T-cell response. Indeed, it was known for a long time that MSI tumors are infiltrated by T cells and

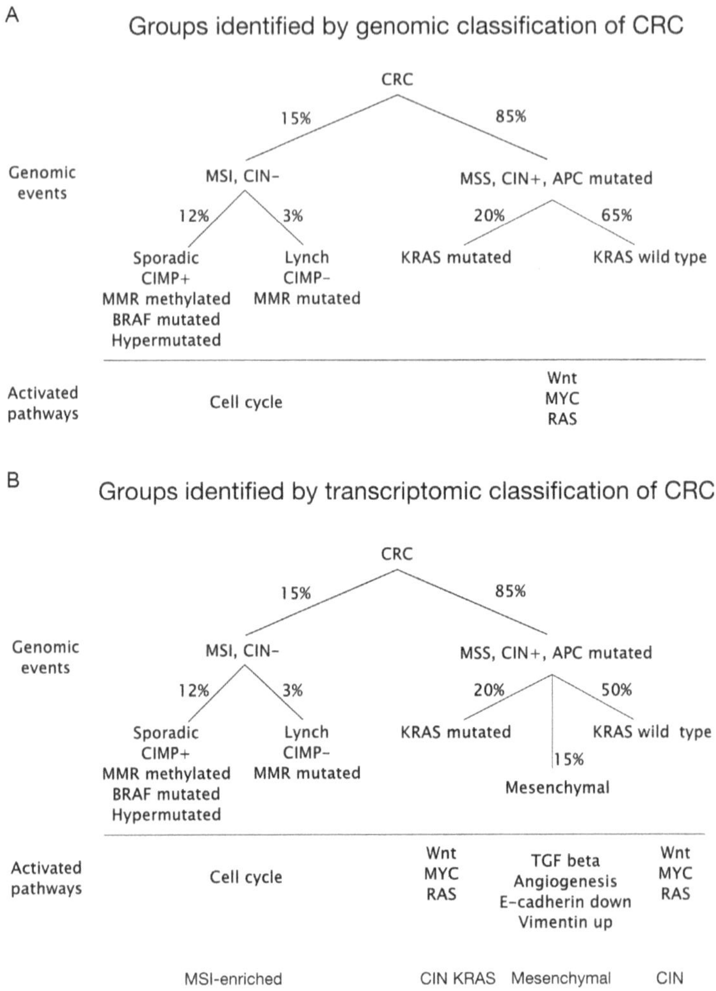

Figure 3 Genomic and transcriptomic subgroups of CRC. (A) Genomic classifications stratify CRC first into microsatellite-stable and microsatellite-instable tumors, the former usually being chromosomal instable. Other mutational events, such as KRAS, can then be used to predict patient's response to particular targeted therapies. (B) Transcriptomic classifications of CRC agree with the existence of four CRC subgroups, including a mesenchymal subgroup that could not be identified on the basis of genomic alterations.

associated with good prognosis in terms of DFS and OS (Dolcetti et al., 1999; Guidoboni et al., 2001). Eventually, CRC metastasizes, particularly in the liver and the lungs where a novel TME is shaped. We have undertaken a comprehensive IHC analysis of CRC tumors of all stages, with or without MSI, in the primary and metastatic sites. The composition of the TME is highly variable. Some tumors exhibit strong angiogenesis and lymphoangiogenesis. Patients with tumor cells detected inside blood or lymphatic vessels or infiltrating along nerves exhibit poor prognosis, the presence of malignant cells in vessels indicating that metastatic processes had begun before surgery of the primary tumor. In these cases, patients even presenting with small tumors with no signs of lymph node invasion and metastatic spread will not be cured by surgery. Variability of the composition and organization of the hematopoietic cell infiltrates is also found within CRC patients. Globally, in tumors with such infiltrates, macrophages are the most abundant cells, followed by lymphocytes, granulocytes, mast cells, and DCs. Among lymphocytes, memory T cells, particularly CD8$^+$, are most represented, followed by B lymphocytes. NK cells, FOXP3$^+$ cells (potential regulatory T cells), and T_{h17} and T_{fh} cells are present at low densities (Bindea et al., 2013). This global picture can be refined by analyzing the organization of the immune landscape. Some tumors exhibit a strong infiltration in their IM without colonization of the tumor core. Some tumors exhibit both IM and CT infiltration. In rare cases, infiltrating cells are found in the center of the tumor with low infiltration detected in the IM. In some tumors, organized lymphoid aggregates are found surrounding tumor nests. Although the densities of the various infiltrating cells are correlated between themselves, their densities differ across locations within the TME. In the center of the tumor, the most represented cells are macrophages and memory CD3$^+$CD8$^+$ T lymphocytes. In the IM, macrophages, memory CD3$^+$CD8$^+$ T lymphocytes, and T_{h17} and NK cells are present. B lymphocytes and T_{fh} are found in the lymphoid aggregates, together with mature and follicular DC, forming potential immunity-generating structures TLSs. FOXP3$^+$ T_{reg} are scarce and found in the IM and the CT (Bindea et al., 2013). Analysis of the clinical impact of these infiltrating cell populations reveals that high densities of memory CD3$^+$CD8$^+$ T cells in the IM and the CT, that of mature DC, T_{fh}, and B cells in the TLS, and, to a lesser extent, that of FOXP3$^+$ and NK cells in the IM and the CT correlate with longer DFS and OS. In contrast, high macrophage, T_{h17} cells, and immature DC infiltration correlate with poor prognosis. Interestingly, the density of innate and proinflammatory cells is higher in high stage tumors compared with low stage tumors, while those of T-cell subsets is lower (Bindea et al., 2013).

Such an analysis provides a first global view of the immune landscape and of the potential clinical impact of its different components (Bindea et al., 2013), leading to the definition of a tumor-associated immune contexture (Fridman et al., 2012) and a routine test called immunoscore (Galon et al., 2012).

2.4 The ccRCC Tumor Microenvironment

A large amount of evidence indicates that RCC displays increased concentrations of almost all types of chemokines and cytokines and often exhibits highly inflammatory features. Indeed, gene expression studies in large cohorts of RCC-bearing patients have suggested that some molecules such as IL-1, IL-6, TNF-α, C1q, and C1r could be implicated in the initiation and promotion of an intratumor inflammatory cascade (Peri, Devarajan, Yang, Knudson, & Balachandran, 2013; Romero et al., 2006; Tan et al., 2011). The RCC tumor cells probably orchestrate this inflammatory environment as suggested by the presence of several cytokines in the supernatant of RCC primary cultures, including IL-8, IL-10, TGF-β, GM-CSF, TNF-α, and VEGF (Gogusev, Augusti, Chrétien, & Droz, 1993; Knoefel et al., 1997; Lahn et al., 1999; Lee, Chun, Nadiminty, Lou, & Gao, 2007; Sievers et al., 2007). Interestingly, some of them can be found in the serum of RCC-bearing patients (Kim, Kim, et al., 2015; Wald et al., 2014), supporting their active production throughout RCC development.

The genetic abnormalities inducing the expression of inflammatory mediators in RCC are still poorly understood. Nonetheless, as the VHL mutation (found in 70% of RCC) causes the overexpression of VEGF, PDGF, and TGF-α, it seems plausible that it could promote the transcription of other inflammatory molecules (Gossage, Eisen, & Maher, 2015). Additionally, some studies suggest that tubular cells (RCC progenitor cell) often acquire a mesenchymal and highly proinflammatory phenotype when stressed (Fragiadaki & Mason, 2011), a characteristic that tumor cells could possibly retain (Conant, Peng, Evans, Naud, & Cooper, 2011; Ho et al., 2012; Huang et al., 2013).

The production of inflammatory cytokines in RCC has proven to be clinically relevant. Thus, the plasma concentration of C-reactive protein in RCC-bearing patients (Hu et al., 2014; Ito et al., 2006; Jagdev et al., 2010; Karakiewicz et al., 2007; Lamb, McMillan, Ramsey, & Aitchison, 2006; Steffens et al., 2012) and the presence of an inflammatory gene signature in RCC (Romero et al., 2006; Tan et al., 2011) are linked to a higher tumor grade and positive metastatic status at diagnosis and poor prognosis. In addition, other inflammatory markers are related to poor survival in RCC-bearing patients, including an augmented erythrocyte sedimentation rate, leukocytosis, thrombocytosis, and the increased plasmatic concentrations of IL-6 and TNF-α (Dosquet, Coudert, Lepage, Cabane, & Richard,

1997; Dosquet et al., 1994; Escudier et al., 2014; Fox et al., 2013; Negrier et al., 2004; Pichler et al., 2013; Wu et al., 2011; Yoshida et al., 2002). In vitro studies reinforce this concept, since several proinflammatory cytokines (eg, IL-6, TNF-α, hypoxia-inducible factor-α, and matrix metalloproteinase-2) are preferentially produced by the RCC cell lines exhibiting the highest malignant potential (Chuang et al., 2008; Ho et al., 2012; König et al., 1999).

Several studies have characterized the function of different immune cellular population in RCC. RCC TAMs are capable of producing a series of immunomodulatory molecules that can both promote inflammation and hamper the cytotoxic antitumor immune response. Some studies suggest that RCC TAMs exhibit an M2-phenotype, characterized by the expression of CD163, interferon regulatory factor 4, fibronectin 1, and IL-10 (Behnes et al., 2014; Dannenmann et al., 2013; Komohara et al., 2011). These cells can induce T_{reg} expansion (Daurkin et al., 2011), in addition to PD-1 and TIM-3 expression on T lymphocytes (Dannenmann et al., 2013). Despite the fact that most of the molecules that induce the accumulation of MDSC are abundant in the RCC microenvironment (Lechner et al., 2011), very few studies have assessed the role of this population in the inhibition of the antitumor immune response in this pathology. Nevertheless, studies indicating that Sunitinib can induce tumor regression trough MDSC depletion suggest that this population might be implicated in RCC pathogenesis. L-Arginine depletion (Ochoa et al., 2007; Rodriguez et al., 2009; Zea et al., 2005) and the overproduction of ROS (Kusmartsev et al., 2008) are only two proven mechanisms of MDSC-induced immunosuppression in human RCC. Several studies have assessed the DC biology in RCC, and some agree that these cells can be found in two main locations: tumor stroma isolated from other immune cells or within lymphocyte aggregates in the IM of the tumor (Giraldo et al., 2015; Troy, Summers, Davidson, Atkinson, & Hart, 1998). Interestingly, these two types of DCs display different phenotypes and capacities of T-cell priming, presumably due to the distinctive microenvironments in which they develop. Isolated DCs display an immature phenotype ($CD80^-$, $CD86^-$, $CD83^-$, and HLA-DR^-) (Gigante et al., 2009; Giraldo et al., 2015), express tumor-promoting molecules (eg, TNF-α and MMP-9) (Figel et al., 2011), and induce a dysfunctional in vitro T-cell activation (Cabillic et al., 2006; Figel et al., 2011). On the other hand, DCs within immune aggregates exhibit a mature phenotype and express activation markers (Giraldo et al., 2015; Middel, Brauneck, Meyer, &

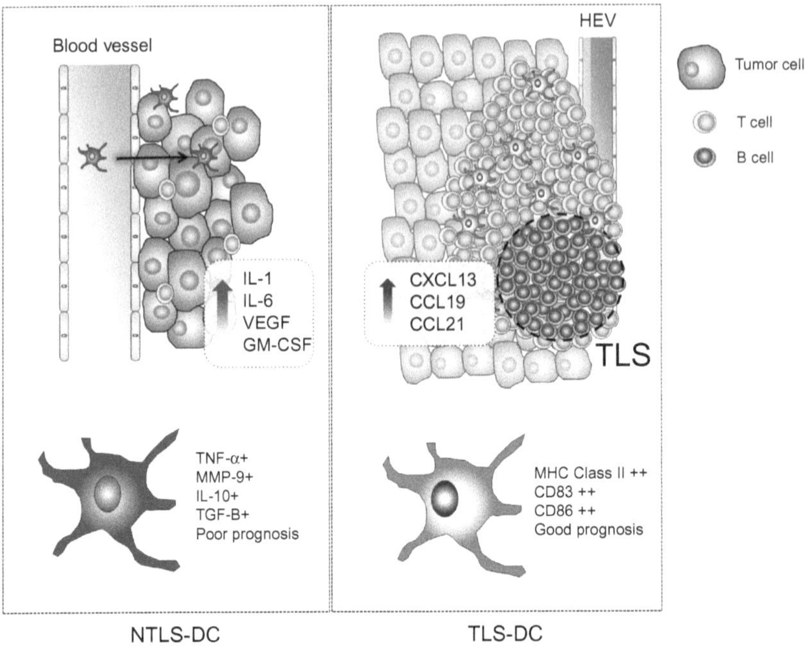

Figure 4 DC compartmentalization and prognosis in RCC. Cartoon depiction of the two major types of DCs in ccRCC. On the left, the expression of inflammatory molecules can induce the stromal accumulation of immature DC, characterized the production of protumoural molecules and poorly effective T-cell priming capacities. On the right, the presence of CXCL13, CCL19, and CCL21 can potentially promote the development of DC within TLS, exhibiting a mature phenotype and capable of inducing T-cell priming. (See the color plate.)

Radzun, 2010; Troy et al., 1998) (Fig. 4). This evidence suggests that RCC microenvironment is enriched in molecules that hamper DC activity and maturation, since their phenotypic abnormalities are restricted to cells embedded in the tumor stroma and can be reversed when they are removed from the tumor. In addition, conditioned media from RCC primary cultures or cell lines can induce in vitro tolerogenic and dysfunctional DCs (HLA-DR^{dim}, $CD80^-/CD86^-$, IL-10^+/TGF-β^+) (Hou et al., 2010; Song, Shurin, Tourkova, Chatta, & Shurin, 2004; Teng et al., 2014) by mechanisms dependent on IL-6, IL-8, and VEGF (Cabillic et al., 2006; Figel et al., 2011).

In addition to express several TAM-attracting molecules, RCC tumor cells secret chemokines that recruit T lymphocytes, including CCL4, CCL5, CXCL9-11, and CXCL16 (Oldham et al., 2012; Polimeno et al., 2013;

Sharpe et al., 2013). Once T cells have breached into the tumor, they become activated, as suggested by the overrepresentation of cells with an effector-memory phenotype expressing activation markers (CD69, HLA-DR, and FAS-L) on TILs, when compared to PBLs (Alexander et al., 1995; Angevin, Kremer, Gaudin, Hercend, & Triebel, 1997; Dannenmann et al., 2013; Elsässer-Beile, Gierschner, Welchner, & Wetterauer, 2003; Kopecký et al., 2007; Kowalczyk, Skorupski, Kwias, & Nowak, 1997; Van den Hove et al., 1997). The biologic consequences of an augmented T-cell recruitment in RCC have been enigmatic, because contrary to most tumors where increased densities Th1 and $CD8^+$ TILs are associated with a favorable clinical outcome (Fridman et al., 2012), a few studies had described the opposite association in RCC. A study in a retrospective cohort of 233 RCC reported that high $CD8^+$ TIL densities were associated with poor prognosis, only by univariate analysis (Nakano et al., 2001). Interestingly, in the same cohort, increased infiltrations with $CD8^+/Ki67^+$ double-positive cells were associated with favorable clinical outcome (Nakano et al., 2001). Although this result needs confirmation, it indicates that when $CD8^+$ RCC TILs are poorly functional (either due to their direct inhibition or due to defects in their priming/activation), they predict an adverse clinical outcome. Other studies have also reported that the increased densities of $CD45RO^+$ cells in RCC are correlated with advanced TNM stages and shorter OS and progression-free survival (Hotta et al., 2011). In the latter and similar studies (Bromwich et al., 2003; Kondo et al., 2006), only the increased densities of $CD4^+$ TILs (and not those of $CD8^+$ TILs) have been associated with poor clinical outcome. Finally, all these studies have consistently reported an association between increased densities of $CD8^+$ TIL and higher Fuhrman grades (Hotta et al., 2011; Igarashi, Murakami, Takahashi, Matsuzaki, & Shimazaki, 1992; Nakano et al., 2001).

Interestingly, few reports have also established that T-cell responses in RCC are characterized by a low amount of expanded clones (Angevin et al., 1997; Gerlinger et al., 2013; Shimabukuro & Naito, 2008; Sittig et al., 2013). The current evidence on the immune microenvironment of RCC suggests that this finding can be associated with dysfunctional T-cell priming (Attig et al., 2009; Remark et al., 2013; Wang, Hanada, Robbins, Li, & Yang, 2012), the inflammatory recruitment of low-affinity T cells (Angevin et al., 1997; Schoof et al., 1993), and the inhibition of T-cell activation/proliferation. In addition, several functional and phenotypic abnormalities on RCC TIL have been reported, and include their diminished proliferative capacity (Rayman et al., 2000; Tartour et al.,

1995), mainly related to alterations in the production of IL-2 and its intracellular signaling pathway (Biswas et al., 2006; Bukowski et al., 1998; Cavalcanti et al., 2010; Knoefel et al., 1997; Kolenko et al., 1997; Kudo et al., 2003; Lahn et al., 1999; Ling et al., 1998; Malinowski et al., 1997; Nakagomi et al., 1995; Onishi, Ohishi, Imagawa, Ohmoto, & Murata, 1999; Rayman et al., 2000, 2004; Rubtsov & Rudensky, 2007; Uzzo et al., 1999), the poor development of polyfunctional responses (Attig et al., 2009), and the increased expression of inhibitory receptors.

Some studies have found that RCC $CD4^+$ TIL are skewed toward a T_{h2}-phenotype and preferentially express IL-10 and IL-4 (Elsässer-Beile et al., 1998; Lauerová et al., 1999; Maeurer et al., 1995; Puccetti et al., 2002; Schoof et al., 1993; Tatsumi et al., 2003, 2002; Van den Hove et al., 1997). Indeed, the high concentrations of COX2-PGE2, gangliosides, and IL-4 and IL-6 in RCC microenvironment have been implicated in the preferential development of T_{h2} responses (Biswas et al., 2006; Smyth et al., 2003; Zhou, Chong, & Littman, 2009). The RCC tumor cells are probably responsible for this phenomenon, since primary RCC tumor supernatants can inhibit the expression of IFN-γ in T cells, and not the production of type 2 cytokines (eg, IL-4, IL-5, and IL-10) (Biswas et al., 2006; Rayman et al., 2004). The impact of the T_{h2} skewing in RCC pathogenesis is currently unknown; nevertheless studies in small cohort of RCC-bearing patients suggested that $CD4^+$ TIL preferentially displays a T_{h2} phenotype in advanced stages of the disease (Cózar et al., 2005). In addition, some studies have implicated the T_{reg} subpopulation in the inhibition of cytotoxic lymphocytes in RCC (Griffiths et al., 2007; Kang et al., 2013; Oldham et al., 2012). Their recruitment seems to be induced by the CCL22/CCR4 and TGF-β/VEGF-A axes (Busse et al., 2011; Desar et al., 2011; Li et al., 2012). The expansion of T_{reg} subpopulation within PBL and TIL in RCC-bearing patients has been associated with advanced TNM stages (Dannenmann et al., 2013; Kim, Kim, et al., 2015; Li et al., 2015), denser tumor microvascularization (Zhan, Gao, Zhou, Pu, & Wang, 2012), and poor survival (Kang et al., 2013; Li et al., 2010; Li, Chu, et al., 2009; Liotta et al., 2011).

NK lymphocytes are the only immune population which correlates with a favorable clinical outcome in RCC (Cózar et al., 2005; Donskov & von der Maase, 2006; Eckl et al., 2012; Remark et al., 2013). These innate immune cells represent a relevant percentage of the RCC TILs (Cózar et al., 2005; Van den Hove et al., 1997), and they often coexpress activation and inhibitory markers (Prinz et al., 2014; Schleypen et al., 2006; Van den Hove et al., 1997).

We have analyzed the immune infiltrate of 135 primary ccRCC tumors and confirmed that high $CD8^+$ T-cell densities correlated with shorter DFS and OS (Giraldo et al., 2015). This study revealed that most intratumoral $CD8^+$ T cells reacted with anti-PD-1 and LAG3 antibodies, suggesting that they may be exhausted. Moreover, tumor cells were often positive for the PD-L1 and/or PD-L2 in many cases. Combination of high densities of PD-1-positive $CD8^+$ T cells and PD-L1- or PD-L2-positive tumor cells identified a group of patients with highly deleterious prognosis (Giraldo et al., 2015). Analysis of The Cancer Genome Atlas (TCGA) database for ccRCC revealed that the immune gene most significantly associated with poor prognosis in ccRCC was IFN-γ (Giraldo et al., 2015), in contrast with many other tumor types (Fridman et al., 2012). This paradoxical observation is reminiscent of the proposed adaptive resistance mechanism which hypothesizes that activated tumor-infiltrating T cells produce IFN-γ which induces PD-L1 (and probably PD-L2) expression on tumor cells, eventually resulting in T cell exhaustion (Eppihimer et al., 2002; Loke & Allison, 2003; Taube et al., 2012). It might explain why $CD8^+$ T cells, expressing PD-1, and other immune checkpoint inhibitors, are associated with poor prognosis in RCC. In contrast, high densities of PD-1-positive $CD8^+$ T cells correlates with favorable prognosis in head and neck (Badoual et al., 2013) and colorectal cancers (Becht et al., 2015). The reasons for these contradicting observations may be due to the differential sensitivity of the malignant cells to the IFN-γ-mediated induction of PD-L1, or other checkpoint inhibitory ligands, or to different organizations of the immune landscape. Indeed, PD-L1 expression can also be induced by other stimuli in the TME, such as hypoxia and VEGF production (Barsoum, Smallwood, Siemens, & Graham, 2014; Noman et al., 2014).

So far ignored from the analyses of the TME, PD-L2 begins to gain importance in the tumor immunology field, since clinical trials with anti-PD-1 agents have pointed at its possible role in the inhibition of T cell response in cancer (Brahmer et al., 2012; Taube et al., 2014; Topalian et al., 2012). Its expression has been reported in several tumors (Hamanishi et al., 2007; Ohigashi et al., 2005; Zhang, Wang, et al., 2014b).

The role of other inhibitory receptors in RCC TIL has not been assessed in detail. Similar to PD-1, few studies haves demonstrated an expansion of the TILs expressing TIM-3 and/or LAG-3 compared to autologous PBLs (Dannenmann et al., 2013; Sittig et al., 2013). The role of other immune checkpoint, including CTLA-4, CD137, OX40, and A2aR in RCC, has not been assessed to date. Nevertheless, a phase II clinical trial with

Ipilimumab (anti-CTLA-4 antibody) in metastatic RCC showed a partial response of approximately 10% of patients (Yang et al., 2007), suggesting the potential role of this molecule in tumor development.

3. CLASSIFICATION OF TUMORS IN THE ERA OF OMIC TECHNIQUES

Tumor classifications have been established and continuously refined by clinicians and researchers. The goals are to integrate knowledge of the biology of different malignancies to improve patient's management. The accurate prediction of the evolution of a cancer is of crucial importance since the advent of cytotoxic chemotherapies, to avoid potentially damaging unnecessary treatments. More recently, the development of targeted therapies, with drugs that interfere with a particular feature of the tumor biology, has opened new therapeutic options and simultaneously pushed for a better characterization of potential responders. In the following section, we recapitulate the evolution of tumor classifications, with a particular attention on the recent molecular classifications.

3.1 Anatomopathological Classifications of Tumors

The most straightforward and currently almost subconsciously accepted way to classify cancers is according to the organ in which the primary tumor originated. As an illustration, the World Health Organization/International Agency for Research on Cancer classification is made of seven books, each covering one or several anatomical compartment (digestive system, breast and female genital organs, soft tissue and bone, skin, urinary and male genital organs, head and neck, and lung, pleura, thymus, and heart) (The International Agency for Research on Cancer, 2000, 2003, 2004a, 2004b, 2005a, 2005b, 2006). Chapters within these books cover cancers of a given organ. This intuitive classification still stands but we now understand that cancer is a disease caused by molecular modifications of normal host cells which disrupt the mechanisms controlling cell proliferation, apoptosis, and motility, among other features, leading to phenotypic modifications and malignant transformation (Hanahan & Weinberg, 2011). Different mutations will lead to either malignant transformation or no phenotypic modification. The phenotype of the tumor cell is therefore the most critical factor in our current understanding of cancer biologies.

Within most malignancies, the current gold standard for prognosis prediction is the Union Internationale Contre le Cancer/TNM staging classification, which is based on the assessment of the size of the primary tumor (T stage), its spread to local or distant lymph nodes (N stage), and its metastatic dissemination to distant organs (M stage) (Sobin, Gospodarowicz, & Wittekind, 2009). The TNM classification is able to accurately depict cancer progression and is therefore strongly associated to prognosis. It is however inapplicable to diffuse cancers such as leukemia.

Other anatomopathological examinations of tumor tissues help clinicians predict the patient's outcome associated with a given tumor. Notably, the tumor grade is a classification based on the appearance of tumor cells. Its definition depends on the malignancy but tries to recapitulate the aggressivity of tumor lesions. One possible grading system is to assess the extent of the differentiation of malignant cells, with low-grade tumors resembling nonmalignant well-differentiated surrounding cells and high-grade tumors looking more anaplastic. In some cancers, such as brain malignancies, the grading system is used instead of the TNM staging system to assess patient's prognosis.

3.2 Molecular Classifications of Tumors

3.2.1 Genomic Classifications

Observations of chromosomal aberrations in cancer cells first hinted at genetic abnormalities as a possible cause for cancer (Calkins, 1914), and it was indeed shown that the transfer of cancer genome in nonmalignant cells provoked their malignant transformation (Krontiris & Cooper, 1981; Shih, Padhy, Murray, & Weinberg, 1981).

Genomic classifications seek to identify genomic modifications of the cancer genome compared to the host's nonmalignant genome and characterize their effect on the tumor cell's biology. The number of known driver mutations ranges in the hundreds (Stratton, Campbell, & Futreal, 2009; Vogelstein et al., 2013), but the true number of possible driver events is likely to range in the thousands (Stratton et al., 2009). Some driver mutations are highly prevalent in many cancers, such as inactivating mutations in the TP53 tumor suppressor gene which occurs in approximately 50% of cancers (Surget, Khoury, & Bourdon, 2013). Others are rare and specific to one or a few malignancies, such as the APC gene which is frequently mutated in colorectal cancer and occasionally mutated in papillary thyroid and adrenocortical cancers (Fearnhead, Britton, & Bodmer, 2001).

The knowledge of mutated genes in a cancer can help establishing molecular classifications, developing targeted therapies, or predicting responders to treatments. For instance, Imatinib is a tyrosine kinase inhibitor specifically targeting the translocation of the ABL oncogene in chronic myeloid leukemia (Druker, 2008). Trastuzumab, a monoclonal antibody targeting the HER2/neu antigen, targets the product of the amplified ERRB2 gene which occurs in 20–30% of breast cancers (Vu & Claret, 2012). Cetuximab, a monoclonal antibody targeting the EGFR receptor, is unable to induce clinical responses in colorectal KRAS-mutated tumors (Lièvre et al., 2006). Such classifications are useful in a clinical setting, but it is difficult to integrate the whole-genome mutation spectrum into unified taxonomies. Driver mutations occur in particular pathways which sustain survival, growth, and other cancer hallmarks, such as resistance to immune elimination. Driver mutations can sometimes be exclusive, but this fact usually indicates functional homology of the encoded proteins, which are usually both critical signaling molecules of a same pathway (Ciriello, Cerami, Sander, & Schultz, 2012; Yeang, McCormick, & Levine, 2008). Creating biologically relevant classification of tumors requires to be able to discriminate between passenger and driver mutations, which is difficult as most somatic mutations in cancers are believed to be passenger (Vogelstein et al., 2013). Driver mutations then need to be sorted according to the pathway they affect, and their effect on the corresponding protein (inactivation, activation or no effect).

3.2.2 Transcriptomic Classifications

The first transcriptomic classifications were established in breast cancer (Perou et al., 2000) and B-cell lymphoma (Alizadeh et al., 2000). These transcriptomic classifications were later confirmed, refined (Ando, Suguro, Kobayashi, Seto, & Honda, 2003; Guedj et al., 2012; Hu et al., 2006; Loi et al., 2007; Mackay et al., 2011; Rosenwald et al., 2002; Sørlie et al., 2001), used in preclinical models to stratify clinical trials (Barton, Hawkes, Wotherspoon, & Cunningham, 2012), and inspired the discovery of clinically and biologically heterogeneous subgroups in many other malignancies, including colorectal cancer (Budinska et al., 2013; De Sousa E Melo et al., 2013; Marisa et al., 2013; Roepman et al., 2014; Sadanandam et al., 2013; Schlicker et al., 2012), renal cell carcinoma (Brannon et al., 2012, 2010), glioma (Nutt et al., 2003; Shai et al., 2003), liver (Boyault et al., 2007; Chiang et al., 2008; Hoshida et al., 2009; Lee et al., 2004), bladder (Biton et al., 2014), prostate (Tomlins et al., 2015), acute myeloid leukemia (de Jonge, Huls, & de Bont, 2011; Mrózek, Radmacher, Bloomfield, & Marcucci, 2009; Silva et al., 2009; Verhaak et al., 2009), and other cancers (Barlin et al., 2015; de Reyniès et al., 2014; Guo et al., 2015).

3.2.3 Multi-omics Classifications

In addition to genomic and transcriptomic characterization of tumors, other high-dimensional techniques are able to identify tumor molecular subtypes. Notably, the methylome, proteome, and miRNome quantify molecular features that result in modifications of biological pathways in tumor cells and their microenvironments. Analog of the techniques used for transcriptome-based tumor classifications has been applied to these different omic techniques. Notably, TCGA consortium aims at the multi-omic characterization of major human cancers.

Their studies, as well as others, show that other omic classifications are significantly correlated with those established using transcriptomics (Assié et al., 2014; Brat et al., 2015; Brennan, Brigl, & Brenner, 2013; Cancer Genome Atlas Network, 2012a, 2012b, 2013a, 2013b, 2014a, 2014b, 2014c, 2014d, 2015a, 2015b; Marisa et al., 2013).

3.3 The Immunoscore

The consensual findings that high densities of $CD8^+$ T cells and a T_{h1} orientation correlate with favorable prognosis in the vast majority of cancers led to propose an immunoscore to classify cancers, and that could be easily used as a routine laboratory prognostic and potentially predictive tool in a clinical setting. The Immunoscore is based on the quantification of two cell populations (CD3/CD45RO or CD8/CD45RO or CD3/CD8) in the IM and the CT. It ranges from 0 to 4, with a low density of a marker being given a score of 0 and a high density of a marker being given a score of 1. This process is repeated in both the IM and the TC. It has been applied to a cohort of 282 early-stage (stage I and stage II) CRC. It appeared very informative since 80% of the patients with a immunoscore of 4 were DFS, 20 years after surgery, while 80% of patients with an immunoscore of 0 had relapsed within 2 years (Galon et al., 2006). Immunoscores of 1, 2, and 3 classified the cohort with increasing DFS (Galon et al., 2006). The immunoscore appeared to be the strongest prognostic factor for DFS and OS, as the classical UICC-TNM classification did not significantly impact clinical outcome when included along with the Immunoscore in multivariate survival analyses of stages I, II, and III patients (Mlecnik et al., 2011). This observation prompted J. Galon to create a worldwide task force grouping over 20 laboratories in over 20 countries, all using the same reagents and standardized technologies to validate the immunoscore in over 10,000 colorectal cancers (Galon et al., 2012). The effort is ongoing. A very interesting application of the immunoscore is its predictive value for response to immunotherapies. The recent years have witnessed the explosion of immunotherapeutic approaches aimed at modulating the immune microenvironment of tumors. The first effective champions of these treatments are the antibodies targeting CTLA-4 and PD-1. To modulate a local immune response, it is necessary that the lymphocytes, even exhausted, are present in the tumors. Indeed, only melanoma patients with $CD8^+$ T cells in their IM respond to anti-PD-1 therapy (Tumeh et al., 2014). Similar findings have been reported for bladder cancer therapy with anti-PD-L1 (Powles et al., 2014). In a more general fashion, therapies that will ultimately result in stimulating local antitumor immune reactions rely on the presence of lymphocytes, particularly potentially cytotoxic $CD8^+$ T cells at the tumor site (Fridman et al., 2012).

The immunoscore allows a standardized, routine test appropriate for clinical use in pathology laboratories. Its strength resides on the fact that the tumor density of memory $CD8^+$ T cells correlates with favorable prognosis in most cancer types and may predict response to certain immunotherapies.

4. THE TUMOR MICROENVIRONMENT AND THE MOLECULARLY DEFINED SUBGROUPS OF HUMAN CANCERS

The immunoscore is a technology usable in routine practice for identifying patients with potential favorable prognosis and responding to selected immunotherapies on the basis of T-cell densities. However, since it quantifies the ultimate result of the very complex tumor–stroma interaction network happening during cancer evolution, the immunoscore does not take into account the myriad of cellular partners that compose the tumor immune contexture. An immunohistochemical approach of such complexity, possibly in research studies (Bindea et al., 2013), cannot be applied in clinical laboratories, due to its price, the time, and the workforce needed. The transcriptomic approach offers rapid, less expensive, and less laborious techniques to quantify cell populations and their functions. They are a way to study the interacting complexities of the immune infiltrating cells and the malignant cells within a tumor, to assess their clinical impact, and to select patients for immunotherapeutic approaches. The availability of databases characterizing patterns of genes expressed in various cell populations, including lymphocytic subsets, myeloid cells, fibroblasts, or endothelial cells, allowed to establish an immunome which defines metagenes specific for a given cell population. It should be highly expressed in the desired population and not expressed, or at very low level, in any other cell type of the tumor sample. Different, largely overlapping immunomes have been established with metagenes specific for T cells, $CD4^+$ T cells, $CD8^+$ T cells, NK cells, cytotoxic cells, T_{h1}, T_{h2}, T_{h17} or T_{fh} cells, B cells, $T_{\gamma\delta}$, DCs of different types, macrophages, granulocytes, endothelial cells, and fibroblasts. Some signatures were validated in quantitative analysis of RNA extracted from purified immune cells, mixed together in controlled proportions and ultimately diluted in cancer cell lines RNA, and were able to accurately predict the corresponding cell type's proportion. This approach allows to robustly quantify immune infiltrates in large collections of tumors and is therefore highly relevant for studying the interplay between complex systems such as the malignant cell and the immune, inflammatory, or vascular and stromal reactions. It has pitfalls, such as the lack of refined functional characterization of the infiltrating cells, which can be completed by specific analysis of genes shaping the functional orientation of the immune response. More importantly, it does not take into account the different locations of the immune infiltration, measuring it at the place where the RNA has been collected, usually in the tumor core. Despite these pitfalls, the use of the immunome allows an immune classification of tumors in relation with molecularly defined cancer subgroups.

In a first simple comparison, a contrasted picture between RCC and CRC appeared. Database analysis of The Cancer Genome Atlas for RCC and the Moffit Cancer Center cohort for CRC revealed that lymphocyte-specific gene expression correlated with poor prognosis in RCC and with favorable prognosis in CRC (Giraldo et al., 2014).

4.1 Integrating Immune and Molecular Classifications in ccRCC

4.1.1 Genomic Classifications of ccRCC

Compared with colorectal cancer, the knowledge of the sequence of genomic events associated with the carcinogenesis of ccRCC is less documented. It is dominated by the disruption of the "cellular response to hypoxia" pathway, while secondary events begin to emerge. In parallel, transcriptomic classifications are being proposed and may help our understanding of the heterogeneity of ccRCC in terms of clinical behavior (Morais, Gobe, Johnson, & Healy, 2011).

4.1.1.1 Genomic Events: Response to Hypoxia Pathway

The first identified germline mutation predisposing to ccRCC is a loss-of-function mutation of the von Hippel-Lindau (VHL) tumor suppressor gene (Cairns, 2010). VHL encodes a protein which associates with products of the TCEB1 (elongin B), TCEB2 (elongin C), and CUL2 (cullin-2) to form a complex with ubiquitin-ligase activity (Sato et al., 2013). In normoxic conditions, this complex ubiquitinates the hypoxia-inducible factor 1a (HIF1a), leading to its degradation.

HIF1a is a transcription factor mediating the cellular response to hypoxia pathway. In hypoxic conditions or when the ubiquitin-ligase complex is inactivated, HIF1a escapes proteasome-mediated degradation and triggers a transcriptomic program which fosters glycolysis and production of VEGF which promotes angiogenesis. VHL inactivation has been reported in around 70–90% of ccRCC tumors (Cancer Genome Atlas Network, 2013b; Sato et al., 2013). Most of these events are recessive loss-of-function mutations, but are associated with a loss of heterozygosity of the 3p21–3p25 chromosomal loci which contains the VHL gene (Cancer Genome Atlas Network, 2013b; Sato et al., 2013). Driver genomic events in VHL wild-type ccRCC tumors are still poorly characterized (Cancer Genome Atlas Network, 2013b), but a study notably reported recurrent mutations in the elongin B gene TCEB1, mutually exclusive with VHL mutations (Sato et al., 2013).

SETD2 mutation appears to be selected as a secondary carcinogenic event among VHL-mutated tumors: SETD2 is located in the same 3p21–3p25 region as VHL (Cancer Genome Atlas Network, 2013b; Sato et al., 2013) and therefore affected by frequent loss-of-heterozygosity (LOH), and mutated SETD2 is detected at lower allelic frequencies than VHL mutations, suggesting a subclonal selection of SETD2 among mutated VHL tumors (Sato et al., 2013). SETD2 seems implicated in widespread DNA hypomethylation (Cancer Genome Atlas Network, 2013b) and associated with faster relapse (Sato et al., 2013).

4.1.1.2 Mutations in the Signaling Pathways

The PI3K–AKT–mTOR pathway, which regulates the cell cycle, ranks third in terms of mutation frequency in ccRCC, with mutations affecting PTEN, PIK3CA, AKT, and mTOR in around one-quarter of tumors (Cancer Genome Atlas Network, 2013b; Sato et al., 2013). Some authors suggested that NF-κB, the main transcriptomic factor implicated in inflammatory transcriptomic programs, but that also regulates cell survival, was overactivated in VHL-mutated tumors (Morais et al., 2011). NF-κB activation was associated to resistance to TNF-α-mediated apoptosis. NF-κB activation also leads to the production of inflammatory cytokines, such as IL-6 and IL-8 (CXCL8) (Morais et al., 2011).

4.1.1.3 Chromatin Remodeling

Epigenetics appears to also play a role in the carcinogenesis of ccRCC. VHL can be inactivated through hypermethylation of its promoter instead of mutation, an event seen in around 7–15% of the cases (Cancer Genome Atlas Network, 2013b; Sato et al., 2013; Varela et al., 2011). A less understood epigenomic event is the disruption of the SWI/SNF chromatin-remodeling complex. Several studies observed a significant rate of somatic mutations affecting this complex, such as the PBRM1 (Cancer Genome Atlas Network, 2013b; Sato et al., 2013; Varela et al., 2011), ARID1A, and SMARCA4 (Cancer Genome Atlas Network, 2013b). It has been proposed that the SWI/SNF complex could be implicated in heterochromatin conversion of chromosomal regions, leading to their epigenetic silencing.

4.1.2 Immune Contexture of ccRCC Molecular Subgroups

Application of the immunome to a molecular classification of ccRCC primary tumors of metastatic patients revealed a high expression of T- and B cell metagenes as well as genes specific for cytotoxic lymphocytes and myeloid cells in the group of patients who did not respond to treatment with the tyrosine kinase inhibitor Sunitinib and had the shorter OS and DFS (Beuselinck et al., 2015). Tumors of this subgroup exhibited a high expression of genes encoding molecules involved in inflammation (TNF, IL-6), myeloid cell activation (CSF1, CSF2, CSF3), T-cell activation (IL-2RA, IL-2RB, IL-2RG, IL-15R), chemotaxis (CXCL9, CXCL10, CXCL11, CXCL13), T_{h1} polarization (IFN-γ), exhaustion (PD-1, LAG3, HAVCR2 [TIM-3], CD274 [PD-L1], PDCD1LG2 [PD-L2]), and immunosuppression (IL-10, TGFB1) (Beuselinck et al., 2015). It is also the group of tumors with the highest infiltration of PD-1-positive lymphocytes and with PD-L1 or PD-L2-expressing tumor cells (Beuselinck et al., 2015). Thus in ccRCC, molecular subtyping identifies an immune high group with high lymphocyte infiltration, high inflammation and local immunosuppression in highly vascularized context, and low TLS density, an immune contexture likely to be involved in the poor prognosis of this group.

4.2 Integrating Molecular and Immune Classifications in CRC

4.2.1 Genomic Classifications of Colorectal Cancer

4.2.1.1 The Adenoma Carcinoma Sequence

The prevailing model for colorectal carcinogenesis was proposed Vogelstein et al. (1988), based on the genetic characterization of four genomic events (mutations in genes of the Ras family, deletions in chromosomes 5, 17, and 18) in early adenomas, advanced adenomas, and colorectal carcinomas. This model describes the course of the genomic events necessary for the progression from benign colorectal adenomas to malignant carcinomas. The first event is a loss of the APC tumor suppressor gene function, which controls the β-catenin pathway. Familial adenomatous polyposis (FAP) is a hereditary condition, most frequently caused by germline inactivation of the APC gene, leading to the development of many adenomatous polyps in the large intestine. Loss of the APC-containing 5p21 genetic locus occurred only in patients without FAP, indicating that either germline or somatic loss of FAP was the main factor associated with the development of adenomas. Second, the authors observed an increase in the incidence of RAS mutations in both advanced (large) adenomas and carcinomas compared to early (small) adenomas, indicating that mutation in the RAS pathway is necessary for the growth of adenomas. Third, loss of the 18q chromosome arm, which carries the SMAD4 oncogene and the still debated tumor suppressor DCC, was most prevalent in carcinomas, followed by advanced adenomas. Finally, Loss of the 17p chromosome arm, which notably hosts the TP53 tumor suppressor gene, was almost exclusively associated to carcinomas.

4.2.1.2 Signaling Pathways Associated with Carcinogenesis

Nearly all colorectal tumors feature dysregulation of the Wnt/β-cathenin pathway (Armaghany, Wilson, Chu, & Mills, 2012). This pathway is important for the migration of colonic stem cell from the bottom to the top of the colon crypt. Inactivation of the Wnt/β-cathenin pathway leads to a proliferation of stem cells which remain in an undifferentiated state in the bottom part of the crypt, resulting in the formation of adenomas. The transition from adenomas to carcinoma necessitate the acquisition of other malignant hallmarks.

Advanced adenoma frequently harbor mutations activating cell-cycle pathways. Notably, mutations in Ras oncogenes (KRAS, NRAS, HRAS), BRAF, or mutations affecting the PI3K pathway (such as activating mutations of PIK3CA or inactivating mutation of PTEN) lead to increase proliferation within the premalignant lesion (Armaghany et al., 2012). Other frequently disregulated pathways include the TP53 pathway, responsible for cell-cycle control and sensing of DNA integrity, and the TGF-β pathway which promotes cell survival, invasion, metastasis, and immunosuppression (Armaghany et al., 2012).

4.2.1.3 Genomic Events Associated with Carcinogenesis

In colorectal cancer, high genomic instability is reported as the main factor contributing to the acquisition of these traits. Three major mechanisms have been implicated. The first one is chromosomal instability. Chromosomal instability (CIN) is observed in up to 85% of colorectal cancers (Fig. 3A). It is characterized by frequent losses or gains of whole of chromosomes during cell replications, which can lead to aneuploidy (inconsistent number of copies of chromosomes), or chromosomal fragments' amplifications or deletions and loss of heterozygosity. The mechanisms underlying this phenomenon are still poorly understood, most likely because of its complexity: hundreds of genes have been shown to induce CIN in yeast, while only around 10 have been identified in humans (Pino & Chung, 2010). Most of these include genes related to chromosome segregation during replication, such as BUB1, BUB1B, and AURKA (Grady & Markowitz, 2015; Pino & Chung, 2010), and genes involved response to DNA damage such as TP53, BRCA1, and BRCA2 (Pino & Chung, 2010).

The second mechanism that favors genomic instability is the CpG-island methylator phenotype (CIMP) (Fig. 3A). The CIMP is defined by hypermethylation of the symmetrical dinucleotide CpG and a global DNA hypomethylation. Many human genes harbor a CpG island in their promoter region, and methylation of the cysteines of the CpG-island has been associated to transcriptional silencing of the gene (Issa, 2004). CIMP is believed to contribute to carcinogenesis through the methylation of promoters of tumor suppressor genes (Jones & Baylin, 2002), and/or by promoting microsatellite instability (MSI), another mechanism involved in genomic instability.

The third mechanism implicated in genomic instability is the MSI. MSI is caused by a defect in the DNA mismatch repair (MMR) machinery. MMR is involved in the correction of DNA replication errors. Microsatellites are short repetitive sequences of DNA, which usually cause DNA-polymerase to "stutter," leading to an increase in replication errors (Armaghany et al., 2012). MSI is therefore the marker of a dysfunctional MMR system. MSI can be due to germline mutations of MMR enzymes, such as MLH1, MSH2, MSH6, and PMS2, which causes a condition known as Lynch syndrome (Armaghany et al., 2012; Gruber & Kohlmann, 2003), associated with an increased risk of developing colorectal, gastric, ovarian, and other cancers (Gruber & Kohlmann, 2003). MSI can also occur in the absence of germline mutations, notably through epigenetic silencing of MMR enzymes in $CIMP^+$ tumors (Armaghany et al., 2012; Issa, 2004). As a result of deficient MMR, MSI^+ tumors are hypermutated, with almost 10 times more nonsynonymous mutations per tumor than mutagen-induced lung cancers and melanoma (Vogelstein et al., 2013). MSI occurs in approximately 15% of cancer cases (Armaghany et al., 2012) (Fig. 3A).

4.2.2 Transcriptomic Subtypes of Colorectal Cancer

Unsupervised transcriptomic classifications of colorectal cancer have been actively studied during the recent years. Six teams have independently reported molecular subtyping studies between 2012 and 2014 (Budinska et al., 2013; De Sousa E Melo et al., 2013; Marisa et al., 2013; Roepman et al., 2014; Sadanandam et al., 2013; Schlicker et al., 2012). These classifications seem to be mostly consistent with each other's: while the reported number of subtypes varies from 3 to 6, some subgroups, identified by their respective associations with molecular and clinical characteristics, appear coherent. We will now briefly introduce the commonly identified subgroups, summarized in Fig. 3B.

4.2.2.1 The MSI-Enriched Subgroup

A subgroup highly enriched for microsatellite-instable tumors has been identified in all proposed classifications (Budinska et al., 2013; De Sousa E Melo et al., 2013; Marisa et al., 2013; Roepman et al., 2014; Sadanandam et al., 2013; Schlicker et al., 2012) and is also highlighted in the TCGA publication (Cancer Genome Atlas Network, 2012a). It is reported to have high CIMP phenotype (Budinska et al., 2013; Cancer Genome Atlas Network, 2012a; De Sousa E Melo et al., 2013; Marisa et al., 2013; Roepman et al., 2014; Sadanandam et al., 2013; Schlicker et al., 2012) consistent with the majority of MSI tumors being sporadic MSI and associated with CIMP. It also highly expresses proliferation-related genes (Budinska et al., 2013; Marisa et al., 2013; Roepman et al., 2014) and is enriched for BRAF-mutated tumors (Budinska et al., 2013; Cancer Genome Atlas Network, 2012a; De Sousa E Melo et al., 2013; Marisa et al., 2013; Roepman et al., 2014).

It is reported to have high immune infiltration (Budinska et al., 2013; Marisa et al., 2013; Sadanandam et al., 2013; Schlicker et al., 2012), including the TCGA study (Cancer Genome Atlas Network, 2012a), but the precise characterization of this infiltration is ambiguous: it was reported to overexpress genes related to antigen processing, HLA Class I and II (Marisa et al., 2013; Schlicker et al., 2012), TLR signaling, NK cells cytotoxicity and TCR signaling (Marisa et al., 2013), interferons and interferon-induced transcripts of the IFI family (Marisa et al., 2013; Sadanandam et al., 2013), chemokines (Sadanandam et al., 2013), and immune system or inflammatory response-related pathways (Cancer Genome Atlas Network, 2012a; Schlicker et al., 2012).

This subgroup is also associated with favorable clinical outcome in five publications (Budinska et al., 2013; De Sousa E Melo et al., 2013; Marisa et al., 2013; Roepman et al., 2014; Sadanandam et al., 2013) and forms a poor prognosis higher-order group when merged with another subgroup in one publication (Schlicker et al., 2012).

4.2.2.2 The Mesenchymal Subgroup

All six publications reporting CRC transcriptomic analyses identified a subgroup with overexpression of mesenchymal markers (Budinska et al., 2013; De Sousa E Melo et al., 2013; Marisa et al., 2013; Roepman et al., 2014; Sadanandam et al., 2013; Schlicker et al., 2012), colon bottom-crypt signature (Marisa et al., 2013; Sadanandam et al., 2013), stem-cell phenotype (Budinska et al., 2013; Marisa et al., 2013; Sadanandam et al., 2013), higher TGF-β pathway signaling (Budinska et al., 2013; De Sousa E Melo et al., 2013; Marisa et al., 2013; Roepman et al., 2014), serrated adenoma signature (De Sousa E Melo et al., 2013; Marisa et al., 2013), desmoplastic histology (Budinska et al., 2013), and low cell proliferation (Budinska et al., 2013; Marisa et al., 2013; Roepman et al., 2014). It is associated with poor prognosis (Budinska et al., 2013; De Sousa E Melo et al., 2013; Marisa et al., 2013; Roepman et al., 2014; Sadanandam et al., 2013; Schlicker et al., 2012) and resistance to therapies, both in vivo to 5-fluouracil (5FU) chemotherapy (De Sousa E Melo et al., 2013; Roepman et al., 2014) and to Cetuximab (a monoclonal antibody targeting the EGFR receptor) (De Sousa E Melo et al., 2013) and in vitro to a panel of targeted therapies (Schlicker et al., 2012). It could however respond to FOLFIRI therapy (a trichemotherapy regimen combining folinic acid, 5FU, and irinotecan) (Sadanandam et al., 2013). Immune infiltration may also be associated with this subgroup, as it has overexpression of genes related to TLR signaling and hematopoietic cell lineage (Marisa et al., 2013) or an immune signature (Budinska et al., 2013). However, other publications did not unveil differentially expressed genes related to immunity in this subgroup (Sadanandam et al., 2013; Schlicker et al., 2012).

4.2.2.3 The Other Subgroups

The other reported subgroups are less consensual based on the reported results. Most studies however report epithelial subgroups with a CIN phenotype (De Sousa E Melo et al., 2013; Marisa et al., 2013) and low immune infiltration (Budinska et al., 2013; Marisa et al., 2013). The Colorectal Cancer Subtyping Consortium is currently investigating the primary data of each reported study, as well as other datasets, to prove the consistency of the various approaches and summarize it by proposing a consensual classification (Dienstmann et al., 2014). In addition to the three above-mentioned subtypes, it appears to identify a fourth subtype, enriched for KRAS and IGFBP2 mutations (Dienstmann et al., 2014).

4.2.3 Immune Infiltration of CRC Transcriptomic Subgroups

A first classification according to the immune metagenes resulted in two clusters of tumors. A cluster with a high expression of cytotoxic cells and IFN-γ, high expression of B cells and T_{fh} cell-specific genes, and high MHC Class I and acute inflammation genes was associated with favorable prognosis (Bindea et al., 2013). It was validated by the corresponding cells in the IM and the CT. The immunome was then applied to molecularly defined subgroups of primary CRC, which revealed differential expression of TME markers across subgroups (Fig. 5A): an immune-down subgroup, with poor immune gene expression including markers of lymphocytic and myeloid subsets, endothelial cells, and fibroblasts. This subgroup was also characterized by low MHC Class I expression (Becht et al., 2015). The CIN/Wnt up and KRAS-mutation-enriched groups exhibit low immune metagene expression, confirming the weak immunogenicity of mutated KRAS. In contrast, the MSI group had a strong expression of lymphocytic, T cells, NK cells, cytotoxic lymphocytes, B cells, and myeloid cells of the monocytic lineage markers, with an intermediate expression of the markers specific for endothelial cells. This subgroup also exhibited a high expression of T-cell activation and chemotaxis markers, such as IFN-γ, IL-15, CXCL9, and CXCL10 as well as of PDCD1 [PD-1], CTLA-4, and LAG3. The MSI-enriched subgroup has the best prognosis (Budinska et al., 2013; De Sousa E Melo et al., 2013; Dienstmann et al., 2014; Marisa et al., 2013; Roepman et al., 2014; Sadanandam et al., 2013; Schlicker et al., 2012), is highly infiltrated with $CD8^+$ T cells expressing

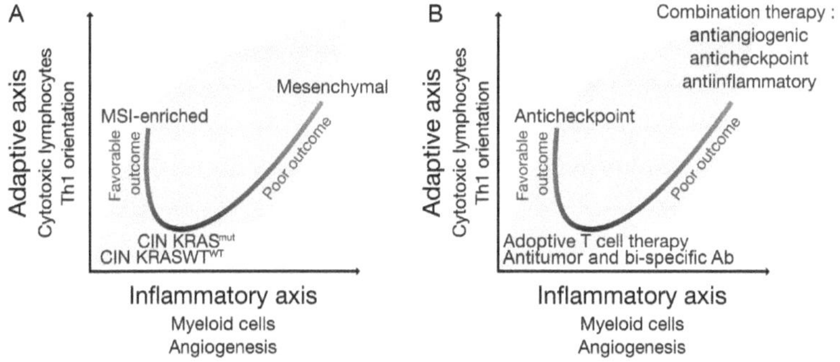

Figure 5 A bi-axis immune stratification identifies CRC molecular subgroups that could benefit from different immunotherapies. (A) Schematic distribution of the four CRC transcriptomic subgroups across the adaptive and inflammatory immune axes, along with their corresponding prognoses. (B) Potential immunotherapeutic approaches that benefit patients depending on the immune components of their tumors. (See the color plate.)

PD-1 as an activation marker (Llosa et al., 2015), and encompasses the few CRC patients who responded to anti-PD-1 therapies (Le et al., 2015). Another molecularly defined subgroup also exhibited a strong immune signature. The cancer stem cell-like group had an intermediate expression of the lymphocytes (T and B cells) and cytotoxic metagenes, together with a strong expression of myeloid, endothelial, and fibroblastic markers (Fig. 5A). This subgroup had the worst prognosis in several cohorts of CRC patients. It has recently been reported that the poor prognosis is in part supported by cancer-associated fibroblasts (CAFs) activated by TGF-ß (Calon et al., 2015; Isella et al., 2015), which is overexpressed in this subgroup where malignant cells may undergo epithelial–mesenchymal transition.

Therefore, in CRC, as in RCC, malignant cells govern their TME by producing factors that recruit and activate different cellular components. Thus, tumors with high load of mutations allow the recruitment of memory $CD8^+$ T cells, educated in TLSs and functionally active against the tumor. Malignant cells may escape the immune attack by strategies that are different across tumors and tumor subtypes. The production of VEGF and other angiogenic factors creates a disrupted vasculature that favors the entry of T and B cells outside of TLS resulting in exhaustion and suppression. Epithelial–mesenchymal transition results in the activation of CAFs that produce growth factors, but also proangiogenic and proinflammatory cytokines that create a milieu hampering potentially effective immune reactions. Integration of molecular classification of tumor cells along with microenvironmental cell metagenes and other genetic modifications will help classifying each tumor for prognostic and theranostic usage.

5. CONCLUSIONS: FROM THE IMMUNE CONTEXTURE OF TUMOR SUBGROUPS TO PRECISION MEDICINE

The concept of immune contexture which takes into account the organization, location, density, and functional orientation of immune cells in the TME has changed the paradigm of cancer/host interactions. It provides a conceptual tool to understand the physiopathological mechanisms that support the clinical impact of various cells of the immune reaction. Moreover, it allowed the definition of a routine, robust, and quantitative "immunoscore" that is a major prognostic factor in the vast majority of human cancers. The immunoscore is also useful to identify patients who may respond to immune therapies.

The immune contexture is the result of interactions between malignant cells and their microenvironments, including not only immune but also inflammatory, angiogenic, and fibroblastic components. Recent data analyzing the immune contexture in primary and metastatic sites indicate that the malignant cells are determinant in shaping their immune contextures. The development of genetic and epigenetic analyses of tumors revealed that different cancers, and different subgroups of a given cancer, correspond to different oncogenic processes. The combination of these analyses with a comprehensive view of the immune contexture refines the prognostic and theranostic power of cancer classifications. More importantly, it opens new avenues to treat subgroups of cancers. For instance, melanoma patients presenting with tumors having high mutational load or CRC patients with MSI are responders to anticheckpoint antibodies (Fig. 5B). However, they represent a minority of these cancers. In CRC, we have identified a molecular subgroup with high lymphocyte infiltration, expressing PD-1 in the context of high angiogenesis and we propose that to treat the corresponding patients with antiangiogenic therapy together with anticheckpoint antibodies (Fig. 5B). In ccRCC, patients with high angiogenesis do respond to antiangiogenic treatments. We characterize a ccRCC subgroup with low angiogenesis but high T-cell infiltration, the T cells expressing exhaustion markers and the tumor cells their ligands. We propose to treat these patients with anticheckpoint antibodies. Finally in CRC and RCC, there are subgroups of tumors with no T-cell infiltration. We propose to treat these patients with adaptive cellular or antibody therapies (Fig. 5B).

A new taxonomy of human cancers is thus emerging in which the combined analysis of the malignant phenotypes and their corresponding immune TME will allow to treat the patients with the most appropriate precision medicine and to adapt therapies by following the dynamic changes of the TME in response to immunotherapies.

ACKNOWLEDGMENTS

We thank the members of teams 10, 13, and 15 of the UMR-S1138 Cordeliers Research Center, of the UMR-1162, of the UMR-S1147, and of the CIT program for their fruitful discussions and performed most of the work cited in this review.

Financial support: This work was supported by the "Institut National de la Santé et de la Recherche Médicale," the University Paris-Descartes, the University Pierre et Marie Curie, the Institut National du Cancer (2009-1-PLBIO-07-INSERM 6-1, 2010-1-PLBIO-03-INSERM 6-1, 2011-1-PLBIO-06-INSERM 6-1), CARPEM (CAncer Research for PErsonalized Medicine), Labex Immuno-Oncology (LAXE62_9UMS872 FRIDMAN, 11LAXE62_9UMS872 FRIDMAN), the Fondation ARC pour la Recherche sur le Cancer (SL220110603483), and the Universidad de los Andes School of Medicine, Colciencias (N.A.G.). E.B. is supported by B3MI doctorate fellowship and N.A.G. by PPATH doctorate fellowship.

REFERENCES

Adachi, Y., Aoki, C., Yoshio-Hoshino, N., Takayama, K., Curiel, D. T., & Nishimoto, N. (2006). Interleukin-6 induces both cell growth and VEGF production in malignant mesotheliomas. *International Journal of Cancer*, *119*, 1303–1311.

Ahmadi, T., Flies, A., Efebera, Y., & Sherr, D. H. (2008). CD40 ligand-activated, antigen-specific B cells are comparable to mature dendritic cells in presenting protein antigens and major histocompatibility complex class I- and class II-binding peptides. *Immunology*, *124*, 129–140.

Al Zaid Siddiquee, K., & Turkson, J. (2008). STAT3 as a target for inducing apoptosis in solid and hematological tumors. *Cell Research*, *18*, 254–267.

Alcalay, J., & Kripke, M. L. (1991). Antigen-presenting activity of draining lymph node cells from mice painted with a contact allergen during ultraviolet carcinogenesis. *Journal of Immunology*, *146*, 1717–1721.

Alexander, R. B., Fitzgerald, E. B., Mixon, A., Carter, C. S., Jakobsen, M., Cohen, P. A., & Rosenberg, S. A. (1995). Helper T cells infiltrating human renal cell carcinomas have the phenotype of activated memory-like T lymphocytes. *Journal of Immunotherapy with Emphasis on Tumor Immunology*, *17*, 39–46.

Algars, A., Irjala, H., Vaittinen, S., Huhtinen, H., Sundström, J., Salmi, M., … Jalkanen, S. (2012). Type and location of tumor-infiltrating macrophages and lymphatic vessels predict survival of colorectal cancer patients. *International Journal of Cancer*, *131*, 864–873.

Algra, A. M., & Rothwell, P. M. (2012). Effects of regular aspirin on long-term cancer incidence and metastasis: A systematic comparison of evidence from observational studies versus randomised trials. *The Lancet Oncology*, *13*, 518–527.

Alizadeh, A. A., Eisen, M. B., Davis, R. E., Ma, C., Lossos, I. S., Rosenwald, A., … Staudt, L. M. (2000). Distinct types of diffuse large B-cell lymphoma identified by gene expression profiling. *Nature*, *403*, 503–511.

Aloisi, F., & Pujol-Borrell, R. (2006). Lymphoid neogenesis in chronic inflammatory diseases. *Nature Reviews Immunology*, *6*, 205–217.

Al-Shibli, K., Al-Saad, S., Donnem, T., Persson, M., Bremnes, R. M., & Busund, L. (2009). The prognostic value of intraepithelial and stromal innate immune system cells in non-small cell lung carcinoma. *Histopathology, 55*, 301–312.

Al-Shibli, K. I., Donnem, T., Al-Saad, S., Persson, M., Bremnes, R. M., & Busund, L. (2008). Prognostic effect of epithelial and stromal lymphocyte infiltration in non-small cell lung cancer. *Clinical Cancer Research, 14*, 5220–5227.

Ammirante, M., Luo, J., Grivennikov, S., Nedospasov, S., & Karin, M. (2010). B-cell-derived lymphotoxin promotes castration-resistant prostate cancer. *Nature, 464*, 302–305.

Amoueian, S., Attaranzadeh, A., & Montazer, M. (2015). Intratumoral CD68-, CD117-, CD56-, and CD1a-positive immune cells and the survival of Iranian patients with non-metastatic intestinal-type gastric carcinoma. *Pathology, Research and Practice, 211*, 326–331.

Ananiev, J., Gulubova, M. V., & Manolova, I. M. (2011). Prognostic significance of CD83 positive tumor-infiltrating dendritic cells and expression of TGF-beta 1 in human gastric cancer. *Hepato-Gastroenterology, 58*, 1834–1840.

Ando, T., Suguro, M., Kobayashi, T., Seto, M., & Honda, H. (2003). Multiple fuzzy neural network system for outcome prediction and classification of 220 lymphoma patients on the basis of molecular profiling. *Cancer Science, 94*, 906–913.

Angell, T. E., Lechner, M. G., Jang, J. K., LoPresti, J. S., & Epstein, A. L. (2014). MHC class I loss is a frequent mechanism of immune escape in papillary thyroid cancer that is reversed by interferon and selumetinib treatment in vitro. *Clinical Cancer Research, 20*, 6034–6044.

Angevin, E., Kremer, F., Gaudin, C., Hercend, T., & Triebel, F. (1997). Analysis of T-cell immune response in renal cell carcinoma: Polarization to type 1-like differentiation pattern, clonal T-cell expansion and tumor-specific cytotoxicity. *International Journal of Cancer, 72*, 431–440.

Anitei, M., Zeitoun, G., Mlecnik, B., Marliot, F., Haicheur, N., Todosi, A., … Pagès, F. (2014). Prognostic and predictive values of the immunoscore in patients with rectal cancer. *Clinical Cancer Research, 20*, 1891–1899.

Armaghany, T., Wilson, J. D., Chu, Q., & Mills, G. (2012). Genetic alterations in colorectal cancer. *Gastrointestinal Cancer Research, 5*, 19–27.

Arzumanyan, A., Reis, H. M. G. P. V., & Feitelson, M. A. (2013). Pathogenic mechanisms in HBV- and HCV-associated hepatocellular carcinoma. *Nature Reviews Cancer, 13*, 123–135.

Assié, G., Letouzé, E., Fassnacht, M., Jouinot, A., Luscap, W., Barreau, O., … Bertherat, J. (2014). Integrated genomic characterization of adrenocortical carcinoma. *Nature Genetics, 46*, 607–612.

Attig, S., Hennenlotter, J., Pawelec, G., Klein, G., Koch, S. D., Pircher, H., … Gouttefangeas, C. (2009). Simultaneous infiltration of polyfunctional effector and suppressor T cells into renal cell carcinomas. *Cancer Research, 69*, 8412–8419.

Ayyoub, M., Taub, R. N., Keohan, M., Hesdorffer, M., Metthez, G., Memeo, L., … Valmori, D. (2004). The frequent expression of cancer/testis antigens provides opportunities for immunotherapeutic targeting of sarcoma. *Cancer Immunity, 4*, 7.

Bachmayr-Heyda, A., Aust, S., Heinze, G., Polterauer, S., Grimm, C., Braicu, E. I., … Castillo-Tong, D. C. (2013). Prognostic impact of tumor infiltrating CD8+ T cells in association with cell proliferation in ovarian cancer patients—A study of the OVCAD consortium. *BMC Cancer, 13*, 422.

Badoual, C., Hans, S., Merillon, N., Van Ryswick, C., Ravel, P., Benhamouda, N., … Tartour, E. (2013). PD-1-expressing tumor-infiltrating T cells are a favorable prognostic biomarker in HPV-associated head and neck cancer. *Cancer Research, 73*, 128–138.

Bais, A. G., Beckmann, I., Lindemans, J., Ewing, P. C., Meijer, C. J. L. M., Snijders, P. J. F., & Helmerhorst, T. J. M. (2005). A shift to a peripheral Th2-type cytokine pattern during the carcinogenesis of cervical cancer becomes manifest in CIN III lesions. *Journal of Clinical Pathology, 58*, 1096–1100.

Balermpas, P., Rödel, F., Weiss, C., Rödel, C., & Fokas, E. (2014). Tumor-infiltrating lymphocytes favor the response to chemoradiotherapy of head and neck cancer. *Oncoimmunology, 3*, e27403.

Balkwill, F. (2009). Tumour necrosis factor and cancer. *Nature Reviews Cancer, 9*, 361–371.

Balkwill, F., Montfort, A., & Capasso, M. (2013). B regulatory cells in cancer. *Trends in Immunology, 34*, 169–173.

Barbera-Guillem, E., May, K. F. J., Nyhus, J. K., & Nelson, M. B. (1999). Promotion of tumor invasion by cooperation of granulocytes and macrophages activated by anti-tumor antibodies. *Neoplasia, 1*, 453–460.

Barbera-Guillem, E., Nelson, M. B., Barr, B., Nyhus, J. K., May, K. F. J., Feng, L., & Sampsel, J. W. (2000). B lymphocyte pathology in human colorectal cancer. Experimental and clinical therapeutic effects of partial B cell depletion. *Cancer Immunology, Immunotherapy, 48*, 541–549.

Barlin, J. N., Zhou, Q. C., Leitao, M. M., Bisogna, M., Olvera, N., Shih, K. K., … Levine, D. A. (2015). Molecular subtypes of uterine leiomyosarcoma and correlation with clinical outcome. *Neoplasia, 17*, 183–189.

Barsoum, I. B., Smallwood, C. A., Siemens, D. R., & Graham, C. H. (2014). A mechanism of hypoxia-mediated escape from adaptive immunity in cancer cells. *Cancer Research, 74*, 665–674.

Barton, S., Hawkes, E. A., Wotherspoon, A., & Cunningham, D. (2012). Are we ready to stratify treatment for diffuse large B-cell lymphoma using molecular hallmarks? *The Oncologist, 17*, 1562–1573.

Becht, E., Giraldo, N. A., Beuselinck, B., Job, S., Marisa, L., Vano, Y., … Sautès-Fridman, C. (2015). Prognostic and theranostic impact of molecular subtypes and immune classifications in Renal Cell Cancer (RCC) and Colorectal Cancer (CRC). *Oncoimmunology, 4*, e1049804.

Becht, E., Goc, J., Germain, C., Giraldo, N. A., Dieu-Nosjean, M., Sautès-Fridman, C., & Fridman, W. (2014). Shaping of an effective immune microenvironment to and by cancer cells. *Cancer Immunology, Immunotherapy, 63*, 991–997.

Becker, J. C., Andersen, M. H., Schrama, D., & Thor Straten, P. (2013). Immune-suppressive properties of the tumor microenvironment. *Cancer Immunology, Immunotherapy, 62*, 1137–1148.

Behnes, C. L., Bremmer, F., Hemmerlein, B., Strauss, A., Ströbel, P., & Radzun, H. (2014). Tumor-associated macrophages are involved in tumor progression in papillary renal cell carcinoma. *Virchows Archiv, 464*, 191–196.

Bernal, M., Ruiz-Cabello, F., Concha, A., Paschen, A., & Garrido, F. (2012). Implication of the β2-microglobulin gene in the generation of tumor escape phenotypes. *Cancer Immunology, Immunotherapy, 61*, 1359–1371.

Beuselinck, B., Job, S., Becht, E., Karadimou, A., Verkarre, V., Couchy, G., … Zucman-Rossi, J. (2015). Molecular subtypes of clear cell renal cell carcinoma are associated with sunitinib response in the metastatic setting. *Clinical Cancer Research, 21*, 1329–1339.

Bindea, G., Mlecnik, B., Tosolini, M., Kirilovsky, A., Waldner, M., Obenauf, A. C., … Galon, J. (2013). Spatiotemporal dynamics of intratumoral immune cells reveal the immune landscape in human cancer. *Immunity, 39*, 782–795.

Bingle, L., Lewis, C. E., Corke, K. P., Reed, M. W. R., & Brown, N. J. (2006). Macrophages promote angiogenesis in human breast tumour spheroids in vivo. *British Journal of Cancer, 94*, 101–107.

Biswas, S. K., & Mantovani, A. (2010). Macrophage plasticity and interaction with lymphocyte subsets: Cancer as a paradigm. *Nature Immunology, 11*, 889–896.

Biswas, K., Richmond, A., Rayman, P., Biswas, S., Thornton, M., Sa, G., … Finke, J. H. (2006). GM2 expression in renal cell carcinoma: Potential role in tumor-induced T-cell dysfunction. *Cancer Research, 66*, 6816–6825.

Biton, A., Bernard-Pierrot, I., Lou, Y., Krucker, C., Chapeaublanc, E., Rubio-Pérez, C., ... Radvanyi, F. (2014). Independent component analysis uncovers the landscape of the bladder tumor transcriptome and reveals insights into luminal and basal subtypes. *Cell Reports*, *9*, 1235–1245.

Blank, C., Brown, I., Peterson, A. C., Spiotto, M., Iwai, Y., Honjo, T., & Gajewski, T. F. (2004). PD-L1/B7H-1 inhibits the effector phase of tumor rejection by T cell receptor (TCR) transgenic CD8+ T cells. *Cancer Research*, *64*, 1140–1145.

Bocelli-Tyndall, C., Barbero, A., Candrian, C., Ceredig, R., Tyndall, A., & Martin, I. (2006). Human articular chondrocytes suppress in vitro proliferation of anti-CD3-activated peripheral blood mononuclear cells. *Journal of Cellular Physiology*, *209*, 732–734.

Boissière-Michot, F., Lazennec, G., Frugier, H., Jarlier, M., Roca, L., Duffour, J., ... Bibeau, F. (2014). Characterization of an adaptive immune response in microsatellite-instable colorectal cancer. *Oncoimmunology*, *3*, e29256.

Bolat, F., Kayaselcuk, F., Nursal, T. Z., Yagmurdur, M. C., Bal, N., & Demirhan, B. (2006). Microvessel density, VEGF expression, and tumor-associated macrophages in breast tumors: Correlations with prognostic parameters. *Journal of Experimental & Clinical Cancer Research*, *25*, 365–372.

Bollrath, J., Phesse, T. J., von Burstin, V. A., Putoczki, T., Bennecke, M., Bateman, T., ... Greten, F. R. (2009). gp130-mediated Stat3 activation in enterocytes regulates cell survival and cell-cycle progression during colitis-associated tumorigenesis. *Cancer Cell*, *15*, 91–102.

Borrello, M. G., Alberti, L., Fischer, A., Degl'innocenti, D., Ferrario, C., Gariboldi, M., ... Pierotti, M. A. (2005). Induction of a proinflammatory program in normal human thyrocytes by the RET/PTC1 oncogene. *Proceedings of the National Academy of Sciences of the United States of America*, *102*, 14825–14830.

Boyault, S., Rickman, D. S., de Reyniès, A., Balabaud, C., Rebouissou, S., Jeannot, E., ... Zucman-Rossi, J. (2007). Transcriptome classification of HCC is related to gene alterations and to new therapeutic targets. *Hepatology*, *45*, 42–52.

Brahmer, J. R., Tykodi, S. S., Chow, L. Q. M., Hwu, W., Topalian, S. L., Hwu, P., ... Wigginton, J. M. (2012). Safety and activity of anti-PD-L1 antibody in patients with advanced cancer. *The New England Journal of Medicine*, *366*, 2455–2465.

Brändle, D., Brasseur, F., Weynants, P., Boon, T., & Van den Eynde, B. (1996). A mutated HLA-A2 molecule recognized by autologous cytotoxic T lymphocytes on a human renal cell carcinoma. *The Journal of Experimental Medicine*, *183*, 2501–2508.

Brannon, A. R., Haake, S. M., Hacker, K. E., Pruthi, R. S., Wallen, E. M., Nielsen, M. E., & Rathmell, W. K. (2012). Meta-analysis of clear cell renal cell carcinoma gene expression defines a variant subgroup and identifies gender influences on tumor biology. *European Urology*, *61*, 258–268.

Brannon, A. R., Reddy, A., Seiler, M., Arreola, A., Moore, D. T., Pruthi, R. S., ... Rathmell, W. K. (2010). Molecular stratification of clear cell renal cell carcinoma by consensus clustering reveals distinct subtypes and survival patterns. *Genes & Cancer*, *1*, 152–163.

Brat, D. J., Verhaak, R. G. W., Aldape, K. D., Yung, W. K. A., Salama, S. R., Cooper, L. A. D., ... Zhang, J. (2015). Comprehensive, integrative genomic analysis of diffuse lower-grade gliomas. *The New England Journal of Medicine*, *372*, 2481–2498.

Breart, B., Lemaître, F., Celli, S., & Bousso, P. (2008). Two-photon imaging of intratumoral CD8+ T cell cytotoxic activity during adoptive T cell therapy in mice. *The Journal of Clinical Investigation*, *118*, 1390–1397.

Brennan, P. J., Brigl, M., & Brenner, M. B. (2013). Invariant natural killer T cells: An innate activation scheme linked to diverse effector functions. *Nature Reviews Immunology*, *13*, 101–117.

Bromberg, J. F., Wrzeszczynska, M. H., Devgan, G., Zhao, Y., Pestell, R. G., Albanese, C., & Darnell, J. E. J. (1999). Stat3 as an oncogene. *Cell*, *98*, 295–303.

Bromwich, E. J., McArdle, P. A., Canna, K., McMillan, D. C., McNicol, A., Brown, M., & Aitchison, M. (2003). The relationship between T-lymphocyte infiltration, stage, tumour grade and survival in patients undergoing curative surgery for renal cell cancer. *British Journal of Cancer*, *89*, 1906–1908.

Bronte-Tinkew, D. M., Terebiznik, M., Franco, A., Ang, M., Ahn, D., Mimuro, H., ... Jones, N. L. (2009). *Helicobacter pylori* cytotoxin-associated gene A activates the signal transducer and activator of transcription 3 pathway in vitro and in vivo. *Cancer Research*, *69*, 632–639.

Brunner, S. M., Kesselring, R., Rubner, C., Martin, M., Jeiter, T., Boerner, T., ... Fichtner-Feigl, S. (2014). Prognosis according to histochemical analysis of liver metastases removed at liver resection. *The British Journal of Surgery*, *101*, 1681–1691.

Budinska, E., Popovici, V., Tejpar, S., D'Ario, G., Lapique, N., Sikora, K. O., ... Delorenzi, M. (2013). Gene expression patterns unveil a new level of molecular heterogeneity in colorectal cancer. *The Journal of Pathology*, *231*, 63–76.

Bukowski, R. M., Rayman, P., Uzzo, R., Bloom, T., Sandstrom, K., Peereboom, D., ... Finke, J. H. (1998). Signal transduction abnormalities in T lymphocytes from patients with advanced renal carcinoma: Clinical relevance and effects of cytokine therapy. *Clinical Cancer Research*, *4*, 2337–2347.

Busse, A., Asemissen, A. M., Nonnenmacher, A., Braun, F., Ochsenreither, S., Stather, D., ... Keilholz, U. (2011). Immunomodulatory effects of sorafenib on peripheral immune effector cells in metastatic renal cell carcinoma. *European Journal of Cancer*, *47*, 690–696.

Cabillic, F., Bouet-Toussaint, F., Toutirais, O., Rioux-Leclercq, N., Fergelot, P., de la Pintière, C. T., ... Catros-Quemener, V. (2006). Interleukin-6 and vascular endothelial growth factor release by renal cell carcinoma cells impedes lymphocyte-dendritic cell cross-talk. *Clinical and Experimental Immunology*, *146*, 518–523.

Cai, X., Gao, Q., Qiu, S., Ye, S., Wu, Z., Fan, J., & Tang, Z. (2006). Dendritic cell infiltration and prognosis of human hepatocellular carcinoma. *Journal of Cancer Research and Clinical Oncology*, *132*, 293–301.

Cairns, P. (2010). Renal cell carcinoma. *Cancer Biomarkers*, *9*, 461–473.

Calkins, G. N. (1914). Zur Frage der Entstehung maligner Tumoren. *Science*, *40*, 857–859.

Calon, A., Lonardo, E., Berenguer-Llergo, A., Espinet, E., Hernando-Momblona, X., Iglesias, M., ... Batlle, E. (2015). Stromal gene expression defines poor-prognosis subtypes in colorectal cancer. *Nature Genetics*, *47*, 320–329.

Campbell, M. J., Tonlaar, N. Y., Garwood, E. R., Huo, D., Moore, D. H., Khramtsov, A. I., ... Esserman, L. J. (2011). Proliferating macrophages associated with high grade, hormone receptor negative breast cancer and poor clinical outcome. *Breast Cancer Research and Treatment*, *128*, 703–711.

Cancer Genome Atlas Network. (2012a). Comprehensive molecular characterization of human colon and rectal cancer. *Nature*, *487*, 330–337.

Cancer Genome Atlas Network. (2012b). Comprehensive genomic characterization of squamous cell lung cancers. *Nature*, *489*, 519–525.

Cancer Genome Atlas Network. (2013a). Genomic and epigenomic landscapes of adult de novo acute myeloid leukemia. *The New England Journal of Medicine*, *368*, 2059–2074.

Cancer Genome Atlas Network. (2013b). Comprehensive molecular characterization of clear cell renal cell carcinoma. *Nature*, *499*, 43–49.

Cancer Genome Atlas Network. (2014a). Comprehensive molecular characterization of urothelial bladder carcinoma. *Nature*, *507*, 315–322.

Cancer Genome Atlas Network. (2014b). Comprehensive molecular characterization of gastric adenocarcinoma. *Nature*, *513*, 202–209.

Cancer Genome Atlas Network. (2014c). Comprehensive molecular profiling of lung adenocarcinoma. *Nature*, *511*, 543–550.

Cancer Genome Atlas Network. (2014d). Integrated genomic characterization of papillary thyroid carcinoma. *Cell*, *159*, 676–690.

Cancer Genome Atlas Network. (2015a). Comprehensive genomic characterization of head and neck squamous cell carcinomas. *Nature*, *517*, 576–582.
Cancer Genome Atlas Network. (2015b). Genomic classification of cutaneous melanoma. *Cell*, *161*, 1681–1696.
Cariani, E., Pilli, M., Zerbini, A., Rota, C., Olivani, A., Pelosi, G., … Missale, G. (2012). Immunological and molecular correlates of disease recurrence after liver resection for hepatocellular carcinoma. *PLoS One*, 7, e32493.
Carmi, Y., Spitzer, M. H., Linde, I. L., Burt, B. M., Prestwood, T. R., Perlman, N., … Engleman, E. G. (2015). Allogeneic IgG combined with dendritic cell stimuli induce antitumour T-cell immunity. *Nature*, *521*, 99–104.
Carrega, P., Morandi, B., Costa, R., Frumento, G., Forte, G., Altavilla, G., … Ferlazzo, G. (2008). Natural killer cells infiltrating human nonsmall-cell lung cancer are enriched in CD56 bright CD16(−) cells and display an impaired capability to kill tumor cells. *Cancer*, *112*, 863–875.
Carter, L., Fouser, L. A., Jussif, J., Fitz, L., Deng, B., Wood, C. R., … Carreno, B. M. (2002). PD-1:PD-L inhibitory pathway affects both CD4(+) and CD8(+) T cells and is overcome by IL-2. *European Journal of Immunology*, *32*, 634–643.
Casares, N., Pequignot, M. O., Tesniere, A., Ghiringhelli, F., Roux, S., Chaput, N., … Kroemer, G. (2005). Caspase-dependent immunogenicity of doxorubicin-induced tumor cell death. *The Journal of Experimental Medicine*, *202*, 1691–1701.
Castle, P. E., Hillier, S. L., Rabe, L. K., Hildesheim, A., Herrero, R., Bratti, M. C., … Schiffman, M. (2001). An association of cervical inflammation with high-grade cervical neoplasia in women infected with oncogenic human papillomavirus (HPV). *Cancer Epidemiology, Biomarkers & Prevention*, *10*, 1021–1027.
Cavalcanti, E., Gigante, M., Mancini, V., Battaglia, M., Ditonno, P., Capobianco, C., … Ranieri, E. (2010). JAK3/STAT5/6 pathway alterations are associated with immune deviation in CD8 T cells in renal cell carcinoma patients. *Journal of Biomedicine & Biotechnology*, 935764.
Čermáková, P., Melichar, B., Tomšová, M., Zoul, Z., Kalábová, H., Spaček, J., & Doležel, M. (2014). Prognostic significance of CD3+ tumor-infiltrating lymphocytes in patients with endometrial carcinoma. *Anticancer Research*, *34*, 5555–5561.
Chahlavi, A., Rayman, P., Richmond, A. L., Biswas, K., Zhang, R., Vogelbaum, M., … Finke, J. H. (2005). Glioblastomas induce T-lymphocyte death by two distinct pathways involving gangliosides and CD70. *Cancer Research*, *65*, 5428–5438.
Chai, C., Chen, W., Hung, W., Kang, W., Huang, Y., Su, Y., & Yang, C. (2008). Hypoxia-inducible factor-1alpha expression correlates with focal macrophage infiltration, angiogenesis and unfavourable prognosis in urothelial carcinoma. *Journal of Clinical Pathology*, *61*, 658–664.
Chan, A. T., Ogino, S., & Fuchs, C. S. (2007). Aspirin and the risk of colorectal cancer in relation to the expression of COX-2. *The New England Journal of Medicine*, *356*, 2131–2142.
Chang, W., Du, Y., Zhao, X., Ma, L., & Cao, G. (2014). Inflammation-related factors predicting prognosis of gastric cancer. *World Journal of Gastroenterology*, *20*, 4586–4596.
Chappell, D. B., & Restifo, N. P. (1998). T cell-tumor cell: A fatal interaction? *Cancer Immunology, Immunotherapy*, *47*, 65–71.
Chaput, N., Svrcek, M., Aupérin, A., Locher, C., Drusch, F., Malka, D., … Boige, V. (2013). Tumour-infiltrating CD68+ and CD57+ cells predict patient outcome in stage II-III colorectal cancer. *British Journal of Cancer*, *109*, 1013–1022.
Chaux, P., Moutet, M., Faivre, J., Martin, F., & Martin, M. (1996). Inflammatory cells infiltrating human colorectal carcinomas express HLA class II but not B7-1 and B7-2 costimulatory molecules of the T-cell activation. *Laboratory Investigation*, *74*, 975–983.
Che, Y., Feng, L., Rong, W., Shen, D., Wang, Q., Yang, L., … Huang, C. (2014). Correlation analysis of peripheral blood T cell subgroups, immunoglobulin and prognosis of

early hepatocellular carcinoma after hepatectomy. *International Journal of Clinical and Experimental Medicine*, 7, 4282–4290.
Chen, J., & Chen, Z. (2014). The effect of immune microenvironment on the progression and prognosis of colorectal cancer. *Medical Oncology*, *31*, 82.
Chen, Z., Chen, X., Zhou, E., Chen, G., Qian, K., Wu, X., … Tang, Z. (2014). Intratumoral $CD8^{+}$ cytotoxic lymphocyte is a favorable prognostic marker in node-negative breast cancer. *PLoS One*, *9*, e95475.
Chen, L., Zhang, Q., Chang, W., Du, Y., Zhang, H., & Cao, G. (2012). Viral and host inflammation-related factors that can predict the prognosis of hepatocellular carcinoma. *European Journal of Cancer*, *48*, 1977–1987.
Chen, L., Zheng, X., Shen, Y., Zhu, Y., Li, Q., Chen, J., … Jiang, J. (2013). Higher numbers of T-bet(+) intratumoral lymphoid cells correlate with better survival in gastric cancer. *Cancer Immunology, Immunotherapy*, *62*, 553–561.
Chew, V., Chen, J., Lee, D., Loh, E., Lee, J., Lim, K. H., … Abastado, J. (2012). Chemokine-driven lymphocyte infiltration: An early intratumoural event determining long-term survival in resectable hepatocellular carcinoma. *Gut*, *61*, 427–438.
Chiang, D. Y., Villanueva, A., Hoshida, Y., Peix, J., Newell, P., Minguez, B., … Llovet, J. M. (2008). Focal gains of VEGFA and molecular classification of hepatocellular carcinoma. *Cancer Research*, *68*, 6779–6788.
Chomarat, P., Banchereau, J., Davoust, J., & Palucka, A. K. (2000). IL-6 switches the differentiation of monocytes from dendritic cells to macrophages. *Nature Immunology*, *1*, 510–514.
Chuang, M., Sun, K., Tang, S., Deng, M., Wu, Y., Sung, J., … Sun, G. (2008). Tumor-derived tumor necrosis factor-alpha promotes progression and epithelial-mesenchymal transition in renal cell carcinoma cells. *Cancer Science*, *99*, 905–913.
Cipponi, A., Mercier, M., Seremet, T., Baurain, J., Théate, I., van den Oord, J., … van Baren, N. (2012). Neogenesis of lymphoid structures and antibody responses occur in human melanoma metastases. *Cancer Research*, *72*, 3997–4007.
Ciriello, G., Cerami, E., Sander, C., & Schultz, N. (2012). Mutual exclusivity analysis identifies oncogenic network modules. *Genome Research*, *22*, 398–406.
Coca, S., Perez-Piqueras, J., Martinez, D., Colmenarejo, A., Saez, M. A., Vallejo, C., … Moreno, M. (1997). The prognostic significance of intratumoral natural killer cells in patients with colorectal carcinoma. *Cancer*, *79*, 2320–2328.
Colotta, F., Allavena, P., Sica, A., Garlanda, C., & Mantovani, A. (2009). Cancer-related inflammation, the seventh hallmark of cancer: Links to genetic instability. *Carcinogenesis*, *30*, 1073–1081.
Conant, J. L., Peng, Z., Evans, M. F., Naud, S., & Cooper, K. (2011). Sarcomatoid renal cell carcinoma is an example of epithelial–mesenchymal transition. *Journal of Clinical Pathology*, *64*, 1088–1092.
Cornelissen, R., Lievense, L. A., Maat, A. P., Hendriks, R. W., Hoogsteden, H. C., Bogers, A. J., … Aerts, J. G. (2014). Ratio of intratumoral macrophage phenotypes is a prognostic factor in epithelioid malignant pleural mesothelioma. *PLoS One*, *9*, e106742.
Correale, P., Rotundo, M. S., Botta, C., Del Vecchio, M. T., Ginanneschi, C., Licchetta, A., … Tagliaferri, P. (2012). Tumor infiltration by T lymphocytes expressing chemokine receptor 7 (CCR7) is predictive of favorable outcome in patients with advanced colorectal carcinoma. *Clinical Cancer Research*, *18*, 850–857.
Correale, P., Walmsley, K., Nieroda, C., Zaremba, S., Zhu, M., Schlom, J., & Tsang, K. Y. (1997). In vitro generation of human cytotoxic T lymphocytes specific for peptides derived from prostate-specific antigen. *Journal of the National Cancer Institute*, *89*, 293–300.
Coulie, P. G., Brichard, V., Van Pel, A., Wölfel, T., Schneider, J., Traversari, C., … Boon, T. (1994). A new gene coding for a differentiation antigen recognized by autologous cytolytic T lymphocytes on HLA-A2 melanomas. *The Journal of Experimental Medicine*, *180*, 35–42.

Coulie, P. G., Van den Eynde, B. J., van der Bruggen, P., & Boon, T. (2014). Tumour antigens recognized by T lymphocytes: At the core of cancer immunotherapy. *Nature Reviews Cancer, 14*, 135–146.

Coussens, L. M., & Werb, Z. (2002). Inflammation and cancer. *Nature, 420*, 860–867.

Cózar, J. M., Canton, J., Tallada, M., Concha, A., Cabrera, T., Garrido, F., & Ruiz-Cabello Osuna, F. (2005). Analysis of NK cells and chemokine receptors in tumor infiltrating CD4 T lymphocytes in human renal carcinomas. *Cancer Immunology, Immunotherapy, 54*, 858–866.

Cunha, L. L., Morari, E. C., Guihen, A. C. T., Razolli, D., Gerhard, R., Nonogaki, S., … Ward, L. S. (2012). Infiltration of a mixture of immune cells may be related to good prognosis in patients with differentiated thyroid carcinoma. *Clinical Endocrinology, 77*, 918–925.

Curiel, T. J., Coukos, G., Zou, L., Alvarez, X., Cheng, P., Mottram, P., … Zou, W. (2004). Specific recruitment of regulatory T cells in ovarian carcinoma fosters immune privilege and predicts reduced survival. *Nature Medicine, 10*, 942–949.

Curtis, R. E., Rowlings, P. A., Deeg, H. J., Shriner, D. A., Socíe, G., Travis, L. B., … Boice, J. D. J. (1997). Solid cancers after bone marrow transplantation. *The New England Journal of Medicine, 336*, 897–904.

Czystowska, M., Gooding, W., Szczepanski, M. J., Lopez-Abaitero, A., Ferris, R. L., Johnson, J. T., & Whiteside, T. L. (2013). The immune signature of CD8(+)CCR7 (+) T cells in the peripheral circulation associates with disease recurrence in patients with HNSCC. *Clinical Cancer Research, 19*, 889–899.

Dai, F., Liu, L., Che, G., Yu, N., Pu, Q., Zhang, S., … You, Z. (2010). The number and microlocalization of tumor-associated immune cells are associated with patient's survival time in non-small cell lung cancer. *BMC Cancer, 10*, 220.

Dannenmann, S. R., Thielicke, J., Stöckli, M., Matter, C., von Boehmer, L., Cecconi, V., … van den Broek, M. (2013). Tumor-associated macrophages subvert T-cell function and correlate with reduced survival in clear cell renal cell carcinoma. *Oncoimmunology, 2*, e23562.

Daurkin, I., Eruslanov, E., Stoffs, T., Perrin, G. Q., Algood, C., Gilbert, S. M., … Kusmartsev, S. (2011). Tumor-associated macrophages mediate immunosuppression in the renal cancer microenvironment by activating the 15-lipoxygenase-2 pathway. *Cancer Research, 71*, 6400–6409.

de Chaisemartin, L., Goc, J., Damotte, D., Validire, P., Magdeleinat, P., Alifano, M., … Dieu-Nosjean, M. (2011). Characterization of chemokines and adhesion molecules associated with T cell presence in tertiary lymphoid structures in human lung cancer. *Cancer Research, 71*, 6391–6399.

de Jonge, H. J. M., Huls, G., & de Bont, E. S. J. M. (2011). Gene expression profiling in acute myeloid leukaemia. *The Netherlands Journal of Medicine, 69*, 167–176.

de Kruijf, E. M., Engels, C. C., van de Water, W., Bastiaannet, E., Smit, V. T. H. B. M., van de Velde, C. J. H., … Kuppen, P. J. K. (2013). Tumor immune subtypes distinguish tumor subclasses with clinical implications in breast cancer patients. *Breast Cancer Research and Treatment, 142*, 355–364.

de Reyniès, A., Jaurand, M., Renier, A., Couchy, G., Hysi, I., Elarouci, N., … Jean, D. (2014). Molecular classification of malignant pleural mesothelioma: Identification of a poor prognosis subgroup linked to the epithelial-to-mesenchymal transition. *Clinical Cancer Research, 20*, 1323–1334.

De Sousa E Melo, F., Wang, X., Jansen, M., Fessler, E., Trinh, A., de Rooij, L. P. M. H., … Vermeulen, L. (2013). Poor-prognosis colon cancer is defined by a molecularly distinct subtype and develops from serrated precursor lesions. *Nature Medicine, 19*, 614–618.

de Visser, K. E., Korets, L. V., & Coussens, L. M. (2005). De novo carcinogenesis promoted by chronic inflammation is B lymphocyte dependent. *Cancer Cell*, 7, 411–423.

de Wit, J., Souwer, Y., Jorritsma, T., Klaasse Bos, H., ten Brinke, A., Neefjes, J., & van Ham, S. M. (2010). Antigen-specific B cells reactivate an effective cytotoxic T cell response against phagocytosed *Salmonella* through cross-presentation. *PLoS One*, *5*, e13016.

DeNardo, D. G., Andreu, P., & Coussens, L. M. (2010). Interactions between lymphocytes and myeloid cells regulate pro-versus anti-tumor immunity. *Cancer Metastasis Reviews*, *29*, 309–316.

DeNardo, D. G., Brennan, D. J., Rexhepaj, E., Ruffell, B., Shiao, S. L., Madden, S. F., … Coussens, L. M. (2011). Leukocyte complexity predicts breast cancer survival and functionally regulates response to chemotherapy. *Cancer Discovery*, *1*, 54–67.

DeNardo, D. G., & Coussens, L. M. (2007). Inflammation and breast cancer. Balancing immune response: Crosstalk between adaptive and innate immune cells during breast cancer progression. *Breast Cancer Research*, *9*(4), 212.

Deola, S., Panelli, M. C., Maric, D., Selleri, S., Dmitrieva, N. I., Voss, C. Y., … Marincola, F. M. (2008). Helper B cells promote cytotoxic T cell survival and proliferation independently of antigen presentation through CD27/CD70 interactions. *Journal of Immunology*, *180*, 1362–1372.

Desar, I. M. E., Jacobs, J. H. F. M., Hulsbergen-vandeKaa, C. A., Oyen, W. J. G., Mulders, P. F. A., van der Graaf, W. T. A., … de Vries, I. J. J. M. (2011). Sorafenib reduces the percentage of tumour infiltrating regulatory T cells in renal cell carcinoma patients. *International Journal of Cancer*, *129*, 507–512.

Di Caro, G., Bergomas, F., Grizzi, F., Doni, A., Bianchi, P., Malesci, A., … Marchesi, F. (2014). Occurrence of tertiary lymphoid tissue is associated with T-cell infiltration and predicts better prognosis in early-stage colorectal cancers. *Clinical Cancer Research*, *20*, 2147–2158.

Dienstmann, R., Guinney, J., Delorenzi, M., de Reynies, A., Roepman, P., Sadanandam, A., … Tejpar, S. (2014). Colorectal Cancer Subtyping Consortium (CRCSC) identification of a consensus of molecular subtypes. *Journal of Clinical Oncology*, *32*, 5s.

Dieu-Nosjean, M., Antoine, M., Danel, C., Heudes, D., Wislez, M., Poulot, V., … Cadranel, J. (2008). Long-term survival for patients with non-small-cell lung cancer with intratumoral lymphoid structures. *Journal of Clinical Oncology*, *26*, 4410–4417.

Dieu-Nosjean, M., Goc, J., Giraldo, N. A., Sautès-Fridman, C., & Fridman, W. H. (2014). Tertiary lymphoid structures in cancer and beyond. *Trends in Immunology*, *35*, 571–580.

DiLillo, D. J., Yanaba, K., & Tedder, T. F. (2010). B cells are required for optimal CD4+ and CD8+ T cell tumor immunity: Therapeutic B cell depletion enhances B16 melanoma growth in mice. *Journal of Immunology*, *184*, 4006–4016.

Ding, T., Xu, J., Wang, F., Shi, M., Zhang, Y., Li, S., & Zheng, L. (2009). High tumor-infiltrating macrophage density predicts poor prognosis in patients with primary hepatocellular carcinoma after resection. *Human Pathology*, *40*, 381–389.

Dolcetti, R., Viel, A., Doglioni, C., Russo, A., Guidoboni, M., Capozzi, E., … Boiocchi, M. (1999). High prevalence of activated intraepithelial cytotoxic T lymphocytes and increased neoplastic cell apoptosis in colorectal carcinomas with microsatellite instability. *The American Journal of Pathology*, *154*, 1805–1813.

Donatelli, S. S., Zhou, J., Gilvary, D. L., Eksioglu, E. A., Chen, X., Cress, W. D., … Djeu, J. Y. (2014). TGF-β-inducible microRNA-183 silences tumor-associated natural killer cells. *Proceedings of the National Academy of Sciences of the United States of America*, *111*, 4203–4208.

Donskov, F. (2007). Interleukin-2 based immunotherapy in patients with metastatic renal cell carcinoma. *Danish Medical Bulletin*, *54*, 249–265.

Donskov, F., & von der Maase, H. (2006). Impact of immune parameters on long-term survival in metastatic renal cell carcinoma. *Journal of Clinical Oncology*, *24*, 1997–2005.

Dosquet, C., Coudert, M. C., Lepage, E., Cabane, J., & Richard, F. (1997). Are angiogenic factors, cytokines, and soluble adhesion molecules prognostic factors in patients with renal cell carcinoma? *Clinical Cancer Research*, *3*, 2451–2458.

Dosquet, C., Schaetz, A., Faucher, C., Lepage, E., Wautier, J. L., Richard, F., & Cabane, J. (1994). Tumour necrosis factor-alpha, interleukin-1 beta and interleukin-6 in patients with renal cell carcinoma. *European Journal of Cancer, 30A*, 162–167.

Druker, B. J. (2008). Translation of the Philadelphia chromosome into therapy for CML. *Blood, 112*, 4808–4817.

Du, R., Lu, K. V., Petritsch, C., Liu, P., Ganss, R., Passegué, E., … Bergers, G. (2008). HIF1alpha induces the recruitment of bone marrow-derived vascular modulatory cells to regulate tumor angiogenesis and invasion. *Cancer Cell, 13*, 206–220.

Dunn, G. P., Old, L. J., & Schreiber, R. D. (2004). The immunobiology of cancer immunosurveillance and immunoediting. *Immunity, 21*, 137–148.

Eckl, J., Buchner, A., Prinz, P. U., Riesenberg, R., Siegert, S. I., Kammerer, R., … Noessner, E. (2012). Transcript signature predicts tissue NK cell content and defines renal cell carcinoma subgroups independent of TNM staging. *Journal of Molecular Medicine (Berlin), 90*, 55–66.

Edin, S., Wikberg, M. L., Dahlin, A. M., Rutegård, J., Öberg, Å., Oldenborg, P., & Palmqvist, R. (2012). The distribution of macrophages with a M1 or M2 phenotype in relation to prognosis and the molecular characteristics of colorectal cancer. *PLoS One*, 7, e47045.

Ehrlich, P. (1909). Uber den jetzigen Stand der Karzinomforschung. *Nederlands Tijdscrift Voor Geneeskunde, 5*, 273–290.

Eisenthal, A., Polyvkin, N., Bramante-Schreiber, L., Misonznik, F., Hassner, A., & Lifschitz-Mercer, B. (2001). Expression of dendritic cells in ovarian tumors correlates with clinical outcome in patients with ovarian cancer. *Human Pathology, 32*, 803–807.

Elinav, E., Nowarski, R., Thaiss, C. A., Hu, B., Jin, C., & Flavell, R. A. (2013). Inflammation-induced cancer: Crosstalk between tumours, immune cells and microorganisms. *Nature Reviews Cancer, 13*, 759–771.

Elsässer-Beile, U., Gierschner, D., Welchner, T., & Wetterauer, U. (2003). Different expression of Fas and Fas ligand in tumor infiltrating and peripheral lymphocytes of patients with renal cell carcinomas. *Anticancer Research, 23*, 433–437.

Elsässer-Beile, U., Kölble, N., Grussenmeyer, T., Schultze-Seemann, W., Wetterauer, U., Gallati, H., … von Kleist, S. (1998). Th1 and Th2 cytokine response patterns in leukocyte cultures of patients with urinary bladder, renal cell and prostate carcinomas. *Tumour Biology, 19*, 470–476.

Engelhardt, J. J., Boldajipour, B., Beemiller, P., Pandurangi, P., Sorensen, C., Werb, Z., … Krummel, M. F. (2012). Marginating dendritic cells of the tumor microenvironment cross-present tumor antigens and stably engage tumor-specific T cells. *Cancer Cell, 21*, 402–417.

Eppihimer, M. J., Gunn, J., Freeman, G. J., Greenfield, E. A., Chernova, T., Erickson, J., & Leonard, J. P. (2002). Expression and regulation of the PD-L1 immunoinhibitory molecule on microvascular endothelial cells. *Microcirculation, 9*, 133–145.

Escudier, B., Porta, C., Schmidinger, M., Algaba, F., Patard, J. J., Khoo, V., … Horwich, A. (2014). Renal cell carcinoma: ESMO Clinical Practice Guidelines for diagnosis, treatment and follow-up. *Annals of Oncology, 25*(Suppl. 3), iii49–iii56.

Facciabene, A., Peng, X., Hagemann, I. S., Balint, K., Barchetti, A., Wang, L., … Coukos, G. (2011). Tumour hypoxia promotes tolerance and angiogenesis via CCL28 and T(reg) cells. *Nature, 475*, 226–230.

Fan, Q., Jing, Y., Yu, G., Kou, X., Ye, F., Gao, L., … Wei, L. (2014). Tumor-associated macrophages promote cancer stem cell-like properties via transforming growth factor-beta1-induced epithelial-mesenchymal transition in hepatocellular carcinoma. *Cancer Letters, 352*, 160–168.

Fan, C., Yang, J., & Engelhardt, J. F. (2002). Temporal pattern of NFkappaB activation influences apoptotic cell fate in a stimuli-dependent fashion. *Journal of Cell Science, 115*, 4843–4853.

Farge, D. (1993). Kaposi's sarcoma in organ transplant recipients. The Collaborative Transplantation Research Group of Ile de France. *The European Journal of Medicine, 2*, 339–343.

Fearnhead, N. S., Britton, M. P., & Bodmer, W. F. (2001). The ABC of APC. *Human Molecular Genetics, 10*, 721–733.

Feichtenbeiner, A., Haas, M., Büttner, M., Grabenbauer, G. G., Fietkau, R., & Distel, L. V. (2014). Critical role of spatial interaction between $CD8^+$ and $Foxp3^+$ cells in human gastric cancer: The distance matters. *Cancer Immunology, Immunotherapy, 63*, 111–119.

Figel, A., Brech, D., Prinz, P. U., Lettenmeyer, U. K., Eckl, J., Turqueti-Neves, A., … Noessner, E. (2011). Human renal cell carcinoma induces a dendritic cell subset that uses T-cell crosstalk for tumor-permissive milieu alterations. *The American Journal of Pathology, 179*, 436–451.

Figenschau, S. L., Fismen, S., Fenton, K. A., Fenton, C., & Mortensen, E. S. (2015). Tertiary lymphoid structures are associated with higher tumor grade in primary operable breast cancer patients. *BMC Cancer, 15*, 101.

Fisk, B., Blevins, T. L., Wharton, J. T., & Ioannides, C. G. (1995). Identification of an immunodominant peptide of HER-2/neu protooncogene recognized by ovarian tumor-specific cytotoxic T lymphocyte lines. *The Journal of Experimental Medicine, 181*, 2109–2117.

Flossmann, E., & Rothwell, P. M. (2007). Effect of aspirin on long-term risk of colorectal cancer: Consistent evidence from randomised and observational studies. *Lancet, 369*, 1603–1613.

Forman, D., Newell, D. G., Fullerton, F., Yarnell, J. W., Stacey, A. R., Wald, N., & Sitas, F. (1991). Association between infection with *Helicobacter pylori* and risk of gastric cancer: Evidence from a prospective investigation. *British Medical Journal, 302*, 1302–1305.

Forssell, J., Oberg, A., Henriksson, M. L., Stenling, R., Jung, A., & Palmqvist, R. (2007). High macrophage infiltration along the tumor front correlates with improved survival in colon cancer. *Clinical Cancer Research, 13*, 1472–1479.

Fox, P., Hudson, M., Brown, C., Lord, S., Gebski, V., De Souza, P., & Lee, C. K. (2013). Markers of systemic inflammation predict survival in patients with advanced renal cell cancer. *British Journal of Cancer, 109*, 147–153.

Fragiadaki, M., & Mason, R. M. (2011). Epithelial-mesenchymal transition in renal fibrosis—Evidence for and against. *International Journal of Experimental Pathology, 92*, 143–150.

Freeman, G. J., Long, A. J., Iwai, Y., Bourque, K., Chernova, T., Nishimura, H., … Honjo, T. (2000). Engagement of the PD-1 immunoinhibitory receptor by a novel B7 family member leads to negative regulation of lymphocyte activation. *The Journal of Experimental Medicine, 192*, 1027–1034.

Fridman, W. H., Pagès, F., Sautès-Fridman, C., & Galon, J. (2012). The immune contexture in human tumours: Impact on clinical outcome. *Nature Reviews Cancer, 12*, 298–306.

Fridman, W. H., Remark, R., Goc, J., Giraldo, N. A., Becht, E., Hammond, S. A., … Sautès-Fridman, C. (2014). The immune microenvironment: A major player in human cancers. *International Archives of Allergy and Immunology, 164*, 13–26.

Fukuda, A., Wang, S. C., Morris, J. P., 4th., Folias, A. E., Liou, A., Kim, G. E., … Hebrok, M. (2011). Stat3 and MMP7 contribute to pancreatic ductal adenocarcinoma initiation and progression. *Cancer Cell, 19*, 441–455.

Furihata, M., Ohtsuki, Y., Ido, E., Iwata, J., Sonobe, H., Araki, K., … Ohmori, K. (1992). HLA-DR antigen- and S-100 protein-positive dendritic cells in esophageal squamous cell carcinoma—Their distribution in relation to prognosis. *Virchows Archiv. B, Cell Pathology Including Molecular Pathology, 61*, 409–414.

Furihata, M., Ono, Y., Ichikawa, K., Tomita, S., Fujimori, T., & Kubota, K. (2005). Prognostic significance of CD83 positive, mature dendritic cells in the gallbladder carcinoma. *Oncology Reports, 14*, 353–356.

Gabrilovich, D. I., Nadaf, S., Corak, J., Berzofsky, J. A., & Carbone, D. P. (1996). Dendritic cells in antitumor immune responses. II. Dendritic cells grown from bone marrow precursors, but not mature DC from tumor-bearing mice, are effective antigen carriers in the therapy of established tumors. *Cellular Immunology*, *170*, 111–119.

Gabrilovich, D. I., & Nagaraj, S. (2009). Myeloid-derived suppressor cells as regulators of the immune system. *Nature Reviews Immunology*, *9*, 162–174.

Galon, J., Costes, A., Sanchez-Cabo, F., Kirilovsky, A., Mlecnik, B., Lagorce-Pagès, C., … Pagès, F. (2006). Type, density, and location of immune cells within human colorectal tumors predict clinical outcome. *Science*, *313*, 1960–1964.

Galon, J., Pagès, F., Marincola, F. M., Angell, H. K., Thurin, M., Lugli, A., … Fox, B. A. (2012). Cancer classification using the Immunoscore: A worldwide task force. *Journal of Translational Medicine*, *10*, 205.

Gao, S. P., Mark, K. G., Leslie, K., Pao, W., Motoi, N., Gerald, W. L., … Bromberg, J. F. (2007). Mutations in the EGFR kinase domain mediate STAT3 activation via IL-6 production in human lung adenocarcinomas. *The Journal of Clinical Investigation*, *117*, 3846–3856.

García-Martínez, E., Gil, G. L., Benito, A. C., González-Billalabeitia, E., Conesa, M. A. V., García García, T., … Ayala de la Peña, F. (2014). Tumor-infiltrating immune cell profiles and their change after neoadjuvant chemotherapy predict response and prognosis of breast cancer. *Breast Cancer Research*, *16*, 488.

Gatto, D., & Brink, R. (2010). The germinal center reaction. *The Journal of Allergy and Clinical Immunology*, *126*, 898–907. quiz 908–909.

Gerber, A. L., Münst, A., Schlapbach, C., Shafighi, M., Kiermeir, D., Hüsler, R., & Hunger, R. E. (2014). High expression of FOXP3 in primary melanoma is associated with tumour progression. *The British Journal of Dermatology*, *170*, 103–109.

Gerlinger, M., Quezada, S. A., Peggs, K. S., Furness, A. J. S., Fisher, R., Marafioti, T., … Swanton, C. (2013). Ultra-deep T cell receptor sequencing reveals the complexity and intratumour heterogeneity of T cell clones in renal cell carcinomas. *The Journal of Pathology*, *231*, 424–432.

Germain, C., Gnjatic, S., & Dieu-Nosjean, M. (2015). Tertiary lymphoid structure-associated B cells are key players in anti-tumor immunity. *Frontiers in Immunology*, *6*, 67.

Germain, C., Gnjatic, S., Tamzalit, F., Knockaert, S., Remark, R., Goc, J., … Dieu-Nosjean, M. (2014). Presence of B cells in tertiary lymphoid structures is associated with a protective immunity in patients with lung cancer. *American Journal of Respiratory and Critical Care Medicine*, *189*, 832–844.

Gierach, G. L., Lacey, J. V. J., Schatzkin, A., Leitzmann, M. F., Richesson, D., Hollenbeck, A. R., & Brinton, L. A. (2008). Nonsteroidal anti-inflammatory drugs and breast cancer risk in the National Institutes of Health-AARP Diet and Health Study. *Breast Cancer Research*, *10*, R38.

Gigante, M., Blasi, A., Loverre, A., Mancini, V., Battaglia, M., Selvaggi, F. P., … Ranieri, E. (2009). Dysfunctional DC subsets in RCC patients: Ex vivo correction to yield an effective anti-cancer vaccine. *Molecular Immunology*, *46*, 893–901.

Giraldo, N. A., Becht, E., Pagès, F., Skliris, G., Verkarre, V., Vano, Y., … Sautès-Fridman, C. (2015). Orchestration and prognostic significance of immune checkpoints in the microenvironment of primary and metastatic renal cell cancer. *Clinical Cancer Research*, *21*, 3031–3040.

Giraldo, N. A., Becht, E., Remark, R., Damotte, D., Sautès-Fridman, C., & Fridman, W. H. (2014). The immune contexture of primary and metastatic human tumours. *Current Opinion in Immunology*, *27*, 8–15.

Giraudo, E., Inoue, M., & Hanahan, D. (2004). An amino-bisphosphonate targets MMP-9-expressing macrophages and angiogenesis to impair cervical carcinogenesis. *The Journal of Clinical Investigation*, *114*, 623–633.

Gnjatic, S., Atanackovic, D., Matsuo, M., Jäger, E., Lee, S. Y., Valmori, D., ... Old, L. J. (2003). Cross-presentation of HLA class I epitopes from exogenous NY-ESO-1 polypeptides by nonprofessional APCs. *Journal of Immunology, 170*, 1191–1196.

Gnjatic, S., Ritter, E., Büchler, M. W., Giese, N. A., Brors, B., Frei, C., ... Jäger, D. (2010). Seromic profiling of ovarian and pancreatic cancer. *Proceedings of the National Academy of Sciences of the United States of America, 107*, 5088–5093.

Goc, J., Germain, C., Vo-Bourgais, T. K. D., Lupo, A., Klein, C., Knockaert, S., ... Dieu-Nosjean, M. (2014). Dendritic cells in tumor-associated tertiary lymphoid structures signal a Th1 cytotoxic immune contexture and license the positive prognostic value of infiltrating CD8+ T cells. *Cancer Research, 74*, 705–715.

Goeppert, B., Frauenschuh, L., Zucknick, M., Stenzinger, A., Andrulis, M., Klauschen, F., ... Weichert, W. (2013). Prognostic impact of tumour-infiltrating immune cells on biliary tract cancer. *British Journal of Cancer, 109*, 2665–2674.

Gogusev, J., Augusti, M., Chrétien, Y., & Droz, D. (1993). Interleukin-6 and TNF alpha production in human renal cell carcinoma. *Kidney International, 44*, 585–592.

Gossage, L., Eisen, T., & Maher, E. R. (2015). VHL, the story of a tumour suppressor gene. *Nature Reviews Cancer, 15*, 55–64.

Gottlin, E. B., Bentley, R. C., Campa, M. J., Pisetsky, D. S., Herndon, J. E., 2nd., & Patz, E. F. J. (2011). The Association of Intratumoral Germinal Centers with early-stage non-small cell lung cancer. *Journal of Thoracic Oncology, 6*, 1687–1690.

Grady, W. M., & Markowitz, S. D. (2015). The molecular pathogenesis of colorectal cancer and its potential application to colorectal cancer screening. *Digestive Diseases and Sciences, 60*, 762–772.

Green, D. R., Ferguson, T., Zitvogel, L., & Kroemer, G. (2009). Immunogenic and tolerogenic cell death. *Nature Reviews Immunology, 9*, 353–363.

Griffiths, R. W., Elkord, E., Gilham, D. E., Ramani, V., Clarke, N., Stern, P. L., & Hawkins, R. E. (2007). Frequency of regulatory T cells in renal cell carcinoma patients and investigation of correlation with survival. *Cancer Immunology, Immunotherapy, 56*, 1743–1753.

Grivennikov, S. I., Greten, F. R., & Karin, M. (2010). Immunity, inflammation, and cancer. *Cell, 140*, 883–899.

Grivennikov, S., Karin, E., Terzic, J., Mucida, D., Yu, G., Vallabhapurapu, S., ... Karin, M. (2009). IL-6 and Stat3 are required for survival of intestinal epithelial cells and development of colitis-associated cancer. *Cancer Cell, 15*, 103–113.

Grivennikov, S. I., Wang, K., Mucida, D., Stewart, C. A., Schnabl, B., Jauch, D., ... Karin, M. (2012). Adenoma-linked barrier defects and microbial products drive IL-23/IL-17-mediated tumour growth. *Nature, 491*, 254–258.

Gruber, S. B., & Kohlmann, W. (2003). The genetics of hereditary non-polyposis colorectal cancer. *Journal of the National Comprehensive Cancer Network, 1*, 137–144.

Guedj, M., Marisa, L., de Reynies, A., Orsetti, B., Schiappa, R., Bibeau, F., ... Theillet, C. (2012). A refined molecular taxonomy of breast cancer. *Oncogene, 31*, 1196–1206.

Guidoboni, M., Gafà, R., Viel, A., Doglioni, C., Russo, A., Santini, A., ... Dolcetti, R. (2001). Microsatellite instability and high content of activated cytotoxic lymphocytes identify colon cancer patients with a favorable prognosis. *The American Journal of Pathology, 159*, 297–304.

Gulubova, M. V., Ananiev, J. R., Vlaykova, T. I., Yovchev, Y., Tsoneva, V., & Manolova, I. M. (2012). Role of dendritic cells in progression and clinical outcome of colon cancer. *International Journal of Colorectal Disease, 27*, 159–169.

Guo, X., Jo, V. Y., Mills, A. M., Zhu, S. X., Lee, C., Espinosa, I., ... van de Rijn, M. (2015). Clinically relevant molecular subtypes in leiomyosarcoma. *Clinical Cancer Research, 21*, 3501–3511.

Guo, S., Lin, D., Li, J., Liu, R., Zhou, C., Wang, D., … Zhang, S. (2007). Tumor-associated macrophages and CD3-zeta expression of tumor-infiltrating lymphocytes in human esophageal squamous-cell carcinoma. *Diseases of the Esophagus*, *20*, 107–116.

Gu-Trantien, C., Loi, S., Garaud, S., Equeter, C., Libin, M., de Wind, A., … Willard-Gallo, K. (2013). $CD4^+$ follicular helper T cell infiltration predicts breast cancer survival. *The Journal of Clinical Investigation*, *123*, 2873–2892.

Haas, M., Dimmler, A., Hohenberger, W., Grabenbauer, G. G., Niedobitek, G., & Distel, L. V. (2009). Stromal regulatory T-cells are associated with a favourable prognosis in gastric cancer of the cardia. *BMC Gastroenterology*, *9*, 65. http://dx.doi.org/10.1186/1471-230X-9-65.

Hahn, M. A., Hahn, T., Lee, D., Esworthy, R. S., Kim, B., Riggs, A. D., … Pfeifer, G. P. (2008). Methylation of polycomb target genes in intestinal cancer is mediated by inflammation. *Cancer Research*, *68*, 10280–10289.

Hahne, M., Rimoldi, D., Schröter, M., Romero, P., Schreier, M., French, L. E., … Tschopp, J. (1996). Melanoma cell expression of Fas(Apo-1/CD95) ligand: Implications for tumor immune escape. *Science*, *274*, 1363–1366.

Hald, S. M., Bremnes, R. M., Al-Shibli, K., Al-Saad, S., Andersen, S., Stenvold, H., … Donnem, T. (2013). CD4/CD8 co-expression shows independent prognostic impact in resected non-small cell lung cancer patients treated with adjuvant radiotherapy. *Lung Cancer*, *80*, 209–215.

Hamaï, A., Duperrier-Amouriaux, K., Pignon, P., Raimbaud, I., Memeo, L., Colarossi, C., … Valmori, D. (2011). Antibody responses to NY-ESO-1 in primary breast cancer identify a subtype target for immunotherapy. *PLoS One*, *6*, e21129.

Hamanishi, J., Mandai, M., Iwasaki, M., Okazaki, T., Tanaka, Y., Yamaguchi, K., … Fujii, S. (2007). Programmed cell death 1 ligand 1 and tumor-infiltrating CD8+ T lymphocytes are prognostic factors of human ovarian cancer. *Proceedings of the National Academy of Sciences of the United States of America*, *104*, 3360–3365.

Hamilton, R., Krauze, M., Romkes, M., Omolo, B., Konstantinopoulos, P., Reinhart, T., … Moschos, S. (2013). Pathologic and gene expression features of metastatic melanomas to the brain. *Cancer*, *119*, 2737–2746.

Han, S., Zhang, C., Li, Q., Dong, J., Liu, Y., Huang, Y., … Wu, A. (2014). Tumour-infiltrating CD4(+) and CD8(+) lymphocytes as predictors of clinical outcome in glioma. *British Journal of Cancer*, *110*, 2560–2568.

Hanada, T., Nakagawa, M., Emoto, A., Nomura, T., Nasu, N., & Nomura, Y. (2000). Prognostic value of tumor-associated macrophage count in human bladder cancer. *International Journal of Urology*, 7, 263–269.

Hanahan, D., & Weinberg, R. A. (2011). Hallmarks of cancer: The next generation. *Cell*, *144*, 646–674.

Haniffa, M. A., Wang, X., Holtick, U., Rae, M., Isaacs, J. D., Dickinson, A. M., … Collin, M. P. (2007). Adult human fibroblasts are potent immunoregulatory cells and functionally equivalent to mesenchymal stem cells. *Journal of Immunology*, *179*, 1595–1604.

Harp, C. T., Lovett-Racke, A. E., Racke, M. K., Frohman, E. M., & Monson, N. L. (2008). Impact of myelin-specific antigen presenting B cells on T cell activation in multiple sclerosis. *Clinical Immunology*, *128*, 382–391.

Haverkos, H. W., & Drotman, D. P. (1985). Prevalence of Kaposi's sarcoma among patients with AIDS. *The New England Journal of Medicine*, *312*, 1518.

Hayashi, K., Nittono, R., Okamoto, N., Tsuji, S., Hara, Y., Goitsuka, R., & Kitamura, D. (2000). The B cell-restricted adaptor BASH is required for normal development and antigen receptor-mediated activation of B cells. *Proceedings of the National Academy of Sciences of the United States of America*, *97*, 2755–2760.

He, K., Zhang, L., Huang, C., Ma, S., Wang, Y., Wang, W., ... Sun, Z. (2014). CD163+ tumor-associated macrophages correlated with poor prognosis and cancer stem cells in oral squamous cell carcinoma. *BioMed Research International, 2014*, 838632.

Heller, D. S., Hameed, M., Cracchiolo, B., Wiederkehr, M., Scott, D., Skurnick, J., ... Lambert, W. C. (2003). Presence and quantification of macrophages in squamous cell carcinoma of the cervix. *International Journal of Gynecological Cancer, 13*, 67–70.

Hernández-Prieto, S., Romera, A., Ferrer, M., Subiza, J. L., López-Asenjo, J. A., Jarabo, J. R., ... Sanz-Ortega, J. (2015). A 50-gene signature is a novel scoring system for tumor-infiltrating immune cells with strong correlation with clinical outcome of stage I/II non-small cell lung cancer. *Clinical & Translational Oncology, 17*, 330–338.

Herrera, M., Herrera, A., Domínguez, G., Silva, J., García, V., García, J. M., ... Peña, C. (2013). Cancer-associated fibroblast and M2 macrophage markers together predict outcome in colorectal cancer patients. *Cancer Science, 104*, 437–444.

Hiraoka, N., Onozato, K., Kosuge, T., & Hirohashi, S. (2006). Prevalence of FOXP3+ regulatory T cells increases during the progression of pancreatic ductal adenocarcinoma and its premalignant lesions. *Clinical Cancer Research, 12*, 5423–5434.

Hiraoka, N., Yamazaki-Itoh, R., Ino, Y., Mizuguchi, Y., Yamada, T., Hirohashi, S., & Kanai, Y. (2011). CXCL17 and ICAM2 are associated with a potential anti-tumor immune response in early intraepithelial stages of human pancreatic carcinogenesis. *Gastroenterology, 140*, 310–321.

Hirayama, S., Ishii, G., Nagai, K., Ono, S., Kojima, M., Yamauchi, C., ... Ochiai, A. (2012). Prognostic impact of CD204-positive macrophages in lung squamous cell carcinoma: Possible contribution of Cd204-positive macrophages to the tumor-promoting microenvironment. *Journal of Thoracic Oncology, 7*, 1790–1797.

Ho, M., Tang, S., Chuang, M., Cha, T., Li, J., Sun, G., & Sun, K. (2012). TNF-α induces epithelial-mesenchymal transition of renal cell carcinoma cells via a GSK3β-dependent mechanism. *Molecular Cancer Research, 10*, 1109–1119.

Honig, A., Schaller, N., Dietl, J., Backe, J., & Kammerer, U. (2005). S100 as an immunohistochemically-detected marker with prognostic significance in endometrial carcinoma. *Anticancer Research, 25*, 1747–1753.

Horikawa, M., Minard-Colin, V., Matsushita, T., & Tedder, T. F. (2011). Regulatory B cell production of IL-10 inhibits lymphoma depletion during CD20 immunotherapy in mice. *The Journal of Clinical Investigation, 121*, 4268–4280.

Hoshida, Y., Nijman, S. M. B., Kobayashi, M., Chan, J. A., Brunet, J., Chiang, D. Y., ... Golub, T. R. (2009). Integrative transcriptome analysis reveals common molecular subclasses of human hepatocellular carcinoma. *Cancer Research, 69*, 7385–7392.

Hotta, K., Sho, M., Fujimoto, K., Shimada, K., Yamato, I., Anai, S., ... Nakajima, Y. (2011). Prognostic significance of CD45RO+ memory T cells in renal cell carcinoma. *British Journal of Cancer, 105*, 1191–1196.

Hou, M., Chang, J. W., Pang, S., Chiang, Y., Shen, Y., Liao, S., ... Chuang, C. (2010). Characterization of the response of dendritic cells and regulatory T cells to tumor antigens in patients with renal cell carcinoma. *Chang Gung Medical Journal, 33*, 25–35.

Hou, Y., Chao, Y., Tung, H., Wang, H., & Shan, Y. (2014). Coexpression of CD44-positive/CD133-positive cancer stem cells and CD204-positive tumor-associated macrophages is a predictor of survival in pancreatic ductal adenocarcinoma. *Cancer, 120*, 2766–2777.

Hsia, J., Chen, J., Chen, C., Hsu, C., Miaw, J., Huang, Y., & Yang, C. (2005). Prognostic significance of intratumoral natural killer cells in primary resected esophageal squamous cell carcinoma. *Chang Gung Medical Journal, 28*, 335–340.

Hu, Z., Fan, C., Oh, D. S., Marron, J. S., He, X., Qaqish, B. F., ... Perou, C. M. (2006). The molecular portraits of breast tumors are conserved across microarray platforms. *BMC Genomics, 7*, 96.

Hu, Q., Gou, Y., Sun, C., Ding, W., Xu, K., Gu, B., … Ding, Q. (2014). The prognostic value of C-reactive protein in renal cell carcinoma: A systematic review and meta-analysis. *Urologic Oncology, 32*, 50.e1–50.e8.

Huang, Y., Wang, F., Wang, T., Wang, Y., Zhu, Z., Gao, Y., & Du, Z. (2012). Tumor-infiltrating FoxP3+ Tregs and CD8+ T cells affect the prognosis of hepatocellular carcinoma patients. *Digestion, 86*, 329–337.

Huang, J., Yao, X., Zhang, J., Dong, B., Chen, Q., Xue, W., … Huang, Y. (2013). Hypoxia-induced downregulation of miR-30c promotes epithelial-mesenchymal transition in human renal cell carcinoma. *Cancer Science, 104*, 1609–1617.

Ichimura, T., Morikawa, T., Kawai, T., Nakagawa, T., Matsushita, H., Kakimi, K., … Fukayama, M. (2014). Prognostic significance of CD204-positive macrophages in upper urinary tract cancer. *Annals of Surgical Oncology, 21*, 2105–2112.

Igarashi, T., Murakami, S., Takahashi, H., Matsuzaki, O., & Shimazaki, J. (1992). Changes on distribution of CD4+/CD45RA- and CD8+/CD11- cells in tumor-infiltrating lymphocytes of renal cell carcinoma associated with tumor progression. *European Urology, 22*, 323–328.

Igney, F. H., Behrens, C. K., & Krammer, P. H. (2000). Tumor counterattack—Concept and reality. *European Journal of Immunology, 30*, 725–731.

Igney, F. H., & Krammer, P. H. (2005). Tumor counterattack: Fact or fiction? *Cancer Immunology, Immunotherapy, 54*, 1127–1136.

Ingels, A., Sanchez Salas, R. E., Ravery, V., Fromont-Hankard, G., Validire, P., Patard, J., … Cathelineau, X. (2014). T-helper 1 immunoreaction influences survival in muscle-invasive bladder cancer: Proof of concept. *Ecancermedicalscience, 8*, 486.

Ino, Y., Yamazaki-Itoh, R., Shimada, K., Iwasaki, M., Kosuge, T., Kanai, Y., & Hiraoka, N. (2013). Immune cell infiltration as an indicator of the immune microenvironment of pancreatic cancer. *British Journal of Cancer, 108*, 914–923.

Inoshima, N., Nakanishi, Y., Minami, T., Izumi, M., Takayama, K., Yoshino, I., & Hara, N. (2002). The influence of dendritic cell infiltration and vascular endothelial growth factor expression on the prognosis of non-small cell lung cancer. *Clinical Cancer Research, 8*, 3480–3486.

Inoue, K., Furihata, M., Ohtsuki, Y., & Fujita, Y. (1993). Distribution of S-100 protein-positive dendritic cells and expression of HLA-DR antigen in transitional cell carcinoma of the urinary bladder in relation to tumour progression and prognosis. *Virchows Archiv. A, Pathological Anatomy and Histopathology, 422*, 351–355.

Isella, C., Terrasi, A., Bellomo, S. E., Petti, C., Galatola, G., Muratore, A., … Medico, E. (2015). Stromal contribution to the colorectal cancer transcriptome. *Nature Genetics, 47*, 312–319.

Ishigami, S., Natsugoe, S., Matsumoto, M., Okumura, H., Sakita, H., Nakashima, S., … Aikou, T. (2003). Clinical implications of intratumoral dendritic cell infiltration in esophageal squamous cell carcinoma. *Oncology Reports, 10*, 1237–1240.

Ishigami, S., Natsugoe, S., Tokuda, K., Nakajo, A., Che, X., Iwashige, H., … Aikou, T. (2000). Prognostic value of intratumoral natural killer cells in gastric carcinoma. *Cancer, 88*, 577–583.

Ishigami, S., Natsugoe, S., Tokuda, K., Nakajo, A., Okumura, H., Matsumoto, M., … Aikou, T. (2003b). Tumor-associated macrophage (TAM) infiltration in gastric cancer. *Anticancer Research, 23*, 4079–4083.

Ishigami, S., Ueno, S., Matsumoto, M., Okumura, H., Arigami, T., Uchikado, Y., … Natsugoe, S. (2010). Prognostic value of CD208-positive cell infiltration in gastric cancer. *Cancer Immunology, Immunotherapy, 59*, 389–395.

Issa, J. (2004). CpG island methylator phenotype in cancer. *Nature Reviews Cancer, 4*, 988–993.

Ito, K., Asano, T., Yoshii, H., Satoh, A., Sumitomo, M., & Hayakawa, M. (2006). Impact of thrombocytosis and C-reactive protein elevation on the prognosis for patients with renal cell carcinoma. *International Journal of Urology, 13*, 1365–1370.

Iwamoto, M., Shinohara, H., Miyamoto, A., Okuzawa, M., Mabuchi, H., Nohara, T., … Tanigawa, N. (2003). Prognostic value of tumor-infiltrating dendritic cells expressing CD83 in human breast carcinomas. *International Journal of Cancer, 104*, 92–97.

Jagdev, S. P. K., Gregory, W., Vasudev, N. S., Harnden, P., Sim, S., Thompson, D., … Banks, R. E. (2010). Improving the accuracy of pre-operative survival prediction in renal cell carcinoma with C-reactive protein. *British Journal of Cancer, 103*, 1649–1656.

Jensen, T. O., Schmidt, H., Møller, H. J., Donskov, F., Høyer, M., Sjoegren, P., … Steiniche, T. (2012). Intratumoral neutrophils and plasmacytoid dendritic cells indicate poor prognosis and are associated with pSTAT3 expression in AJCC stage I/II melanoma. *Cancer, 118*, 2476–2485.

Jensen, T. O., Schmidt, H., Møller, H. J., Høyer, M., Maniecki, M. B., Sjoegren, P., … Steiniche, T. (2009). Macrophage markers in serum and tumor have prognostic impact in American Joint Committee on Cancer stage I/II melanoma. *Journal of Clinical Oncology, 27*, 3330–3337.

João, C., Ogle, B. M., Gay-Rabinstein, C., Platt, J. L., & Cascalho, M. (2004). B cell-dependent TCR diversification. *Journal of Immunology, 172*, 4709–4716.

Johnson, S. K., Kerr, K. M., Chapman, A. D., Kennedy, M. M., King, G., Cockburn, J. S., & Jeffrey, R. R. (2000). Immune cell infiltrates and prognosis in primary carcinoma of the lung. *Lung Cancer, 27*, 27–35.

Jones, P. A., & Baylin, S. B. (2002). The fundamental role of epigenetic events in cancer. *Nature Reviews Genetics, 3*, 415–428.

Jones, S., Horwood, N., Cope, A., & Dazzi, F. (2007). The antiproliferative effect of mesenchymal stem cells is a fundamental property shared by all stromal cells. *Journal of Immunology, 179*, 2824–2831.

Jumaa, H., Wollscheid, B., Mitterer, M., Wienands, J., Reth, M., & Nielsen, P. J. (1999). Abnormal development and function of B lymphocytes in mice deficient for the signaling adaptor protein SLP-65. *Immunity, 11*, 547–554.

Jung, A. C., Guihard, S., Krugell, S., Ledrappier, S., Brochot, A., Dalstein, V., … Abecassis, J. (2013). CD8-alpha T-cell infiltration in human papillomavirus-related oropharyngeal carcinoma correlates with improved patient prognosis. *International Journal of Cancer, 132*, E26–E36.

Kang, M. J., Kim, K. M., Bae, J. S., Park, H. S., Lee, H., Chung, M. J., … Jang, K. Y. (2013). Tumor-infiltrating PD1-positive lymphocytes and FoxP3-positive regulatory T cells predict distant metastatic relapse and survival of clear cell renal cell carcinoma. *Translational Oncology, 6*, 282–289.

Karakiewicz, P. I., Hutterer, G. C., Trinh, Q., Jeldres, C., Perrotte, P., Gallina, A., … Patard, J. (2007). C-reactive protein is an informative predictor of renal cell carcinoma-specific mortality: A European study of 313 patients. *Cancer, 110*, 1241–1247.

Kashimura, S., Saze, Z., Terashima, M., Soeta, N., Ohtani, S., Osuka, F., … Gotoh, M. (2012). CD83(+) dendritic cells and Foxp3(+) regulatory T cells in primary lesions and regional lymph nodes are inversely correlated with prognosis of gastric cancer. *Gastric Cancer, 15*, 144–153.

Kasiske, B. L., Snyder, J. J., Gilbertson, D. T., & Wang, C. (2004). Cancer after kidney transplantation in the United States. *American Journal of Transplantation, 4*, 905–913.

Katz, S. C., Bamboat, Z. M., Maker, A. V., Shia, J., Pillarisetty, V. G., Yopp, A. C., … DeMatteo, R. P. (2013). Regulatory T cell infiltration predicts outcome following resection of colorectal cancer liver metastases. *Annals of Surgical Oncology, 20*, 946–955.

Kawahara, A., Hattori, S., Akiba, J., Nakashima, K., Taira, T., Watari, K., ... Ono, M. (2010). Infiltration of thymidine phosphorylase-positive macrophages is closely associated with tumor angiogenesis and survival in intestinal type gastric cancer. *Oncology Reports*, *24*, 405–415.

Kawai, O., Ishii, G., Kubota, K., Murata, Y., Naito, Y., Mizuno, T., ... Ochiai, A. (2008). Predominant infiltration of macrophages and CD8(+) T cells in cancer nests is a significant predictor of survival in stage IV nonsmall cell lung cancer. *Cancer*, *113*, 1387–1395.

Kawakami, Y., Eliyahu, S., Delgado, C. H., Robbins, P. F., Rivoltini, L., Topalian, S. L., ... Rosenberg, S. A. (1994a). Cloning of the gene coding for a shared human melanoma antigen recognized by autologous T cells infiltrating into tumor. *Proceedings of the National Academy of Sciences of the United States of America*, *91*, 3515–3519.

Kawakami, Y., Eliyahu, S., Sakaguchi, K., Robbins, P. F., Rivoltini, L., Yannelli, J. R., ... Rosenberg, S. A. (1994b). Identification of the immunodominant peptides of the MART-1 human melanoma antigen recognized by the majority of HLA-A2-restricted tumor infiltrating lymphocytes. *The Journal of Experimental Medicine*, *180*, 347–352.

Khorana, A. A., Ryan, C. K., Cox, C., Eberly, S., & Sahasrabudhe, D. M. (2003). Vascular endothelial growth factor, CD68, and epidermal growth factor receptor expression and survival in patients with Stage II and Stage III colon carcinoma: A role for the host response in prognosis. *Cancer*, *97*, 960–968.

Kim, Y., Bae, J. M., Li, G., Cho, N. Y., & Kang, G. H. (2015). Image analyzer-based assessment of tumor-infiltrating T cell subsets and their prognostic values in colorectal carcinomas. *PLoS One*, *10*, e0122183.

Kim, S. T., Jeong, H., Woo, O. H., Seo, J. H., Kim, A., Lee, E. S., ... Park, K. H. (2013). Tumor-infiltrating lymphocytes, tumor characteristics, and recurrence in patients with early breast cancer. *American Journal of Clinical Oncology*, *36*, 224–231.

Kim, C., Kim, Y., Kwon, T., Yoon, J. H., Kim, K. H., You, D., ... Jeong, I. G. (2015). Regulatory T cells and TGF-β1 in clinically localized renal cell carcinoma: Comparison with age-matched healthy controls. *Urologic Oncology*, *33*, 113.e19–113.e25.

Kim, K., Lee, K. S., Cho, H. J., Kim, Y. H., Yang, H. K., Kim, W. H., & Kang, G. H. (2014). Prognostic implications of tumor-infiltrating FoxP3+ regulatory T cells and CD8+ cytotoxic T cells in microsatellite-unstable gastric cancers. *Human Pathology*, *45*, 285–293.

Klein, J. L., Nguyen, T. T., Bien-Willner, G. A., Chen, L., Foyil, K. V., Bartlett, N. L., ... Kreisel, F. (2014). CD163 immunohistochemistry is superior to CD68 in predicting outcome in classical Hodgkin lymphoma. *American Journal of Clinical Pathology*, *141*, 381–387.

Kluger, H. M., Zito, C. R., Barr, M. L., Baine, M. K., Chiang, V. L. S., Sznol, M., ... Jilaveanu, L. B. (2015). Characterization of PD-L1 expression and associated T-cell infiltrates in metastatic melanoma samples from variable anatomic sites. *Clinical Cancer Research*, *21*, 3052–3060.

Knoefel, B., Nuske, K., Steiner, T., Junker, K., Kosmehl, H., Rebstock, K., ... Junker, U. (1997). Renal cell carcinomas produce IL-6, IL-10, IL-11, and TGF-beta 1 in primary cultures and modulate T lymphocyte blast transformation. *Journal of Interferon & Cytokine Research*, *17*, 95–102.

Kobayashi, T., Hamaguchi, Y., Hasegawa, M., Fujimoto, M., Takehara, K., & Matsushita, T. (2014). B cells promote tumor immunity against B16F10 melanoma. *The American Journal of Pathology*, *184*, 3120–3129.

Kolenko, V., Wang, Q., Riedy, M. C., O'Shea, J., Ritz, J., Cathcart, M. K., ... Finke, J. (1997). Tumor-induced suppression of T lymphocyte proliferation coincides with inhibition of Jak3 expression and IL-2 receptor signaling: Role of soluble products from human renal cell carcinomas. *Journal of Immunology*, *159*, 3057–3067.

Komohara, Y., Hasita, H., Ohnishi, K., Fujiwara, Y., Suzu, S., Eto, M., & Takeya, M. (2011). Macrophage infiltration and its prognostic relevance in clear cell renal cell carcinoma. *Cancer Science*, *102*, 1424–1431.

Kondo, T., Nakazawa, H., Ito, F., Hashimoto, Y., Osaka, Y., Futatsuyama, K., ... Tanabe, K. (2006). Favorable prognosis of renal cell carcinoma with increased expression of chemokines associated with a Th1-type immune response. *Cancer Science, 97*, 780–786.

Kong, L., Zhu, X., Xu, H., Zhang, J., Lu, L., Wang, W., ... Sun, H. (2013). The clinical significance of the CD163+ and CD68+ macrophages in patients with hepatocellular carcinoma. *PLoS One, 8*, e59771.

König, B., Steinbach, F., Janocha, B., Drynda, A., Stumm, M., Philipp, C., ... König, W. (1999). The differential expression of proinflammatory cytokines IL-6, IL-8 and TNF-alpha in renal cell carcinoma. *Anticancer Research, 19*, 1519–1524.

Kopecký, O., Lukesová, S., Vroblová, V., Vokurková, D., Morávek, P., Safránek, H., ... Soucek, P. (2007). Phenotype analysis of tumour-infiltrating lymphocytes and lymphocytes in peripheral blood in patients with renal carcinoma. *Acta Medica (Hradec Kralove), 50*, 207–212.

Kotlan, B., Simsa, P., Teillaud, J., Fridman, W. H., Toth, J., McKnight, M., & Glassy, M. C. (2005). Novel ganglioside antigen identified by B cells in human medullary breast carcinomas: The proof of principle concerning the tumor-infiltrating B lymphocytes. *Journal of Immunology, 175*, 2278–2285.

Kowalczyk, D., Skorupski, W., Kwias, Z., & Nowak, J. (1997). Flow cytometric analysis of tumour-infiltrating lymphocytes in patients with renal cell carcinoma. *British Journal of Urology, 80*, 543–547.

Kraman, M., Bambrough, P. J., Arnold, J. N., Roberts, E. W., Magiera, L., Jones, J. O., ... Fearon, D. T. (2010). Suppression of antitumor immunity by stromal cells expressing fibroblast activation protein-alpha. *Science, 330*, 827–830.

Kristensen, V. N., Vaske, C. J., Ursini-Siegel, J., Van Loo, P., Nordgard, S. H., Sachidanandam, R., ... Børresen-Dale, A. (2012). Integrated molecular profiles of invasive breast tumors and ductal carcinoma in situ (DCIS) reveal differential vascular and interleukin signaling. *Proceedings of the National Academy of Sciences of the United States of America, 109*, 2802–2807.

Krontiris, T. G., & Cooper, G. M. (1981). Transforming activity of human tumor DNAs. *Proceedings of the National Academy of Sciences of the United States of America, 78*, 1181–1184.

Kübler, K., Ayub, T. H., Weber, S. K., Zivanovic, O., Abramian, A., Keyver-Paik, M., ... Rudlowski, C. (2014). Prognostic significance of tumor-associated macrophages in endometrial adenocarcinoma. *Gynecologic Oncology, 135*, 176–183.

Kudo, D., Rayman, P., Horton, C., Cathcart, M. K., Bukowski, R. M., Thornton, M., ... Finke, J. H. (2003). Gangliosides expressed by the renal cell carcinoma cell line SK-RC-45 are involved in tumor-induced apoptosis of T cells. *Cancer Research, 63*, 1676–1683.

Kurahara, H., Shinchi, H., Mataki, Y., Maemura, K., Noma, H., Kubo, F., ... Takao, S. (2011). Significance of M2-polarized tumor-associated macrophage in pancreatic cancer. *The Journal of Surgical Research, 167*, e211–e219.

Kusmartsev, S., Su, Z., Heiser, A., Dannull, J., Eruslanov, E., Kübler, H., ... Vieweg, J. (2008). Reversal of myeloid cell-mediated immunosuppression in patients with metastatic renal cell carcinoma. *Clinical Cancer Research, 14*, 8270–8278.

La Rocca, G., Anzalone, R., Corrao, S., Magno, F., Rappa, F., Marasà, S., ... Cappello, F. (2008). CD1a down-regulation in primary invasive ductal breast carcinoma may predict regional lymph node invasion and patient outcome. *Histopathology, 52*, 203–212.

Ladányi, A., Kiss, J., Mohos, A., Somlai, B., Liszkay, G., Gilde, K., ... Tímár, J. (2011). Prognostic impact of B-cell density in cutaneous melanoma. *Cancer Immunology, Immunotherapy, 60*, 1729–1738.

Ladányi, A., Kiss, J., Somlai, B., Gilde, K., Fejos, Z., Mohos, A., ... Tímár, J. (2007). Density of DC-LAMP(+) mature dendritic cells in combination with activated T lymphocytes infiltrating primary cutaneous melanoma is a strong independent prognostic factor. *Cancer Immunology, Immunotherapy, 56*, 1459–1469.

Lahn, M., Fisch, P., Köhler, G., Kunzmann, R., Hentrich, I., Jesuiter, H., … Lindemann, A. (1999). Pro-inflammatory and T cell inhibitory cytokines are secreted at high levels in tumor cell cultures of human renal cell carcinoma. *European Urology*, *35*, 70–80.

Lamb, G. W. A., McMillan, D. C., Ramsey, S., & Aitchison, M. (2006). The relationship between the preoperative systemic inflammatory response and cancer-specific survival in patients undergoing potentially curative resection for renal clear cell cancer. *British Journal of Cancer*, *94*, 781–784.

Lan, C., Huang, X., Lin, S., Huang, H., Cai, Q., Wan, T., … Liu, J. (2013). Expression of M2-polarized macrophages is associated with poor prognosis for advanced epithelial ovarian cancer. *Technology in Cancer Research & Treatment*, *12*, 259–267.

Lauerová, L., Dusek, L., Simícková, M., Rovný, F., Spurný, V., Rovný, A., … Kovarík, J. (1999). Renal cell carcinoma-associated immune impairment that may interfere with the response to cytokine therapy. *Neoplasma*, *46*, 141–149.

Le, D. T., Uram, J. N., Wang, H., Bartlett, B. R., Kemberling, H., Eyring, A. D., … Diaz, L. A. J. (2015). PD-1 blockade in tumors with mismatch-repair deficiency. *The New England Journal of Medicine*, *372*, 2509–2520.

Lechner, M. G., Megiel, C., Russell, S. M., Bingham, B., Arger, N., Woo, T., & Epstein, A. L. (2011). Functional characterization of human Cd33+ and Cd11b+ myeloid-derived suppressor cell subsets induced from peripheral blood mononuclear cells co-cultured with a diverse set of human tumor cell lines. *Journal of Translational Medicine*, *9*, 90.

Lee, J., Chu, I., Heo, J., Calvisi, D. F., Sun, Z., Roskams, T., … Thorgeirsson, S. S. (2004). Classification and prediction of survival in hepatocellular carcinoma by gene expression profiling. *Hepatology*, *40*, 667–676.

Lee, S. O., Chun, J. Y., Nadiminty, N., Lou, W., & Gao, A. C. (2007). Interleukin-6 undergoes transition from growth inhibitor associated with neuroendocrine differentiation to stimulator accompanied by androgen receptor activation during LNCaP prostate cancer cell progression. *Prostate*, *67*, 764–773.

Lee, S., Seo, S., Kim, B., Kim, C., Lee, J., Kang, J., … Lim, J. (2005). IFN-gamma regulates the expression of B7-H1 in dermal fibroblast cells. *Journal of Dermatological Science*, *40*, 95–103.

Leek, R. D., Lewis, C. E., Whitehouse, R., Greenall, M., Clarke, J., & Harris, A. L. (1996). Association of macrophage infiltration with angiogenesis and prognosis in invasive breast carcinoma. *Cancer Research*, *56*, 4625–4629.

Lesina, M., Kurkowski, M. U., Ludes, K., Rose-John, S., Treiber, M., Klöppel, G., … Algül, H. (2011). Stat3/Socs3 activation by IL-6 trans-signaling promotes progression of pancreatic intraepithelial neoplasia and development of pancreatic cancer. *Cancer Cell*, *19*, 456–469.

Lespagnard, L., Gancberg, D., Rouas, G., Leclercq, G., de Saint-Aubain Somerhausen, N., Di Leo, A., … Larsimont, D. (1999). Tumor-infiltrating dendritic cells in adenocarcinomas of the breast: A study of 143 neoplasms with a correlation to usual prognostic factors and to clinical outcome. *International Journal of Cancer*, *84*, 309–314.

Li, J. F., Chu, Y. W., Wang, G. M., Zhu, T. Y., Rong, R. M., Hou, J., & Xu, M. (2009). The prognostic value of peritumoral regulatory T cells and its correlation with intratumoral cyclooxygenase-2 expression in clear cell renal cell carcinoma. *BJU International*, *103*, 399–405.

Li, J., Feng, G., Liu, J., Rong, R., Luo, F., Guo, L., … Chu, Y. (2010). Renal cell carcinoma may evade the immune system by converting CD4+Foxp3-T cells into CD4+CD25+Foxp3+ regulatory T cells: Role of tumor COX-2-derived PGE2. *Molecular Medicine Reports*, *3*, 959–963.

Li, Y., Guo, Z., Liu, C. F., Xing, W. G., Si, T. G., Liu, F., … Xing, J. Z. (2012). Effect of transcatheter renal arterial embolization combined with cryoablation on regulatory CD4+CD25+ T lymphocytes in the peripheral blood of patients with advanced renal carcinoma. *Cryobiology*, *65*, 56–59.

Li, Y., Qiu, S., Fan, J., Gao, Q., Zhou, J., Xiao, Y., ... Huang, X. (2009). Tumor-infiltrating macrophages can predict favorable prognosis in hepatocellular carcinoma after resection. *Journal of Cancer Research and Clinical Oncology, 135*, 439–449.

Li, L., Yang, C., Zhao, Z., Xu, B., Zheng, M., Zhang, C., ... Rong, R. (2015). Skewed T-helper (Th)1/2- and Th17/T regulatory-cell balances in patients with renal cell carcinoma. *Molecular Medicine Reports, 11*, 947–953.

Lièvre, A., Bachet, J., Le Corre, D., Boige, V., Landi, B., Emile, J., ... Laurent-Puig, P. (2006). KRAS mutation status is predictive of response to cetuximab therapy in colorectal cancer. *Cancer Research, 66*, 3992–3995.

Lijun, Z., Xin, Z., Danhua, S., Xiaoping, L., Jianliu, W., Huilan, W., & Lihui, W. (2012). Tumor-infiltrating dendritic cells may be used as clinicopathologic prognostic factors in endometrial carcinoma. *International Journal of Gynecological Cancer, 22*, 836–841.

Lim, S. H., Chua, W., Cheng, C., Descallar, J., Ng, W., Solomon, M., ... Lee, C. S. (2014). Effect of neoadjuvant chemoradiation on tumor-infiltrating/associated lymphocytes in locally advanced rectal cancers. *Anticancer Research, 34*(11), 6505–6513.

Lin, E. Y., Li, J., Bricard, G., Wang, W., Deng, Y., Sellers, R., ... Pollard, J. W. (2007). Vascular endothelial growth factor restores delayed tumor progression in tumors depleted of macrophages. *Molecular Oncology, 1*, 288–302.

Lin, E. Y., Li, J., Gnatovskiy, L., Deng, Y., Zhu, L., Grzesik, D. A., ... Pollard, J. W. (2006). Macrophages regulate the angiogenic switch in a mouse model of breast cancer. *Cancer Research, 66*, 11238–11246.

Ling, A., Edin, S., Wikberg, M. L., Öberg, Å., & Palmqvist, R. (2014). The intratumoural subsite and relation of CD8(+) and FOXP3(+) T lymphocytes in colorectal cancer provide important prognostic clues. *British Journal of Cancer, 110*, 2551–2559.

Ling, W., Rayman, P., Uzzo, R., Clark, P., Kim, H. J., Tubbs, R., ... Finke, J. (1998). Impaired activation of NFkappaB in T cells from a subset of renal cell carcinoma patients is mediated by inhibition of phosphorylation and degradation of the inhibitor, IkappaBalpha. *Blood, 92*, 1334–1341.

Liotta, F., Gacci, M., Frosali, F., Querci, V., Vittori, G., Lapini, A., ... Annunziato, F. (2011). Frequency of regulatory T cells in peripheral blood and in tumour-infiltrating lymphocytes correlates with poor prognosis in renal cell carcinoma. *BJU International, 107*, 1500–1506.

Liu, S., Chang, L., Pan, L., Hung, Y., Lee, C., & Shieh, Y. (2008). Clinicopathologic significance of tumor cell-lined vessel and microenvironment in oral squamous cell carcinoma. *Oral Oncology, 44*, 277–285.

Liu, S., Lachapelle, J., Leung, S., Gao, D., Foulkes, W. D., & Nielsen, T. O. (2012). CD8+ lymphocyte infiltration is an independent favorable prognostic indicator in basal-like breast cancer. *Breast Cancer Research, 14*, R48.

Llosa, N. J., Cruise, M., Tam, A., Wicks, E. C., Hechenbleikner, E. M., Taube, J. M., ... Housseau, F. (2015). The vigorous immune microenvironment of microsatellite instable colon cancer is balanced by multiple counter-inhibitory checkpoints. *Cancer Discovery, 5*, 43–51.

Lohr, M., Edlund, K., Botling, J., Hammad, S., Hellwig, B., Othman, A., ... Micke, P. (2013). The prognostic relevance of tumour-infiltrating plasma cells and immunoglobulin kappa C indicates an important role of the humoral immune response in non-small cell lung cancer. *Cancer Letters, 333*, 222–228.

Loi, S., Haibe-Kains, B., Desmedt, C., Lallemand, F., Tutt, A. M., Gillet, C., ... Sotiriou, C. (2007). Definition of clinically distinct molecular subtypes in estrogen receptor-positive breast carcinomas through genomic grade. *Journal of Clinical Oncology, 25*, 1239–1246.

Loke, P., & Allison, J. P. (2003). PD-L1 and PD-L2 are differentially regulated by Th1 and Th2 cells. *Proceedings of the National Academy of Sciences of the United States of America, 100*, 5336–5341.

Long, H. M., Parsonage, G., Fox, C. P., & Lee, S. P. (2010). Immunotherapy for Epstein-Barr virus-associated malignancies. *Drug News & Perspectives*, *23*, 221–228.

Lu, L., Pan, K., Zheng, H., Li, J., Qiu, H., Zhao, J., … Xia, J. (2013). IL-17A promotes immune cell recruitment in human esophageal cancers and the infiltrating dendritic cells represent a positive prognostic marker for patient survival. *Journal of Immunotherapy*, *36*, 451–458.

Ma, J., Liu, L., Che, G., Yu, N., Dai, F., & You, Z. (2010). The M1 form of tumor-associated macrophages in non-small cell lung cancer is positively associated with survival time. *BMC Cancer*, *10*, 112.

Mackay, A., Weigelt, B., Grigoriadis, A., Kreike, B., Natrajan, R., A'Hern, R., … Reis-Filho, J. S. (2011). Microarray-based class discovery for molecular classification of breast cancer: Analysis of interobserver agreement. *Journal of the National Cancer Institute*, *103*, 662–673.

Maeurer, M. J., Martin, D. M., Castelli, C., Elder, E., Leder, G., Storkus, W. J., & Lotze, M. T. (1995). Host immune response in renal cell cancer: Interleukin-4 (IL-4) and IL-10 mRNA are frequently detected in freshly collected tumor-infiltrating lymphocytes. *Cancer Immunology, Immunotherapy*, *41*, 111–121.

Mahmoud, S. M. A., Lee, A. H. S., Paish, E. C., Macmillan, R. D., Ellis, I. O., & Green, A. R. (2012a). The prognostic significance of B lymphocytes in invasive carcinoma of the breast. *Breast Cancer Research and Treatment*, *132*, 545–553.

Mahmoud, S. M. A., Lee, A. H. S., Paish, E. C., Macmillan, R. D., Ellis, I. O., & Green, A. R. (2012b). Tumour-infiltrating macrophages and clinical outcome in breast cancer. *Journal of Clinical Pathology*, *65*, 159–163.

Malinowski, K., Kono, K., Takayama, T., Terashima, T., Tsukuda, K., Waltzer, W., & Rapaport, F. T. (1997). Inhibition of lymphocyte proliferative responses by renal cell carcinoma extract. *Transplantation Proceedings*, *29*, 839–841.

Mamessier, E., Sylvain, A., Thibult, M., Houvenaeghel, G., Jacquemier, J., Castellano, R., … Olive, D. (2011). Human breast cancer cells enhance self tolerance by promoting evasion from NK cell antitumor immunity. *The Journal of Clinical Investigation*, *121*, 3609–3622.

Mantovani, A., Allavena, P., Sica, A., & Balkwill, F. (2008). Cancer-related inflammation. *Nature*, *454*, 436–444.

Marisa, L., de Reyniès, A., Duval, A., Selves, J., Gaub, M. P., Vescovo, L., … Boige, V. (2013). Gene expression classification of colon cancer into molecular subtypes: Characterization, validation, and prognostic value. *PLoS Medicine*, *10*, e1001453.

Martinet, L., Garrido, I., Filleron, T., Le Guellec, S., Bellard, E., Fournie, J., … Girard, J. (2011). Human solid tumors contain high endothelial venules: Association with T- and B-lymphocyte infiltration and favorable prognosis in breast cancer. *Cancer Research*, *71*, 5678–5687.

Mazanet, M. M., & Hughes, C. C. W. (2002). B7-H1 is expressed by human endothelial cells and suppresses T cell cytokine synthesis. *Journal of Immunology*, *169*, 3581–3588.

McMullen, T. P. W., Lai, R., Dabbagh, L., Wallace, T. M., & de Gara, C. J. (2010). Survival in rectal cancer is predicted by T cell infiltration of tumour-associated lymphoid nodules. *Clinical and Experimental Immunology*, *161*, 81–88.

Medrek, C., Pontén, F., Jirström, K., & Leandersson, K. (2012). The presence of tumor associated macrophages in tumor stroma as a prognostic marker for breast cancer patients. *BMC Cancer*, *12*, 306.

Menetrier-Caux, C., Montmain, G., Dieu, M. C., Bain, C., Favrot, M. C., Caux, C., & Blay, J. Y. (1998). Inhibition of the differentiation of dendritic cells from CD34(+) progenitors by tumor cells: Role of interleukin-6 and macrophage colony-stimulating factor. *Blood*, *92*, 4778–4791.

Menon, A. G., Janssen-van Rhijn, C. M., Morreau, H., Putter, H., Tollenaar, R. A. E. M., van de Velde, C. J. H., … Kuppen, P. J. K. (2004). Immune system and prognosis in colorectal cancer: A detailed immunohistochemical analysis. *Laboratory Investigation*, *84*, 493–501.

Meshcheryakova, A., Tamandl, D., Bajna, E., Stift, J., Mittlboeck, M., Svoboda, M., … Mechtcheriakova, D. (2014). B cells and ectopic follicular structures: Novel players in anti-tumor programming with prognostic power for patients with metastatic colorectal cancer. *PLoS One*, *9*, e99008.

Micheau, O., & Tschopp, J. (2003). Induction of TNF receptor I-mediated apoptosis via two sequential signaling complexes. *Cell*, *114*, 181–190.

Middel, P., Brauneck, S., Meyer, W., & Radzun, H. (2010). Chemokine-mediated distribution of dendritic cell subsets in renal cell carcinoma. *BMC Cancer*, *10*, 578.

Millrud, C. R., Månsson Kvarnhammar, A., Uddman, R., Björnsson, S., Riesbeck, K., & Cardell, L. O. (2012). The activation pattern of blood leukocytes in head and neck squamous cell carcinoma is correlated to survival. *PLoS One*, *7*, e51120.

Minegishi, Y., Rohrer, J., Coustan-Smith, E., Lederman, H. M., Pappu, R., Campana, D., … Conley, M. E. (1999). An essential role for BLNK in human B cell development. *Science*, *286*, 1954–1957.

Mlecnik, B., Tosolini, M., Kirilovsky, A., Berger, A., Bindea, G., Meatchi, T., … Galon, J. (2011). Histopathologic-based prognostic factors of colorectal cancers are associated with the state of the local immune reaction. *Journal of Clinical Oncology*, *29*, 610–618.

Mohammed, Z. M. A., Going, J. J., Edwards, J., Elsberger, B., & McMillan, D. C. (2013). The relationship between lymphocyte subsets and clinico-pathological determinants of survival in patients with primary operable invasive ductal breast cancer. *British Journal of Cancer*, *109*, 1676–1684.

Morais, C., Gobe, G., Johnson, D. W., & Healy, H. (2011). The emerging role of nuclear factor kappa B in renal cell carcinoma. *The International Journal of Biochemistry & Cell Biology*, *43*, 1537–1549.

Moulin, V., Andris, F., Thielemans, K., Maliszewski, C., Urbain, J., & Moser, M. (2000). B lymphocytes regulate dendritic cell (DC) function in vivo: Increased interleukin 12 production by DCs from B cell-deficient mice results in T helper cell type 1 deviation. *The Journal of Experimental Medicine*, *192*, 475–482.

Movahedi, K., Laoui, D., Gysemans, C., Baeten, M., Stangé, G., Van den Bossche, J., … Van Ginderachter, J. A. (2010). Different tumor microenvironments contain functionally distinct subsets of macrophages derived from Ly6C(high) monocytes. *Cancer Research*, *70*, 5728–5739.

Mrózek, K., Radmacher, M. D., Bloomfield, C. D., & Marcucci, G. (2009). Molecular signatures in acute myeloid leukemia. *Current Opinion in Hematology*, *16*, 64–69.

Nakagomi, H., Pisa, P., Pisa, E. K., Yamamoto, Y., Halapi, E., Backlin, K., … Kiessling, R. (1995). Lack of interleukin-2 (IL-2) expression and selective expression of IL-10 mRNA in human renal cell carcinoma. *International Journal of Cancer*, *63*, 366–371.

Nakakubo, Y., Miyamoto, M., Cho, Y., Hida, Y., Oshikiri, T., Suzuoki, M., … Katoh, H. (2003). Clinical significance of immune cell infiltration within gallbladder cancer. *British Journal of Cancer*, *89*, 1736–1742.

Nakano, O., Sato, M., Naito, Y., Suzuki, K., Orikasa, S., Aizawa, M., … Ohtani, H. (2001). Proliferative activity of intratumoral CD8(+) T-lymphocytes as a prognostic factor in human renal cell carcinoma: Clinicopathologic demonstration of antitumor immunity. *Cancer Research*, *61*, 5132–5136.

Nakayama, Y., Inoue, Y., Minagawa, N., Katsuki, T., Nagashima, N., Onitsuka, K., … Itoh, H. (2003). Relationships between S-100 protein-positive cells and clinicopathological factors in patients with colorectal cancer. *Anticancer Research*, *23*, 4423–4426.

Näsman, A., Romanitan, M., Nordfors, C., Grün, N., Johansson, H., Hammarstedt, L., … Ramqvist, T. (2012). Tumor infiltrating CD8+ and Foxp3+ lymphocytes correlate to clinical outcome and human papillomavirus (HPV) status in tonsillar cancer. *PLoS One*, *7*, e38711.

Neagu, M., Constantin, C., & Zurac, S. (2013). Immune parameters in the prognosis and therapy monitoring of cutaneous melanoma patients: Experience, role, and limitations. *BioMed Research International, 2013*, 107940.

Negrier, S., Perol, D., Menetrier-Caux, C., Escudier, B., Pallardy, M., Ravaud, A., … Blay, J. (2004). Interleukin-6, interleukin-10, and vascular endothelial growth factor in metastatic renal cell carcinoma: Prognostic value of interleukin-6—From the Groupe Francais d'Immunotherapie. *Journal of Clinical Oncology, 22*, 2371–2378.

Ness, N., Andersen, S., Valkov, A., Nordby, Y., Donnem, T., Al-Saad, S., … Richardsen, E. (2014). Infiltration of CD8+ lymphocytes is an independent prognostic factor of biochemical failure-free survival in prostate cancer. *Prostate, 74*, 1452–1461.

Nielsen, J. S., Sahota, R. A., Milne, K., Kost, S. E., Nesslinger, N. J., Watson, P. H., & Nelson, B. H. (2012). CD20+ tumor-infiltrating lymphocytes have an atypical CD27- memory phenotype and together with CD8+ T cells promote favorable prognosis in ovarian cancer. *Clinical Cancer Research, 18*, 3281–3292.

Noguchi, T., Kato, T., Wang, L., Maeda, Y., Ikeda, H., Sato, E., … Nishikawa, H. (2012). Intracellular tumor-associated antigens represent effective targets for passive immunotherapy. *Cancer Research, 72*, 1672–1682.

Noman, M. Z., Desantis, G., Janji, B., Hasmim, M., Karray, S., Dessen, P., … Chouaib, S. (2014). PD-L1 is a novel direct target of HIF-1α, and its blockade under hypoxia enhanced MDSC-mediated T cell activation. *The Journal of Experimental Medicine, 211*, 781–790.

Nomura, A., Stemmermann, G. N., Chyou, P. H., Kato, I., Perez-Perez, G. I., & Blaser, M. J. (1991). *Helicobacter pylori* infection and gastric carcinoma among Japanese Americans in Hawaii. *The New England Journal of Medicine, 325*, 1132–1136.

Nutt, C. L., Mani, D. R., Betensky, R. A., Tamayo, P., Cairncross, J. G., Ladd, C., … Louis, D. N. (2003). Gene expression-based classification of malignant gliomas correlates better with survival than histological classification. *Cancer Research, 63*, 1602–1607.

Nzula, S., Going, J. J., & Stott, D. I. (2003). Antigen-driven clonal proliferation, somatic hypermutation, and selection of B lymphocytes infiltrating human ductal breast carcinomas. *Cancer Research, 63*, 3275–3280.

Ochoa, A. C., Zea, A. H., Hernandez, C., & Rodriguez, P. C. (2007). Arginase, prostaglandins, and myeloid-derived suppressor cells in renal cell carcinoma. *Clinical Cancer Research, 13*, 721s–726s.

O'Connell, J., O'Sullivan, G. C., Collins, J. K., & Shanahan, F. (1996). The Fas counterattack: Fas-mediated T cell killing by colon cancer cells expressing Fas ligand. *The Journal of Experimental Medicine, 184*, 1075–1082.

Ohigashi, Y., Sho, M., Yamada, Y., Tsurui, Y., Hamada, K., Ikeda, N., … Nakajima, Y. (2005). Clinical significance of programmed death-1 ligand-1 and programmed death-1 ligand-2 expression in human esophageal cancer. *Clinical Cancer Research, 11*, 2947–2953.

Ohno, S., Inagawa, H., Dhar, D. K., Fujii, T., Ueda, S., Tachibana, M., … Nagasue, N. (2003). The degree of macrophage infiltration into the cancer cell nest is a significant predictor of survival in gastric cancer patients. *Anticancer Research, 23*, 5015–5022.

Ohno, S., Ohno, Y., Suzuki, N., Kamei, T., Koike, K., Inagawa, H., … Inoue, M. (2004). Correlation of histological localization of tumor-associated macrophages with clinicopathological features in endometrial cancer. *Anticancer Research, 24*, 3335–3342.

Ohri, C. M., Shikotra, A., Green, R. H., Waller, D. A., & Bradding, P. (2009). Macrophages within NSCLC tumour islets are predominantly of a cytotoxic M1 phenotype associated with extended survival. *The European Respiratory Journal, 33*, 118–126.

Ohtaki, Y., Ishii, G., Nagai, K., Ashimine, S., Kuwata, T., Hishida, T., … Ochiai, A. (2010). Stromal macrophage expressing CD204 is associated with tumor aggressiveness in lung adenocarcinoma. *Journal of Thoracic Oncology, 5*, 1507–1515.

Ohue, Y., Eikawa, S., Okazaki, N., Mizote, Y., Isobe, M., Uenaka, A., ... Nakayama, E. (2012). Spontaneous antibody, and CD4 and CD8 T-cell responses against XAGE-1b (GAGED2a) in non-small cell lung cancer patients. *International Journal of Cancer, 131*, E649–E658.

Okita, Y., Tanaka, H., Ohira, M., Muguruma, K., Kubo, N., Watanabe, M., ... Hirakawa, K. (2014). Role of tumor-infiltrating $CD11b^{+}$ antigen-presenting cells in the progression of gastric cancer. *The Journal of Surgical Research, 186*(1), 192–200. http://dx.doi.org/10.1016/j.jss.2013.08.024. Epub 2013 Sep 24.

Oldham, K. A., Parsonage, G., Bhatt, R. I., Wallace, D. M. A., Deshmukh, N., Chaudhri, S., ... Lee, S. P. (2012). T lymphocyte recruitment into renal cell carcinoma tissue: A role for chemokine receptors CXCR3, CXCR6, CCR5, and CCR6. *European Urology, 61*, 385–394.

Olkhanud, P. B., Damdinsuren, B., Bodogai, M., Gress, R. E., Sen, R., Wejksza, K., ... Biragyn, A. (2011). Tumor-evoked regulatory B cells promote breast cancer metastasis by converting resting $CD4^{+}$ T cells to T-regulatory cells. *Cancer Research, 71*, 3505–3515.

Olson, B. M., Frye, T. P., Johnson, L. E., Fong, L., Knutson, K. L., Disis, M. L., & McNeel, D. G. (2010). HLA-A2-restricted T-cell epitopes specific for prostatic acid phosphatase. *Cancer Immunology, Immunotherapy, 59*, 943–953.

Ong, C. W., Chong, P. Y., McArt, D. G., Chan, J. Y., Tan, H. T., Kumar, A. P., ... Salto-Tellez, M. (2015). The prognostic value of the stem-like group in colorectal cancer using a panel of immunohistochemistry markers. *Oncotarget, 6*, 12763–12773.

Onishi, T., Ohishi, Y., Imagawa, K., Ohmoto, Y., & Murata, K. (1999). An assessment of the immunological environment based on intratumoral cytokine production in renal cell carcinoma. *BJU International, 83*, 488–492.

Origoni, M., Parma, M., Dell'Antonio, G., Gelardi, C., Stefani, C., Salvatore, S., & Candiani, M. (2013). Prognostic significance of immunohistochemical phenotypes in patients treated for high-grade cervical intraepithelial neoplasia. *BioMed Research International, 2013*, 831907.

Osinsky, S., Bubnovskaya, L., Ganusevich, I., Kovelskaya, A., Gumenyuk, L., Olijnichenko, G., & Merentsev, S. (2011). Hypoxia, tumour-associated macrophages, microvessel density, VEGF and matrix metalloproteinases in human gastric cancer: Interaction and impact on survival. *Clinical and Translational Oncology, 13*, 133–138.

Oyama, T., Ran, S., Ishida, T., Nadaf, S., Kerr, L., Carbone, D. P., & Gabrilovich, D. I. (1998). Vascular endothelial growth factor affects dendritic cell maturation through the inhibition of nuclear factor-kappa B activation in hemopoietic progenitor cells. *Journal of Immunology, 160*, 1224–1232.

Palucka, K., & Banchereau, J. (2012). Cancer immunotherapy via dendritic cells. *Nature Reviews Cancer, 12*, 265–277.

Pantano, F., Berti, P., Guida, F. M., Perrone, G., Vincenzi, B., Amato, M. M. C., ... Santini, D. (2013). The role of macrophages polarization in predicting prognosis of radically resected gastric cancer patients. *Journal of Cellular and Molecular Medicine, 17*, 1415–1421.

Pape, J. W., Liautaud, B., Thomas, F., Mathurin, J. R., St Amand, M. M., Boncy, M., ... Johnson, W. D. J. (1983). Characteristics of the acquired immunodeficiency syndrome (AIDS) in Haiti. *The New England Journal of Medicine, 309*, 945–950.

Pappu, R., Cheng, A. M., Li, B., Gong, Q., Chiu, C., Griffin, N., ... Chan, A. C. (1999). Requirement for B cell linker protein (BLNK) in B cell development. *Science, 286*, 1949–1954.

Park, M. H., Lee, J. S., & Yoon, J. H. (2012). High expression of CX3CL1 by tumor cells correlates with a good prognosis and increased tumor-infiltrating CD8+ T cells, natural killer cells, and dendritic cells in breast carcinoma. *Journal of Surgical Oncology, 106*, 386–392.

Parker Harp, C. R., Archambault, A. S., Sim, J., Ferris, S. T., Mikesell, R. J., Koni, P. A., ... Wu, G. F. (2015). B cell antigen presentation is sufficient to drive neuroinflammation in an animal model of multiple sclerosis. *Journal of Immunology, 194*, 5077–5084.

Parsonnet, J., Friedman, G. D., Vandersteen, D. P., Chang, Y., Vogelman, J. H., Orentreich, N., & Sibley, R. K. (1991). *Helicobacter pylori* infection and the risk of gastric carcinoma. *The New England Journal of Medicine, 325*, 1127–1131.

Parsonnet, J., Hansen, S., Rodriguez, L., Gelb, A. B., Warnke, R. A., Jellum, E., ... Friedman, G. D. (1994). *Helicobacter pylori* infection and gastric lymphoma. *The New England Journal of Medicine, 330*, 1267–1271.

Paulson, K. G., Iyer, J. G., Simonson, W. T., Blom, A., Thibodeau, R. M., Schmidt, M., ... Nghiem, P. (2014). CD8+ lymphocyte intratumoral infiltration as a stage-independent predictor of Merkel cell carcinoma survival: A population-based study. *American Journal of Clinical Pathology, 142*, 452–458.

Pelletier, M. P., Edwardes, M. D., Michel, R. P., Halwani, F., & Morin, J. E. (2001). Prognostic markers in resectable non-small cell lung cancer: A multivariate analysis. *Canadian Journal of Surgery, 44*, 180–188.

Peri, S., Devarajan, K., Yang, D., Knudson, A. G., & Balachandran, S. (2013). Meta-analysis identifies NF-κB as a therapeutic target in renal cancer. *PLoS One, 8*, e76746.

Perou, C. M., Sørlie, T., Eisen, M. B., van de Rijn, M., Jeffrey, S. S., Rees, C. A., ... Botstein, D. (2000). Molecular portraits of human breast tumours. *Nature, 406*, 747–752.

Pichler, M., Hutterer, G. C., Stoeckigt, C., Chromecki, T. F., Stojakovic, T., Golbeck, S., ... Zigeuner, R. (2013). Validation of the pre-treatment neutrophil-lymphocyte ratio as a prognostic factor in a large European cohort of renal cell carcinoma patients. *British Journal of Cancer, 108*, 901–907.

Pietra, G., Manzini, C., Rivara, S., Vitale, M., Cantoni, C., Petretto, A., ... Mingari, M. C. (2012). Melanoma cells inhibit natural killer cell function by modulating the expression of activating receptors and cytolytic activity. *Cancer Research, 72*, 1407–1415.

Pinchuk, I. V., Saada, J. I., Beswick, E. J., Boya, G., Qiu, S. M., Mifflin, R. C., ... Powell, D. W. (2008). PD-1 ligand expression by human colonic myofibroblasts/fibroblasts regulates CD4+ T-cell activity. *Gastroenterology, 135*, 1228–1237. 1237.e1–2.

Pino, M. S., & Chung, D. C. (2010). The chromosomal instability pathway in colon cancer. *Gastroenterology, 138*, 2059–2072.

Platonova, S., Cherfils-Vicini, J., Damotte, D., Crozet, L., Vieillard, V., Validire, P., ... Cremer, I. (2011). Profound coordinated alterations of intratumoral NK cell phenotype and function in lung carcinoma. *Cancer Research, 71*, 5412–5422.

Polimeno, M., Napolitano, M., Costantini, S., Portella, L., Esposito, A., Capone, F., ... Scala, S. (2013). Regulatory T cells, interleukin (IL)-6, IL-8, vascular endothelial growth factor (VEGF), CXCL10, CXCL11, epidermal growth factor (EGF) and hepatocyte growth factor (HGF) as surrogate markers of host immunity in patients with renal cell carcinoma. *BJU International, 112*, 686–696.

Powles, T., Eder, J. P., Fine, G. D., Braiteh, F. S., Loriot, Y., Cruz, C., ... Vogelzang, N. J. (2014). MPDL3280A (anti-PD-L1) treatment leads to clinical activity in metastatic bladder cancer. *Nature, 515*, 558–562.

Preston, C. C., Maurer, M. J., Oberg, A. L., Visscher, D. W., Kalli, K. R., Hartmann, L. C., ... Knutson, K. L. (2013). The ratios of CD8+ T cells to CD4+CD25+ FOXP3+ and FOXP3- T cells correlate with poor clinical outcome in human serous ovarian cancer. *PLoS One, 8*, e80063.

Prinz, P. U., Mendler, A. N., Brech, D., Masouris, I., Oberneder, R., & Noessner, E. (2014). NK-cell dysfunction in human renal carcinoma reveals diacylglycerol kinase as key regulator and target for therapeutic intervention. *International Journal of Cancer, 135*, 1832–1841.

Puccetti, L., Manetti, R., Parronchi, P., Piccinni, M., Mavilia, C., Carini, M., ... Maggi, E. (2002). Role of low nuclear grading of renal carcinoma cells in the functional profile of tumor-infiltrating T cells. *International Journal of Cancer, 98*, 674–681.

Pyonteck, S. M., Akkari, L., Schuhmacher, A. J., Bowman, R. L., Sevenich, L., Quail, D. F., ... Joyce, J. A. (2013). CSF-1R inhibition alters macrophage polarization and blocks glioma progression. *Nature Medicine, 19*, 1264–1272.

Rabinovich, G. A., Gabrilovich, D., & Sotomayor, E. M. (2007). Immunosuppressive strategies that are mediated by tumor cells. *Annual Review of Immunology, 25*, 267–296.

Rathore, A. S., Kumar, S., Konwar, R., Makker, A., Negi, M. P. S., & Goel, M. M. (2014). CD3+, CD4+ & CD8+ tumour infiltrating lymphocytes (TILs) are predictors of favourable survival outcome in infiltrating ductal carcinoma of breast. *The Indian Journal of Medical Research, 140*, 361–369.

Rayman, P., Uzzo, R. G., Kolenko, V., Bloom, T., Cathcart, M. K., Molto, L., ... Finke, J. H. (2000). Tumor-induced dysfunction in interleukin-2 production and interleukin-2 receptor signaling: A mechanism of immune escape. *The Cancer Journal from Scientific American, 6*(Suppl. 1), S81–S87.

Rayman, P., Wesa, A. K., Richmond, A. L., Das, T., Biswas, K., Raval, G., ... Finke, J. (2004). Effect of renal cell carcinomas on the development of type 1 T-cell responses. *Clinical Cancer Research, 10*, 6360S–6366S.

Rebouissou, S., Amessou, M., Couchy, G., Poussin, K., Imbeaud, S., Pilati, C., ... Zucman-Rossi, J. (2009). Frequent in-frame somatic deletions activate gp130 in inflammatory hepatocellular tumours. *Nature, 457*, 200–204.

Reichert, T. E., Scheuer, C., Day, R., Wagner, W., & Whiteside, T. L. (2001). The number of intratumoral dendritic cells and zeta-chain expression in T cells as prognostic and survival biomarkers in patients with oral carcinoma. *Cancer, 91*, 2136–2147.

Remark, R., Alifano, M., Cremer, I., Lupo, A., Dieu-Nosjean, M., Riquet, M., ... Damotte, D. (2013). Characteristics and clinical impacts of the immune environments in colorectal and renal cell carcinoma lung metastases: Influence of tumor origin. *Clinical Cancer Research, 19*, 4079–4091.

Reuschenbach, M., von Knebel Doeberitz, M., & Wentzensen, N. (2009). A systematic review of humoral immune responses against tumor antigens. *Cancer Immunology, Immunotherapy, 58*, 1535–1544.

Riker, A., Cormier, J., Panelli, M., Kammula, U., Wang, E., Abati, A., ... Marincola, F. (1999). Immune selection after antigen-specific immunotherapy of melanoma. *Surgery, 126*, 112–120.

Rivera, L. B., & Bergers, G. (2015). Intertwined regulation of angiogenesis and immunity by myeloid cells. *Trends in Immunology, 36*, 240–249.

Robbins, P. F., El-Gamil, M., Li, Y. F., Kawakami, Y., Loftus, D., Appella, E., & Rosenberg, S. A. (1996). A mutated beta-catenin gene encodes a melanoma-specific antigen recognized by tumor infiltrating lymphocytes. *The Journal of Experimental Medicine, 183*, 1185–1192.

Rodriguez, P. C., Ernstoff, M. S., Hernandez, C., Atkins, M., Zabaleta, J., Sierra, R., & Ochoa, A. C. (2009). Arginase I-producing myeloid-derived suppressor cells in renal cell carcinoma are a subpopulation of activated granulocytes. *Cancer Research, 69*, 1553–1560.

Rodriguez, P. C., Hernandez, C. P., Quiceno, D., Dubinett, S. M., Zabaleta, J., Ochoa, J. B., ... Ochoa, A. C. (2005). Arginase I in myeloid suppressor cells is induced by COX-2 in lung carcinoma. *The Journal of Experimental Medicine, 202*, 931–939.

Roepman, P., Schlicker, A., Tabernero, J., Majewski, I., Tian, S., Moreno, V., ... Simon, I. M. (2014). Colorectal cancer intrinsic subtypes predict chemotherapy benefit, deficient mismatch repair and epithelial-to-mesenchymal transition. *International Journal of Cancer, 134*, 552–562.

Romero, J. M., Aptsiauri, N., Vazquez, F., Cozar, J. M., Canton, J., Cabrera, T., ... Ruiz-Cabello, F. (2006). Analysis of the expression of HLA class I, proinflammatory cytokines and chemokines in primary tumors from patients with localized and metastatic renal cell carcinoma. *Tissue Antigens, 68*, 303–310.

Rosenwald, A., Wright, G., Chan, W. C., Connors, J. M., Campo, E., Fisher, R. I., ... Staudt, L. M. (2002). The use of molecular profiling to predict survival after chemotherapy for diffuse large-B-cell lymphoma. *The New England Journal of Medicine, 346*, 1937–1947.

Rosser, E. C., & Mauri, C. (2015). Regulatory B cells: Origin, phenotype, and function. *Immunity, 42*, 607–612.

Rubtsov, Y. P., & Rudensky, A. Y. (2007). TGFbeta signalling in control of T-cell-mediated self-reactivity. *Nature Reviews Immunology, 7*, 443–453.

Rusakiewicz, S., Semeraro, M., Sarabi, M., Desbois, M., Locher, C., Mendez, R., ... Zitvogel, L. (2013). Immune infiltrates are prognostic factors in localized gastrointestinal stromal tumors. *Cancer Research, 73*, 3499–3510.

Ryder, M., Ghossein, R. A., Ricarte-Filho, J. C. M., Knauf, J. A., & Fagin, J. A. (2008). Increased density of tumor-associated macrophages is associated with decreased survival in advanced thyroid cancer. *Endocrine-Related Cancer, 15*, 1069–1074.

Sadanandam, A., Lyssiotis, C. A., Homicsko, K., Collisson, E. A., Gibb, W. J., Wullschleger, S., ... Hanahan, D. (2013). A colorectal cancer classification system that associates cellular phenotype and responses to therapy. *Nature Medicine, 19*, 619–625.

Salama, N. R., Hartung, M. L., & Müller, A. (2013). Life in the human stomach: Persistence strategies of the bacterial pathogen *Helicobacter pylori*. *Nature Reviews Microbiology, 11*, 385–399.

Salmon, H., Franciszkiewicz, K., Damotte, D., Dieu-Nosjean, M., Validire, P., Trautmann, A., ... Donnadieu, E. (2012). Matrix architecture defines the preferential localization and migration of T cells into the stroma of human lung tumors. *The Journal of Clinical Investigation, 122*, 899–910.

Salvesen, H. B., & Akslen, L. A. (1999). Significance of tumour-associated macrophages, vascular endothelial growth factor and thrombospondin-1 expression for tumour angiogenesis and prognosis in endometrial carcinomas. *International Journal of Cancer, 84*, 538–543.

Sandel, M. H., Dadabayev, A. R., Menon, A. G., Morreau, H., Melief, C. J. M., Offringa, R., ... Kuppen, P. J. K. (2005). Prognostic value of tumor-infiltrating dendritic cells in colorectal cancer: Role of maturation status and intratumoral localization. *Clinical Cancer Research, 11*, 2576–2582.

Sato, E., Olson, S. H., Ahn, J., Bundy, B., Nishikawa, H., Qian, F., ... Odunsi, K. (2005). Intraepithelial CD8+ tumor-infiltrating lymphocytes and a high CD8+/regulatory T cell ratio are associated with favorable prognosis in ovarian cancer. *Proceedings of the National Academy of Sciences of the United States of America, 102*, 18538–18543.

Sato, Y., Yoshizato, T., Shiraishi, Y., Maekawa, S., Okuno, Y., Kamura, T., ... Ogawa, S. (2013). Integrated molecular analysis of clear-cell renal cell carcinoma. *Nature Genetics, 45*, 860–867.

Schleypen, J. S., Baur, N., Kammerer, R., Nelson, P. J., Rohrmann, K., Gröne, E. F., ... Noessner, E. (2006). Cytotoxic markers and frequency predict functional capacity of natural killer cells infiltrating renal cell carcinoma. *Clinical Cancer Research, 12*, 718–725.

Schlicker, A., Beran, G., Chresta, C. M., McWalter, G., Pritchard, A., Weston, S., ... Wessels, L. F. A. (2012). Subtypes of primary colorectal tumors correlate with response to targeted treatment in colorectal cell lines. *BMC Medical Genomics, 5*, 66.

Schmidt, M., Böhm, D., von Törne, C., Steiner, E., Puhl, A., Pilch, H., ... Gehrmann, M. (2008). The humoral immune system has a key prognostic impact in node-negative breast cancer. *Cancer Research, 68*, 5405–5413.

Schoof, D. D., Terashima, Y., Peoples, G. E., Goedegebuure, P. S., Andrews, J. V., Richie, J. P., & Eberlein, T. J. (1993). CD4+ T cell clones isolated from human renal cell carcinoma possess the functional characteristics of Th2 helper cells. *Cellular Immunology, 150*, 114–123.

Schreiber, T. H., & Podack, E. R. (2009). A critical analysis of the tumour immunosurveillance controversy for 3-MCA-induced sarcomas. *British Journal of Cancer, 101*, 381–386.

Séguier, S., Tartour, E., Guérin, C., Couty, L., Lemitre, M., Lallement, L., … Coulomb, B. (2013). Inhibition of the differentiation of monocyte-derived dendritic cells by human gingival fibroblasts. *PLoS One, 8*, e70937.

Seo, A. N., Lee, H. J., Kim, E. J., Kim, H. J., Jang, M. H., Lee, H. E., … Park, S. Y. (2013). Tumour-infiltrating CD8+ lymphocytes as an independent predictive factor for pathological complete response to primary systemic therapy in breast cancer. *British Journal of Cancer, 109*, 2705–2713.

Serreze, D. V., Fleming, S. A., Chapman, H. D., Richard, S. D., Leiter, E. H., & Tisch, R. M. (1998). B lymphocytes are critical antigen-presenting cells for the initiation of T cell-mediated autoimmune diabetes in nonobese diabetic mice. *Journal of Immunology, 161*, 3912–3918.

Shai, R., Shi, T., Kremen, T. J., Horvath, S., Liau, L. M., Cloughesy, T. F., … Nelson, S. F. (2003). Gene expression profiling identifies molecular subtypes of gliomas. *Oncogene, 22*, 4918–4923.

Shalapour, S., Font-Burgada, J., Di Caro, G., Zhong, Z., Sanchez-Lopez, E., Dhar, D., … Karin, M. (2015). Immunosuppressive plasma cells impede T-cell-dependent immunogenic chemotherapy. *Nature, 521*, 94–98.

Shankaran, V., Ikeda, H., Bruce, A. T., White, J. M., Swanson, P. E., Old, L. J., & Schreiber, R. D. (2001). IFNgamma and lymphocytes prevent primary tumour development and shape tumour immunogenicity. *Nature, 410*, 1107–1111.

Sharpe, K., Stewart, G. D., Mackay, A., Van Neste, C., Rofe, C., Berney, D., … Powles, T. (2013). The effect of VEGF-targeted therapy on biomarker expression in sequential tissue from patients with metastatic clear cell renal cancer. *Clinical Cancer Research, 19*, 6924–6934.

Shchors, K., Shchors, E., Rostker, F., Lawlor, E. R., Brown-Swigart, L., & Evan, G. I. (2006). The Myc-dependent angiogenic switch in tumors is mediated by interleukin 1beta. *Genes & Development, 20*, 2527–2538.

Shi, J., Gao, Q., Wang, Z., Zhou, J., Wang, X., Min, Z., … Fan, J. (2013). Margin-infiltrating CD20(+) B cells display an atypical memory phenotype and correlate with favorable prognosis in hepatocellular carcinoma. *Clinical Cancer Research, 19*, 5994–6005.

Shigeoka, M., Urakawa, N., Nakamura, T., Nishio, M., Watajima, T., Kuroda, D., … Yokozaki, H. (2013). Tumor associated macrophage expressing CD204 is associated with tumor aggressiveness of esophageal squamous cell carcinoma. *Cancer Science, 104*, 1112–1119.

Shih, C., Padhy, L. C., Murray, M., & Weinberg, R. A. (1981). Transforming genes of carcinomas and neuroblastomas introduced into mouse fibroblasts. *Nature, 290*, 261–264.

Shimabukuro, T., & Naito, K. (2008). Tumor-infiltrating lymphocytes derived from human renal cell carcinoma: Clonal analysis of its characteristics. *International Journal of Urology, 15*, 241–244.

Shimabukuro-Vornhagen, A., Schlößer, H. A., Gryschok, L., Malcher, J., Wennhold, K., Garcia-Marquez, M., … von Bergwelt-Baildon, M. S. (2014). Characterization of tumor-associated B-cell subsets in patients with colorectal cancer. *Oncotarget, 5*, 4651–4664.

Shimura, S., Yang, G., Ebara, S., Wheeler, T. M., Frolov, A., & Thompson, T. C. (2000). Reduced infiltration of tumor-associated macrophages in human prostate cancer: Association with cancer progression. *Cancer Research, 60*, 5857–5861.

Shinto, E., Hase, K., Hashiguchi, Y., Sekizawa, A., Ueno, H., Shikina, A., … Yamamoto, J. (2014). CD8+ and FOXP3+ tumor-infiltrating T cells before and after chemoradiotherapy for rectal cancer. *Annals of Surgical Oncology, 21*(Suppl. 3), S414–S421.

Shojaei, F., Wu, X., Zhong, C., Yu, L., Liang, X., Yao, J., … Ferrara, N. (2007). Bv8 regulates myeloid-cell-dependent tumour angiogenesis. *Nature, 450*, 825–831.

Sievers, E., Dreimuller, P., Haferkamp, A., Schmidt-Wolf, I. G. H., Buchler, M. W., Schmidt, J., & Marten, A. (2007). Characterization of primary renal carcinoma cultures. *Urologia Internationalis, 79*, 235–243.

Silva, F. P. G., Swagemakers, S. M. A., Erpelinck-Verschueren, C., Wouters, B. J., Delwel, R., Vrieling, H., … Giphart-Gassler, M. (2009). Gene expression profiling of minimally differentiated acute myeloid leukemia: M0 is a distinct entity subdivided by RUNX1 mutation status. *Blood, 114*, 3001–3007.

Simonetti, O., Goteri, G., Lucarini, G., Rubini, C., Stramazzotti, D., Lo Muzio, L., … Offidani, A. (2007). In melanoma changes of immature and mature dendritic cell expression correlate with tumor thickness: An immunohistochemical study. *International Journal of Immunopathology and Pharmacology, 20*, 325–333.

Simsa, P., Teillaud, J., Stott, D. I., Tóth, J., & Kotlan, B. (2005). Tumor-infiltrating B cell immunoglobulin variable region gene usage in invasive ductal breast carcinoma. *Pathology Oncology Research, 11*, 92–97.

Sittig, S. P., Køllgaard, T., Grønbæk, K., Idorn, M., Hennenlotter, J., Stenzl, A., … Thor Straten, P. (2013). Clonal expansion of renal cell carcinoma-infiltrating T lymphocytes. *Oncoimmunology, 2*, e26014.

Sjödahl, G., Lövgren, K., Lauss, M., Chebil, G., Patschan, O., Gudjonsson, S., … Höglund, M. (2014). Infiltration of $CD3^+$ and $CD68^+$ cells in bladder cancer is subtype specific and affects the outcome of patients with muscle-invasive tumors. *Urologic Oncology, 32*, 791–797.

Smyth, G. P., Stapleton, P. P., Barden, C. B., Mestre, J. R., Freeman, T. A., Duff, M. D., … Daly, J. M. (2003). Renal cell carcinoma induces prostaglandin E2 and T-helper type 2 cytokine production in peripheral blood mononuclear cells. *Annals of Surgical Oncology, 10*, 455–462.

Sobin, L. H., Gospodarowicz, M. K., & Wittekind, C. (2009). *TNM classification of malignant tumours*. Wiley-Blackwell.

Song, E. Y., Shurin, M. R., Tourkova, I. L., Chatta, G., & Shurin, G. V. (2004). Human renal cell carcinoma inhibits dendritic cell maturation and functions. *Der Urologe. Ausg. A, 43*(Suppl. 3), S128–S130.

Sørlie, T., Perou, C. M., Tibshirani, R., Aas, T., Geisler, S., Johnsen, H., … Børresen-Dale, A. L. (2001). Gene expression patterns of breast carcinomas distinguish tumor subclasses with clinical implications. *Proceedings of the National Academy of Sciences of the United States of America, 98*, 10869–10874.

Soucek, L., Lawlor, E. R., Soto, D., Shchors, K., Swigart, L. B., & Evan, G. I. (2007). Mast cells are required for angiogenesis and macroscopic expansion of Myc-induced pancreatic islet tumors. *Nature Medicine, 13*, 1211–1218.

Sparmann, A., & Bar-Sagi, D. (2004). Ras-induced interleukin-8 expression plays a critical role in tumor growth and angiogenesis. *Cancer Cell, 6*, 447–458.

Sripa, B., Brindley, P. J., Mulvenna, J., Laha, T., Smout, M. J., Mairiang, E., … Loukas, A. (2012). The tumorigenic liver fluke Opisthorchis viverrini—Multiple pathways to cancer. *Trends in Parasitology, 28*, 395–407.

Steffens, S., Köhler, A., Rudolph, R., Eggers, H., Seidel, C., Janssen, M., ... Schrader, A. J. (2012). Validation of CRP as prognostic marker for renal cell carcinoma in a large series of patients. *BMC Cancer, 12*, 399.

Stratton, M. R., Campbell, P. J., & Futreal, P. A. (2009). The cancer genome. *Nature, 458*, 719–724.

Stutman, O. (1974). Tumor development after 3-methylcholanthrene in immunologically deficient athymic-nude mice. *Science, 183*, 534–536.

Stutman, O. (1979). Chemical carcinogenesis in nude mice: Comparison between nude mice from homozygous matings and heterozygous matings and effect of age and carcinogen dose. *Journal of the National Cancer Institute, 62*, 353–358.

Sugimoto, M., Mitsunaga, S., Yoshikawa, K., Kato, Y., Gotohda, N., Takahashi, S., ... Kaneko, H. (2014). Prognostic impact of M2 macrophages at neural invasion in patients with invasive ductal carcinoma of the pancreas. *European Journal of Cancer, 50*, 1900–1908.

Sugimura, K., Miyata, H., Tanaka, K., Takahashi, T., Kurokawa, Y., Yamasaki, M., ... Doki, Y. (2015). High infiltration of tumor-associated macrophages is associated with a poor response to chemotherapy and poor prognosis of patients undergoing neoadjuvant chemotherapy for esophageal cancer. *Journal of Surgical Oncology, 111*, 752–759.

Surget, S., Khoury, M. P., & Bourdon, J. (2013). Uncovering the role of p53 splice variants in human malignancy: A clinical perspective. *OncoTargets and Therapy*, 7, 57–68.

Sznurkowski, J. J., Zawrocki, A., & Biernat, W. (2014). Subtypes of cytotoxic lymphocytes and natural killer cells infiltrating cancer nests correlate with prognosis in patients with vulvar squamous cell carcinoma. *Cancer Immunology, Immunotherapy, 63*, 297–303.

Takahashi, H., Ogata, H., Nishigaki, R., Broide, D. H., & Karin, M. (2010). Tobacco smoke promotes lung tumorigenesis by triggering IKKbeta- and JNK1-dependent inflammation. *Cancer Cell, 17*, 89–97.

Takanami, I., Takeuchi, K., & Giga, M. (2001). The prognostic value of natural killer cell infiltration in resected pulmonary adenocarcinoma. *The Journal of Thoracic and Cardiovascular Surgery, 121*, 1058–1063.

Taketomi, A., Shimada, M., Shirabe, K., Kajiyama, K., Gion, T., & Sugimachi, K. (1998). Natural killer cell activity in patients with hepatocellular carcinoma: A new prognostic indicator after hepatectomy. *Cancer, 83*, 58–63.

Takeuchi, H., Maehara, Y., Tokunaga, E., Koga, T., Kakeji, Y., & Sugimachi, K. (2001). Prognostic significance of natural killer cell activity in patients with gastric carcinoma: A multivariate analysis. *The American Journal of Gastroenterology, 96*, 574–578.

Tan, T., & Coussens, L. M. (2007). Humoral immunity, inflammation and cancer. *Current Opinion in Immunology, 19*, 209–216.

Tan, W., Hildebrandt, M. A. T., Pu, X., Huang, M., Lin, J., Matin, S. F., ... Wu, X. (2011). Role of inflammatory related gene expression in clear cell renal cell carcinoma development and clinical outcomes. *The Journal of Urology, 186*, 2071–2077.

Tang, Y., Xu, X., Guo, S., Zhang, C., Tang, Y., Tian, Y., ... Wang, H. (2014). An increased abundance of tumor-infiltrating regulatory T cells is correlated with the progression and prognosis of pancreatic ductal adenocarcinoma. *PLoS One, 9*, e91551.

Tao, H., Lu, L., Xia, Y., Dai, F., Wang, Y., Bao, Y., ... Chang, A. E. (2015). Antitumor effector B cells directly kill tumor cells via the Fas/FasL pathway and are regulated by IL-10. *European Journal of Immunology, 45*, 999–1009.

Tartour, E., Gey, A., Sastre-Garau, X., Lombard Surin, I., Mosseri, V., & Fridman, W. H. (1998). Prognostic value of intratumoral interferon gamma messenger RNA expression in invasive cervical carcinomas. *Journal of the National Cancer Institute, 90*, 287–294.

Tartour, E., Latour, S., Mathiot, C., Thiounn, N., Mosseri, V., Joyeux, I., ... Fridman, W. H. (1995). Variable expression of CD3-zeta chain in tumor-infiltrating

lymphocytes (TIL) derived from renal-cell carcinoma: Relationship with TIL phenotype and function. *International Journal of Cancer, 63*, 205–212.
Tas, M. P., Simons, P. J., Balm, F. J., & Drexhage, H. A. (1993). Depressed monocyte polarization and clustering of dendritic cells in patients with head and neck cancer: In vitro restoration of this immunosuppression by thymic hormones. *Cancer Immunology, Immunotherapy, 36*, 108–114.
Tatsumi, T., Herrem, C. J., Olson, W. C., Finke, J. H., Bukowski, R. M., Kinch, M. S., … Storkus, W. J. (2003). Disease stage variation in CD4+ and CD8+ T-cell reactivity to the receptor tyrosine kinase EphA2 in patients with renal cell carcinoma. *Cancer Research, 63*, 4481–4489.
Tatsumi, T., Kierstead, L. S., Ranieri, E., Gesualdo, L., Schena, F. P., Finke, J. H., … Storkus, W. J. (2002). Disease-associated bias in T helper type 1 (Th1)/Th2 CD4 (+) T cell responses against MAGE-6 in HLA-DRB10401(+) patients with renal cell carcinoma or melanoma. *The Journal of Experimental Medicine, 196*, 619–628.
Taube, J. M., Anders, R. A., Young, G. D., Xu, H., Sharma, R., McMiller, T. L., … Chen, L. (2012). Colocalization of inflammatory response with B7-h1 expression in human melanocytic lesions supports an adaptive resistance mechanism of immune escape. *Science Translational Medicine, 4*, 127ra37.
Taube, J. M., Klein, A., Brahmer, J. R., Xu, H., Pan, X., Kim, J. H., … Anders, R. A. (2014). Association of PD-1, PD-1 Ligands, and Other Features of the Tumor Immune Microenvironment with Response to Anti-PD-1 Therapy. *Clinical Cancer Research, 20*, 5064–5074.
Teng, L., Chen, Y., Ding, D., Dai, H., Liu, G., & Li, C. (2014). Immunosuppressive effect of renal cell carcinoma on phenotype and function of dendritic cells. *International Urology and Nephrology, 46*, 915–920.
Teng, M. W. L., Galon, J., Fridman, W., & Smyth, M. J. (2015). From mice to humans: Developments in cancer immunoediting. *The Journal of Clinical Investigation, 125*, 3338–3346.
Terme, M., Pernot, S., Marcheteau, E., Sandoval, F., Benhamouda, N., Colussi, O., … Taieb, J. (2013). VEGFA-VEGFR pathway blockade inhibits tumor-induced regulatory T-cell proliferation in colorectal cancer. *Cancer Research, 73*, 539–549.
Tewari, N., Zaitoun, A. M., Arora, A., Madhusudan, S., Ilyas, M., & Lobo, D. N. (2013). The presence of tumour-associated lymphocytes confers a good prognosis in pancreatic ductal adenocarcinoma: An immunohistochemical study of tissue microarrays. *BMC Cancer, 13*, 436. http://dx.doi.org/10.1186/1471-2407-13-436.
The International Agency for Research on Cancer. (2000). *Pathology and genetics of tumours of the digestive system*. Lyon: IARC Press.
The International Agency for Research on Cancer. (2003). In IARC WHO Classification of Tumours (Ed.), *Pathology and genetics of tumours of the breast and female genital organs*: Lyon, France: World Health Organization; IARC Press.
The International Agency for Research on Cancer. (2004a). In IARC WHO Classification of Tumours (Ed.), *Pathology and genetics of tumours of the lung, pleura, thymus and heart*: Lyon, France: World Health Organization; IARC Press.
The International Agency for Research on Cancer. (2004b). In IARC WHO Classification of Tumours (Ed.), *Pathology and genetics of tumours of the urinary system and male genital organs*: Lyon, France: World Health Organization; IARC Press.
The International Agency for Research on Cancer. (2005a). In IARC WHO Classification of Tumours (Ed.), *Pathology and genetics of head and neck tumours*: Lyon, France: World Health Organization; IARC Press.
The International Agency for Research on Cancer. (2005b). In IARC WHO Classification of Tumours (Ed.), *Pathology and genetics of skin tumours*: Lyon, France: World Health Organization; IARC Press.

The International Agency for Research on Cancer. (2006). In IARC WHO Classification of Tumours (Ed.), *Pathology and genetics of tumours of soft tissue and bone*: Lyon, France: World Health Organization; IARC Press.

Thorat, M. A., & Cuzick, J. (2013). Role of aspirin in cancer prevention. *Current Oncology Reports*, *15*, 533–540.

Tian, C., Lu, S., Fan, Q., Zhang, W., Jiao, S., Zhao, X., … Wang, L. (2015). Prognostic significance of tumor-infiltrating $CD8^+$ or $CD3^+$ T lymphocytes and interleukin-2 expression in radically resected non-small cell lung cancer. *Chinese Medical Journal*, *128*, 105–110.

Tomlins, S. A., Alshalalfa, M., Davicioni, E., Erho, N., Yousefi, K., Zhao, S., … Feng, F. Y. (2015). Characterization of 1577 primary prostate cancers reveals novel biological and clinicopathologic insights into molecular subtypes. *European Urology*, *68*, 555–567.

Topalian, S. L., Hodi, F. S., Brahmer, J. R., Gettinger, S. N., Smith, D. C., McDermott, D. F., … Sznol, M. (2012). Safety, activity, and immune correlates of anti-PD-1 antibody in cancer. *The New England Journal of Medicine*, *366*, 2443–2454.

Treilleux, I., Blay, J., Bendriss-Vermare, N., Ray-Coquard, I., Bachelot, T., Guastalla, J., … Lebecque, S. (2004). Dendritic cell infiltration and prognosis of early stage breast cancer. *Clinical Cancer Research*, *10*, 7466–7474.

Troy, A. J., Summers, K. L., Davidson, P. J., Atkinson, C. H., & Hart, D. N. (1998). Minimal recruitment and activation of dendritic cells within renal cell carcinoma. *Clinical Cancer Research*, *4*, 585–593.

Tsujimoto, Y., Finger, L. R., Yunis, J., Nowell, P. C., & Croce, C. M. (1984). Cloning of the chromosome breakpoint of neoplastic B cells with the t(14;18) chromosome translocation. *Science*, *226*, 1097–1099.

Tsukayama, S., Omura, K., Yoshida, K., Tanaka, Y., & Watanabe, G. (2005). Prognostic value of CD83-positive mature dendritic cells and their relation to vascular endothelial growth factor in advanced human gastric cancer. *Oncology Reports*, *14*, 369–375.

Tumeh, P. C., Harview, C. L., Yearley, J. H., Shintaku, I. P., Taylor, E. J. M., Robert, L., … Ribas, A. (2014). PD-1 blockade induces responses by inhibiting adaptive immune resistance. *Nature*, *515*, 568–571.

Uzzo, R. G., Rayman, P., Kolenko, V., Clark, P. E., Cathcart, M. K., Bloom, T., … Finke, J. H. (1999). Renal cell carcinoma-derived gangliosides suppress nuclear factor-kappaB activation in T cells. *The Journal of Clinical Investigation*, *104*, 769–776.

Valković, T., Dobrila, F., Melato, M., Sasso, F., Rizzardi, C., & Jonjić, N. (2002). Correlation between vascular endothelial growth factor, angiogenesis, and tumor-associated macrophages in invasive ductal breast carcinoma. *Virchows Archiv*, *440*, 583–588.

Van den Hove, L. E., Van Gool, S. W., Van Poppel, H., Baert, L., Coorevits, L., Van Damme, B., & Ceuppens, J. L. (1997). Phenotype, cytokine production and cytolytic capacity of fresh (uncultured) tumour-infiltrating T lymphocytes in human renal cell carcinoma. *Clinical and Experimental Immunology*, *109*, 501–509.

van der Bruggen, P., Traversari, C., Chomez, P., Lurquin, C., De Plaen, E., Van den Eynde, B., … Boon, T. (1991). A gene encoding an antigen recognized by cytolytic T lymphocytes on a human melanoma. *Science*, *254*, 1643–1647.

van der Burg, S. H., & Melief, C. J. M. (2011). Therapeutic vaccination against human papilloma virus induced malignancies. *Current Opinion in Immunology*, *23*, 252–257.

van Herpen, C. M. L., van der Voort, R., van der Laak, J. A. W. M., Klasen, I. S., de Graaf, A. O., van Kempen, L. C. L., … De Mulder, P. H. M. (2008). Intratumoral rhIL-12 administration in head and neck squamous cell carcinoma patients induces B cell activation. *International Journal of Cancer*, *123*, 2354–2361.

Varela, I., Tarpey, P., Raine, K., Huang, D., Ong, C. K., Stephens, P., … Futreal, P. A. (2011). Exome sequencing identifies frequent mutation of the SWI/SNF complex gene PBRM1 in renal carcinoma. *Nature*, *469*, 539–542.

Väyrynen, J. P., Sajanti, S. A., Klintrup, K., Mäkelä, J., Herzig, K., Karttunen, T. J., … Mäkinen, M. J. (2014). Characteristics and significance of colorectal cancer associated lymphoid reaction. *International Journal of Cancer, 134*, 2126–2135.

Väyrynen, J. P., Tuomisto, A., Klintrup, K., Mäkelä, J., Karttunen, T. J., & Mäkinen, M. J. (2013). Detailed analysis of inflammatory cell infiltration in colorectal cancer. *British Journal of Cancer, 109*(7), 1839–1847. http://dx.doi.org/10.1038/bjc.2013.508. Epub 2013 Sep 5.

Vennervald, B. J., & Polman, K. (2009). Helminths and malignancy. *Parasite Immunology, 31*, 686–696.

Verhaak, R. G. W., Wouters, B. J., Erpelinck, C. A. J., Abbas, S., Beverloo, H. B., Lugthart, S., … Valk, P. J. M. (2009). Prediction of molecular subtypes in acute myeloid leukemia based on gene expression profiling. *Haematologica, 94*, 131–134.

Villegas, F. R., Coca, S., Villarrubia, V. G., Jiménez, R., Chillón, M. J., Jareño, J., … Callol, L. (2002). Prognostic significance of tumor infiltrating natural killer cells subset CD57 in patients with squamous cell lung cancer. *Lung Cancer, 35*, 23–28.

Vivier, E., Ugolini, S., Blaise, D., Chabannon, C., & Brossay, L. (2012). Targeting natural killer cells and natural killer T cells in cancer. *Nature Reviews Immunology, 12*, 239–252.

Vogelstein, B., Fearon, E. R., Hamilton, S. R., Kern, S. E., Preisinger, A. C., Leppert, M., … Bos, J. L. (1988). Genetic alterations during colorectal-tumor development. *The New England Journal of Medicine, 319*, 525–532.

Vogelstein, B., Papadopoulos, N., Velculescu, V. E., Zhou, S., Diaz, L. A. J., & Kinzler, K. W. (2013). Cancer genome landscapes. *Science, 339*, 1546–1558.

Voron, T., Colussi, O., Marcheteau, E., Pernot, S., Nizard, M., Pointet, A., … Terme, M. (2015). VEGF-A modulates expression of inhibitory checkpoints on CD8+ T cells in tumors. *The Journal of Experimental Medicine, 212*, 139–148.

Vu, T., & Claret, F. X. (2012). Trastuzumab: Updated mechanisms of action and resistance in breast cancer. *Frontiers in Oncology, 2*, 62.

Wald, G., Barnes, K. T., Bing, M. T., Kresowik, T. P., Tomanek-Chalkley, A., Kucaba, T. A., … Norian, L. A. (2014). Minimal changes in the systemic immune response after nephrectomy of localized renal masses. *Urologic Oncology, 32*, 589–600.

Wallis, S. P., Stafford, N. D., & Greenman, J. (2015). Clinical relevance of immune parameters in the tumor microenvironment of head and neck cancers. *Head & Neck, 37*, 449–459.

Wang, Q. J., Hanada, K., Robbins, P. F., Li, Y. F., & Yang, J. C. (2012). Distinctive features of the differentiated phenotype and infiltration of tumor-reactive lymphocytes in clear cell renal cell carcinoma. *Cancer Research, 72*, 6119–6129.

Wansom, D., Light, E., Thomas, D., Worden, F., Prince, M., Urba, S., … Wolf, G. (2012). Infiltrating lymphocytes and human papillomavirus-16—Associated oropharyngeal cancer. *Laryngoscope, 122*, 121–127.

Welsh, T. J., Green, R. H., Richardson, D., Waller, D. A., O'Byrne, K. J., & Bradding, P. (2005). Macrophage and mast-cell invasion of tumor cell islets confers a marked survival advantage in non-small-cell lung cancer. *Journal of Clinical Oncology, 23*, 8959–8967.

Wesolowski, R., Markowitz, J., & Carson, W. E., 3rd. (2013). Myeloid derived suppressor cells—A new therapeutic target in the treatment of cancer. *Journal for ImmunoTherapy of Cancer, 1*, 10.

West, N. R., Kost, S. E., Martin, S. D., Milne, K., Deleeuw, R. J., Nelson, B. H., & Watson, P. H. (2013). Tumour-infiltrating FOXP3(+) lymphocytes are associated with cytotoxic immune responses and good clinical outcome in oestrogen receptor-negative breast cancer. *British Journal of Cancer, 108*, 155–162.

Wirsing, A. M., Rikardsen, O. G., Steigen, S. E., Uhlin-Hansen, L., & Hadler-Olsen, E. (2014). Characterisation and prognostic value of tertiary lymphoid structures in oral squamous cell carcinoma. *BMC Clinical Pathology, 14*, 38.

Wolf, G. T., Chepeha, D. B., Bellile, E., Nguyen, A., Thomas, D., & McHugh, J. (2015). Tumor infiltrating lymphocytes (TIL) and prognosis in oral cavity squamous carcinoma: A preliminary study. *Oral Oncology, 51*, 90–95.

Wölfel, T., Hauer, M., Schneider, J., Serrano, M., Wölfel, C., Klehmann-Hieb, E., ... Beach, D. (1995). A p16INK4a-insensitive CDK4 mutant targeted by cytolytic T lymphocytes in a human melanoma. *Science, 269*, 1281–1284.

Wotherspoon, A. C., Doglioni, C., Diss, T. C., Pan, L., Moschini, A., de Boni, M., & Isaacson, P. G. (1993). Regression of primary low-grade B-cell gastric lymphoma of mucosa-associated lymphoid tissue type after eradication of *Helicobacter pylori*. *Lancet, 342*, 575–577.

Wotherspoon, A. C., Ortiz-Hidalgo, C., Falzon, M. R., & Isaacson, P. G. (1991). *Helicobacter pylori*-associated gastritis and primary B-cell gastric lymphoma. *Lancet, 338*, 1175–1176.

Wu, Y., Fu, X., Zhu, X., He, X., Zou, C., Han, Y., ... Zhao, Y. (2011). Prognostic role of systemic inflammatory response in renal cell carcinoma: A systematic review and meta-analysis. *Journal of Cancer Research and Clinical Oncology, 137*, 887–896.

Xu, L., Zhu, Y., Chen, L., An, H., Zhang, W., Wang, G., ... Xu, J. (2014). Prognostic value of diametrically polarized tumor-associated macrophages in renal cell carcinoma. *Annals of Surgical Oncology, 21*, 3142–3150.

Yang, J. C., Hughes, M., Kammula, U., Royal, R., Sherry, R. M., Topalian, S. L., ... Rosenberg, S. A. (2007). Ipilimumab (anti-CTLA4 antibody) causes regression of metastatic renal cell cancer associated with enteritis and hypophysitis. *Journal of Immunotherapy, 30*, 825–830.

Yasuda, M., Takenoyama, M., Obata, Y., Sugaya, M., So, T., Hanagiri, T., ... Yasumoto, K. (2002). Tumor-infiltrating B lymphocytes as a potential source of identifying tumor antigen in human lung cancer. *Cancer Research, 62*, 1751–1756.

Yeang, C., McCormick, F., & Levine, A. (2008). Combinatorial patterns of somatic gene mutations in cancer. *The FASEB Journal, 22*, 2605–2622.

Yin, X., Lu, M., Lai, Y., Liang, L., & Huang, J. (2003). Prognostic significances of tumor-infiltrating S-100 positive dendritic cells and lymphocytes in patients with hepatocellular carcinoma. *Hepato-Gastroenterology, 50*, 1281–1284.

Yoon, H. H., Orrock, J. M., Foster, N. R., Sargent, D. J., Smyrk, T. C., & Sinicrope, F. A. (2012). Prognostic impact of FoxP3+ regulatory T cells in relation to CD8+ T lymphocyte density in human colon carcinomas. *PLoS One*, 7, e42274.

Yoshida, N., Ikemoto, S., Narita, K., Sugimura, K., Wada, S., Yasumoto, R., ... Nakatani, T. (2002). Interleukin-6, tumour necrosis factor alpha and interleukin-1beta in patients with renal cell carcinoma. *British Journal of Cancer, 86*, 1396–1400.

Yue, Q., Zhang, X., Ye, H. X., Wang, Y., Du, Z. G., Yao, Y., & Mao, Y. (2014). The prognostic value of Foxp3+ tumor-infiltrating lymphocytes in patients with glioblastoma. *Journal of Neuro-oncology, 116*(2), 251–259. http://dx.doi.org/10.1007/s11060-013-1314-0. Epub 2013 Nov 26.

Yuseff, M., Pierobon, P., Reversat, A., & Lennon-Duménil, A. (2013). How B cells capture, process and present antigens: A crucial role for cell polarity. *Nature Reviews Immunology, 13*, 475–486.

Zea, A. H., Rodriguez, P. C., Atkins, M. B., Hernandez, C., Signoretti, S., Zabaleta, J., ... Ochoa, A. C. (2005). Arginase-producing myeloid suppressor cells in renal cell carcinoma patients: A mechanism of tumor evasion. *Cancer Research, 65*, 3044–3048.

Zeng, L., Guo, Y., Liang, J., Chen, S., Peng, P., Zhang, Q., ... Huang, K. (2014). Perineural invasion and TAMs in pancreatic ductal adenocarcinomas: Review of the original pathology reports using immunohistochemical enhancement and relationships with clinicopathological features. *Journal of Cancer, 5*, 754–760.

Zeni, E., Mazzetti, L., Miotto, D., Lo Cascio, N., Maestrelli, P., Querzoli, P., ... Boschetto, P. (2007). Macrophage expression of interleukin-10 is a prognostic factor in nonsmall cell lung cancer. *The European Respiratory Journal, 30*, 627–632.

Zhan, H., Gao, X., Zhou, X., Pu, X., & Wang, D. (2012). Presence of tumour-infiltrating FOXP3+ lymphocytes correlates with immature tumour angiogenesis in renal cell carcinomas. *Asian Pacific Journal of Cancer Prevention, 13*, 867–872.

Zhang, Y., Cheng, S., Zhang, M., Zhen, L., Pang, D., Zhang, Q., & Li, Z. (2013). High-infiltration of tumor-associated macrophages predicts unfavorable clinical outcome for node-negative breast cancer. *PLoS One, 8*, e76147.

Zhang, L., Conejo-Garcia, J. R., Katsaros, D., Gimotty, P. A., Massobrio, M., Regnani, G., ... Coukos, G. (2003). Intratumoral T cells, recurrence, and survival in epithelial ovarian cancer. *The New England Journal of Medicine, 348*, 203–213.

Zhang, B. C., Gao, J., Wang, J., Rao, Z. G., Wang, B. C., & Gao, J. F. (2011). Tumor-associated macrophages infiltration is associated with peritumoral lymphangiogenesis and poor prognosis in lung adenocarcinoma. *Medical Oncology, 28*, 1447–1452.

Zhang, M., He, Y., Sun, X., Li, Q., Wang, W., Zhao, A., & Di, W. (2014). A high M1/M2 ratio of tumor-associated macrophages is associated with extended survival in ovarian cancer patients. *Journal of Ovarian Research*, 7, 19.

Zhang, Z., Huang, J., Zhang, C., Yang, H., Qiu, H., Li, J., ... Shan, B. (2015). Infiltration of dendritic cells and T lymphocytes predicts favorable outcome in epithelial ovarian cancer. *Cancer Gene Therapy, 22*, 198–206.

Zhang, Q., Liu, L., Gong, C., Shi, H., Zeng, Y., Wang, X., ... Wei, Y. (2012). Prognostic significance of tumor-associated macrophages in solid tumor: A meta-analysis of the literature. *PLoS One*, 7, e50946.

Zhang, Y., Wang, L., Li, Y., Pan, Y., Wang, R., Hu, H., ... Chen, H. (2014b). Protein expression of programmed death 1 ligand 1 and ligand 2 independently predict poor prognosis in surgically resected lung adenocarcinoma. *OncoTargets and Therapy*, 7, 567–573.

Zhang, H., Wang, X., Shen, Z., Xu, J., Qin, J., & Sun, Y. (2014a). Infiltration of diametrically polarized macrophages predicts overall survival of patients with gastric cancer after surgical resection. *Gastric Cancer, 18*, 740–750.

Zhou, L., Chong, M. M. W., & Littman, D. R. (2009). Plasticity of CD4+ T cell lineage differentiation. *Immunity, 30*, 646–655.

Zhou, Q., Peng, R., Wu, X., Xia, Q., Hou, J., Ding, Y., ... Zhang, X. (2010). The density of macrophages in the invasive front is inversely correlated to liver metastasis in colon cancer. *Journal of Translational Medicine, 8*, 13.

Zhu, X., Zhang, J., Zhuang, P., Zhu, H., Zhang, W., Xiong, Y., ... Sun, H. (2008). High expression of macrophage colony-stimulating factor in peritumoral liver tissue is associated with poor survival after curative resection of hepatocellular carcinoma. *Journal of Clinical Oncology, 26*, 2707–2716.

CHAPTER FIVE

Advances in Therapeutic Cancer Vaccines

Karrie K. Wong*, WeiWei Aileen Li[†,‡], David J. Mooney[†,‡], Glenn Dranoff[*,1]

*Novartis Institutes for Biomedical Research, Cambridge, MA, United States
†Wyss Institute for Bioinspired Engineering, Harvard University, Boston, MA, United States
‡School of Engineering and Applied Sciences, Harvard University, Cambridge, MA, United States
[1]Corresponding author: e-mail address: glenn_dranoff@dfci.harvard.edu

Contents

Abstract

Therapeutic cancer vaccines aim to induce durable antitumor immunity that is capable of systemic protection against tumor recurrence or metastatic disease. Many approaches to therapeutic cancer vaccines have been explored, with varying levels of success. However, with the exception of Sipuleucel T, an ex vivo dendritic cell vaccine for prostate cancer, no therapeutic cancer vaccine has yet shown clinical efficacy in phase 3 randomized trials. Though disappointing, lessons learned from these studies have suggested new strategies to improve cancer vaccines. The clinical success of

Advances in Immunology, Volume 130
ISSN 0065-2776
http://dx.doi.org/10.1016/bs.ai.2015.12.001

checkpoint blockade has underscored the role of peripheral tolerance mechanisms in limiting vaccine responses and highlighted the potential for combination therapies. Recent advances in transcriptome sequencing, computational modeling, and material engineering further suggest new opportunities to intensify cancer vaccines. This review will discuss the major approaches to therapeutic cancer vaccination and explore recent advances that inform the design of the next generation of cancer vaccines.

1. INTRODUCTION

1.1 Cancer Immunotherapy Coming of Age

A role for the immune system in cancer was first postulated in the dawn of the 20th century; however, it was not until recently, with the publication of seminal work by Schreiber and colleagues uncovering the control of tumor progression by the immune system and the modulation of such responses by the tumor in turn, that the concept of cancer immunoediting became widely accepted as a hallmark of cancer (Schreiber, Old, & Smyth, 2011). Tumor infiltration of effector immune cells and the presence of suppressor populations are now established to have significant prognostic value (Galon, Angell, Bedognetti, & Marincola, 2013). The recognition of a role for the immune system in cancer has catalyzed exciting developments in cancer vaccinology. The immune system, equipped with an adaptive arm characterized by specificity and immunological memory, might provide a solution for two major problems in cancer treatment: specificity that reduces off-target effects and memory that controls tumor relapse.

In theory at least, a therapeutic cancer vaccine has the potential to stimulate specific immunity against tumors while sparing normal tissues, leading not only to tumor lysis but also to the induction of a long lasting, systemic immunological memory that protects against recurrent disease and metastasis. The generation of an efficacious antitumor immune response requires that a number of important steps are achieved: (a) the presence of immunogenic tumor antigens; (b) the recruitment and maturation of antigen-presenting cells—primarily dendritic cells (DCs); (c) the uptake and processing of tumor antigens by mature DCs; and (d) the presence of costimulatory signals and stimulatory cytokines and the absence of inhibitory signals that together favor the induction of T helper-1 (Th1) cells, CD8+T cytotoxic lymphocytes (CTLs), and high-titer, high-affinity antibodies.

Thus far a vaccination approach for cancer that meets all of these criteria has yet to be developed. Indeed, the field has suffered major disappointments with the failure of several therapeutic vaccines in phase 3 studies due to a lack of clinical efficacy. Nevertheless, valuable lessons have been learned from these efforts, and an improved understanding of the key factors required for immune activation presents new opportunities to improve cancer vaccine development. This review aims to provide an overview of some of the most important lessons learned from past experiences and the exciting recent progress that is reenergizing the crafting of therapeutic cancer vaccines.

2. CANCER ANTIGENS

2.1 Identification of Tumor Antigens

The presence of tumor-specific CTLs in human cancer was first reported in the early 1980s by Vose and Bonnard, who showed that T lymphocytes expanded ex vivo in the presence of IL-2-mediated lysis of autologous tumor cells (Vose & Bonnard, 1982a, 1982b). Methods to expand tumor-infiltrating lymphocytes ex vivo were subsequently developed, and the adoptive transfer of these T cells mediated tumor regression, demonstrating their functional importance (Maryanski et al., 1991; Rosenberg et al., 1988). Together, these data supported the hypothesis that tumor cells express antigens that are distinguishable from normal cells, and effective adaptive immune responses can be mounted against such antigens.

A protective antitumor immune response requires activation of specific T effector cells that recognize tumor antigen-derived peptides presented by MHC molecules. Tumor cell proteins are catabolized to peptide fragments in the proteasome, shuttled into the endoplasmic reticulum (ER) through the transporter associated with antigen processing (TAP) where they are complexed with nascent MHC class I molecules, and then transported through the Golgi apparatus to the tumor cell surface (Hansen & Bouvier, 2009; Leone et al., 2013). Cancer antigens can be broadly categorized into four main classes: oncofetal antigens, which are fetal proteins that are normally silenced in adult tissues but then are aberrantly reexpressed in tumors due to epigenetic alterations; viral antigens (in cancers of viral origin); neoantigens that result from somatic mutations; and differentiation antigens that are normal cellular proteins overexpressed in cancers. The first three classes of antigens are almost exclusively found only in tumor cells and are thus ideal as vaccine targets. Differentiation antigens, on the other hand, are self-antigens that can be found at low levels in normal tissues. Efficacious immunity against this latter class of antigens has to overcome immune tolerance, rendering them challenging vaccine targets. In spite of this inherent difficulty, functional CTL responses against differentiation antigens can sometimes arise without therapy in melanoma patients, although they can be associated with damage to normal melanocytes and result in depigmentation (Coulie et al., 1994; Cox et al., 1994; Kawakami et al., 1994).

Boon and colleagues pioneered the identification of tumor antigens using a genetic approach. Their strategy involved the screening of tumor-reactive autologous CTL clones against a tumor-derived cDNA expression library in order to determine the sequences of the epitopes recognized by the CTL clones (reviewed in detail by Coulie, Van den Eynde, van der Bruggen, & Boon, 2014). Using this method, they characterized the first human tumor antigen in melanoma, MAGE-A1, which was the founding member of a family of germline encoded genes called cancer/testis (CT) antigens. This discovery was shortly followed, using the same approach, with the identification of the differentiation antigen MART-1 as the second melanoma T cell antigens (Coulie et al., 1994; van der Bruggen et al., 1991). Several years later, using a biochemical approach with tandem mass spectrometry for screening of antigenic peptides bound to HLA-A2.1, Cox et al. described an antigenic peptide derived from the melanocyte differentiation protein gp100 that was recognized by CTL clones from five different melanoma patients (Cox et al., 1994). Both methods provided valuable information regarding the nature of peptide–MHC interactions in tumor cells and the induction of spontaneous antitumor CTL responses in patents.

A third, less cumbersome approach to tumor antigen identification, serological analysis of recombinant cDNA expression libraries (SEREX), involves the screening of cancer patient sera against tumor-derived cDNA expression libraries (Sahin et al., 1995). This strategy is significantly easier to perform and results in the delineation of tumor targets that elicit antibody responses. By this approach, Chen et al. identified the cancer germline-encoded antigen NY-ESO-1, which is highly expressed on multiple cancers from melanoma to hepatocellular carcinoma, but absent on normal adult tissues except testis and ovary (Chen et al., 1997). Thus far, over 2000 cancer genes that encode proteins capable of eliciting antibody responses have been identified using this approach, many of which belong to the cancer/testis (CT) antigen family, such as MAGE-A1 (Chen, 2000). Interestingly, despite immune tolerance mechanisms, the majority of antigens that elicit antibodies in cancer patients are self-antigens. The presence of a humoral response is not always associated with a corresponding CTL response, which may reflect differences in the mechanisms of tumor antigen presentation or of peripheral tolerance. Nonetheless, the capability of a tumor antigen to generate spontaneous humoral responses suggests inherent immunogenicity; thus, SEREX has been instrumental in defining antigens of interest for further pursuit in the search for T cell epitopes.

2.2 The Search for T Cell Epitopes

Two important determinants for an effective T cell response are antigen processing and peptide–MHC interactions—both critical steps for antigen presentation and hence effector T cell activation. Biochemical analyses and studies on the crystal structures of human MHC class I molecules show that each allotype binds a specific, nonoverlapping pattern of amino acids, termed the peptide motif (Carreno, Anderson, Coligan, & Biddison, 1990; Falk, Rotzschke, Stevanovic, Jung, & Rammensee, 1991). In vitro analyses of the differentiation antigen gp100 showed a direct association between peptide immunogenicity and MHC binding affinity (Salgaller, Marincola, Cormier, & Rosenberg, 1996). The accumulated knowledge on peptide motifs led to the development of prediction algorithms for peptide–MHC binding affinity (Davenport et al., 1997; Peters, Tong, Sidney, Sette, & Weng, 2003; Rammensee, Bachmann, Emmerich, Bachor, & Stevanovic, 1999). With advances in bioinformatics and the increasing accessibility of powerful computational tools in recent years, research on cancer vaccine targets has moved into a new era of high-throughput discoveries. The potential cancer antigen repertoire can be established by a comparison of tumor to corresponding normal tissue in high-throughput, transcriptome-based analyses (Viatte, Alves, & Romero, 2006). In silico epitope prediction algorithms can then be applied for the entire antigen repertoire or for antigens derived from the SEREX database to screen for CTL epitopes. The accuracy of prediction models continues to improve with an increased availability of large training datasets, which should advance target discovery for cancer vaccine development (Kuksa, Min, Dugar, & Gerstein, 2015).

3. APPROACHES TO CANCER VACCINES: WHAT WE HAVE LEARNED

3.1 Peptide/Protein Vaccines

Prior to the identification of tumor-specific antigens, the idea to use inactivated autologous tumor cells as vaccines to elicit antitumor immunity had been explored since the early 20th century, though with little clinical success. Recent advances in cancer vaccines were enabled through an improved understanding of tumor antigen presentation and the mechanisms of T cell activation. These insights included the concept of antigen presentation by DCs via MHC molecules; the identification of TCRs recognizing tumor antigenic epitopes; and the engineering of tumor cells to produce immunostimulatory cytokines that could enhance their immunogenicity in syngeneic murine hosts (Colombo et al., 1991; Coulie et al., 1994; van der Bruggen et al., 1991). Based on these developments, three major vaccine strategies emerged. Defined tumor antigen vaccines are based on specific gene products identified through the techniques described in the last section, and can come in the form of peptides, full-length proteins, or genetically encoded vectors. Whole-cell cancer vaccines employ the entire tumor cell and, in the cases of autologous tumors, the associated tumor stroma and vasculature, to potentiate immune activation. A third approach involves the direct loading of whole tumor material or defined antigen to autologous DCs ex vivo followed by inoculation into patients.

With the exception of neoantigens and viral antigens, cancer antigens are by nature less immunogenic in comparison to foreign pathogens, and thus a necessary component for an effective vaccine is an adjuvant that functions to activate, mature, and recruit cells of the innate immunity, which in turn provide the stimulatory signals essential for an optimal adaptive immune response. Due to the diversity and plasticity of innate immune cells, the choice of adjuvant exerts considerable influence on immunological outcomes. Aluminum salts (alum), a component of a number of licensed human vaccines for common viruses, are known to induce Th2 CD4+T helper response and high-titer antibodies, but are less effective on promoting CTL responses (Dubensky & Reed, 2010). The pleotropic cytokine GM-CSF is commonly used as an adjuvant in various cancer vaccines because of its impact on the maturation and recruitment of DCs, which are critical for the induction of CD4+T cells and the cross-priming of CD8+T cells (Bowne et al., 1999; Disis et al., 1996). Recently, GM-CSF as an adjuvant was also shown to induce expression of two homing molecules on vaccine-specific CD8+T cells, thus enhancing tumor infiltration of vaccine-induced effector cells (Clancy-Thompson et al., 2013). Cholesterol, phospholipid, and saponin, preformed with antigen or antigen-free, constitute immunostimulatory complexes (ISCOM) that induce both cellular and humoral responses against a broad range of antigens (Sun, Xie, & Ye, 2009). Incomplete Freund's adjuvant (IFA) and other oil-based adjuvants such as Montanide combined with an agonist for pattern recognition receptors (PRRs) expressed on innate immune cells can trigger pro-inflammatory cytokine production and elicit CD8+T cell responses (Billiau & Matthys, 2001). The addition of CpG ODN, a synthetic agonist to TLR-9, to a NY-ESO-1 peptide vaccine formulated in Montanide ISA-51 was shown in a small clinical study to improve induction of NY-ESO-1-specific immune responses and clinical outcome compared to Montanide ISA-51 alone (Karbach et al., 2010). The combination of poly-ICLC, the agonist for TLR-3, in Montanide ISA-51 for a NY-ESO-1 vaccine with long overlapping peptides was shown in another study to enhance Th1 CD4+T cell responses, which is in turn important for a robust CTL response (Tsuji et al., 2013). Importantly, this adjuvant combination also suppressed the induction of IL4-producing Th2 cells. Peptide vaccines formulated in IFA with agonists for other TLRs, such as for TLR-4 (MPL) and TLR-7 (imiquimod), are similarly being investigated (Dubensky & Reed, 2010). Since expression of the different TLRs varies in DC and myeloid cell subsets, the choice of which TLR agonist to incorporate into a vaccine might influence the type of ensuing immune response. A recent study showed that the application of the TLR-7 agonist resiquimod as a topical adjuvant with a NY-ESO-1 protein vaccine was insufficient for inducing potent CTL responses (Sabado et al., 2015).

Cognate activation of Th1 CD4+T cells, which produce IFNγ and IL-21 that help CD8+T cell expansion and differentiation into CTLs, is another important component for an optimal vaccine response (Casey & Mescher, 2007; Frohlich et al., 2009; Whitmire, Tan, & Whitton, 2005; Zeng et al., 2005). In a small study involving patients with high-risk vulvar intraepithelial neoplasia that is caused by human papillomavirus type-16 (HPV-16), vaccination with a mix of long peptides derived from the viral oncoproteins E6 and E7 formulated with Montanide ISA-51 induced both CD4+T cell and CD8+T cell responses and resulted in tumor regression in 94% of patients, including complete responses in 47% (Kenter et al., 2009). This impressive clinical response rate, rare among cancer vaccines, was associated with not only vaccine-induced CTL responses but also expansion of IFNγ producing CD4+T cells, indicating the importance of T helper response in potentiating CTL effector functions. The potent clinical efficacy seen with this vaccine is also likely related to its targeting of viral, and thus non-self, proteins. In the case of NY-ESO-1, a peptide vaccine containing a single NY-ESO epitope in IFA that is the target of both CD4+T cells and CD8+T cells was tested in NY-ESO-1 + ovarian cancer patients with minimal disease (Odunsi et al., 2007). Following immunization, vaccine-specific CD4+ T cells and CD8+T cells were both found to recognize NY-ESO-1 expressing tumor cells; importantly, both responses persisted to up to 12 months after immunization. The long-lasting antitumor responses by both CD4+T cells and CD8+T cells seen in this trial supports the notion that provision of cognate CD4+T cell help can enhance CTL responses and is an important consideration for vaccine design.

Other investigators have explored a different approach to NY-ESO-1 vaccination, using full-length recombinant NY-ESO-1 protein with specific adjuvants designed to induce both antibody and CTL responses. In a clinical trial where the full-length protein was used with ISCOMATRIX as adjuvant, patients with resected NY-ESO-1 + tumors generated detectable levels of circulating CD4+T cells and CD8+T cells reactive to a broad range of NY-ESO-1 epitopes (Chen et al., 2004; Davis et al., 2004; Valmori et al., 2007). Under certain contexts, however, vaccines that target CD4+T cells may induce antigen-specific regulatory T (Treg) cells. Indeed, the lack of clinical responses to the full-length NY-ESO-1 ISCOMATRIX vaccine in patients with advanced melanoma in another study was attributed to induction of NY-ESO-1-specific Treg cells (Ebert et al., 2012; Nicholaou et al., 2009). An improved clinical response to the vaccine in these patients was seen when a Treg depletion agent, cyclophosphamide, was included (Klein et al., 2015).

Induction of tumor-specific Treg cells was also observed in a MAGE-A3 peptide vaccine using a peptide epitope capable of inducing both CD4+T cell and CD8+T cell responses (Francois et al., 2009). Vaccination using a full-length recombinant MAGE-A3 protein with an immunostimulant adjuvant (GSK), in contrast, showed superior induction of humoral and cellular responses in patients with resected non-small cell lung cancer (NSCLC) and possibly clinical activity in patients with metastatic melanoma in phase 2 studies (Atanackovic et al., 2008; Kruit et al., 2013; Vansteenkiste et al., 2013). However, the phase 3 MAGRIT trial of MAGE-A3 peptide vaccine for NSCLC patients with MAGE-3 + and resected tumor failed to show an increase in the primary endpoint of disease-free survival compared with controls and was terminated in 2014, once again highlighting the clinical challenges of cancer vaccines (Vansteenkiste et al., 2014). It is not clear whether induction of regulatory cells or other inhibitory elements at play hindered vaccine efficacy in the MAGE-A3 trial; however, microarray analysis of tumor samples from earlier phase 2 trials identified that a prevaccine IFNγ- and STAT1-driven gene signature, with expression of chemokines that are important for immune cell recruitment in the tumor, might be associated with positive treatment response (Ulloa-Montoya et al., 2013).

Compelling evidence for a functional role of Tregs in diminishing vaccine responses is further provided by a recent report of a phase 2 study on IMA901 peptide vaccine for renal cell carcinoma patients, which provided one of the most thorough immunological analyses to date of pre- and postvaccine treatment (Walter et al., 2012). IMA901 is distinguished from other peptide vaccines in being the first multipeptide vaccine consisting of multiple HLA-restricted novel tumor antigen peptides discovered with a combination of mass spectrometry analysis and a computer-based antigen discovery platform (XPRESIDENT), which involves analysis of the immunopeptidome (Weinschenk et al., 2002). Retrospective analysis from this study showed that a low frequency of prevaccination Tregs was associated with multiple T cell responses, which, in turn, was correlated with disease control. Pretreatment with the Treg-depleting agent cyclophosphamide downregulated Tregs as early as 3 days after treatment and was associated with prolonged overall survival in patients. The numbers of circulating myeloid-derived suppressor cells (MDSCs) prevaccination were also associated with reduced vaccine efficacy. MDSC infiltration in tumors is often associated with immunosuppression, and in this study, there was a strong correlation between the presence of MDSCs and T cell dysfunction.

Unlike NY-ESO-1 and MAGE-A3, CTL epitopes for self-differentiation antigens tend to have intermediate MHC binding affinity and typically require amino acid substitutions to enhance immunogenicity (Houghton & Guevara-Patino, 2004). For gp100, a vaccine using two immunogenic peptides in IFA plus IL-2 showed induction of peptide and melanoma-specific effector responses and a 42% objective tumor response rate, suggesting that vaccination against a self-antigen might be possible in some cases (Rosenberg et al., 1998). For MART-1, another differentiation antigen, a peptide vaccine containing the modified, immunogenic peptide formulated in a saponin-based adjuvant was shown to induce MART-1-specific CTLs that were able to kill melanoma cells in vitro (Ayyoub et al., 2003). The addition of CpG ODN to the vaccine formulation further enhanced induction of MART-1 reactive CD8+T cells with effector memory phenotype (Speiser et al., 2005). The effectiveness of CTL response to this peptide vaccine was inversely associated with expression of the immunomodulatory molecule BTLA on MART-1-specific CD8+T cells. BTLA mediates T cell suppression through binding the cognate ligand HVEM; the inclusion of CpG ODN improved CTL response in part by downregulating BTLA expression on effector cells (Derre et al., 2010). However, when compared with virus-specific T cells from healthy donors, MART-1-specific CD8+T cells from peptide-CpG ODN-vaccinated patients still showed a gene signature of exhaustion, with expression of other inhibitory receptors including CTLA-4, TIM-3, and LAG-3 (Baitsch et al., 2011). These results illustrate cell-intrinsic inhibitory mechanisms that prevent cancer vaccines from unleashing full therapeutic power. Further, in another study, CD4+T cells from patients vaccinated with a MART-1 peptide included antigen-specific Foxp3+ Treg cells, which likely attenuated vaccine efficacy (Jandus et al., 2009).

In a randomized phase 3 clinical trial involving 183 patients with advanced melanoma, the clinical activity of a gp100 peptide vaccine +IL-2 was evaluated against IL-2 alone. Patients who received vaccine +IL-2 showed an increase in clinical response and progression free survival compared to those who received IL2 alone, but there was no impact on overall survival (Schwartzentruber et al., 2011). Vaccinated patients who showed a clinical response also harbored higher numbers of circulating CD4+Foxp3+ Treg cells, but the role of this increase in the vaccine response was unclear. Interestingly, in the seminal phase 3 study of ipilimumab—a monoclonal antibody against the inhibitory receptor CTLA-4 which is commonly expressed on Tregs and effector T cells in the tumor microenvironment—Hodi et al. reported that while ipilimumab improved overall survival of patients with metastatic melanoma, gp100 peptide vaccine with ipilimumab showed no advantage compared to ipilimumab alone (Hodi et al., 2010). In another report, mice that received a gp100 melanoma short peptide vaccine formulated in IFA showed that CD8+ T cells accumulated at vaccine injection sites where antigens persisted, and subsequently underwent FasL induced apoptosis (Hailemichael et al., 2013). This problem was only circumvented by a different vaccine formulation that reduces antigen persistence or employs the use of long peptides, illustrating yet another nuance in vaccine design that could affect efficacy. In another study, IFNα was shown to be an effective adjuvant for a gp100 peptide vaccine in mice (Sikora et al., 2009).

Indeed, for peptide/protein vaccines, formulation can significantly affect vaccine efficacy through modulating antigen availability, persistence, and conformation. Muc1, a mucin that is often aberrantly glycosylated in a variety of tumor types, manifests truncated O-linked saccharides. Due to its ubiquitous expression across tumors, Muc1 has been intensively studied as a vaccine target. In the case of Muc1, formulation critically affects vaccine response: vaccination with nonglycosylated MUC1 sequences, which differ from tumor Muc1 in conformation, failed to induce anti-Muc1 responses against Muc1+ tumor cells. A formulation of Muc1 peptides in liposomes provided preliminary evidence for the induction of Muc1-specific CTLs in patients with NSCLC (Agrawal, Krantz, Reddish, & Longenecker, 1998; Butts et al., 2005). However, a randomized phase 3 multicenter study of a lipopeptide vaccine (L-BLP25 or Tecemotide) for stage III patients reported no survival benefits, despite the use of cyclophosphamide for Treg depletion (Butts et al., 2014). Other Muc1 vaccine formulations have also been explored, such as the use of a glycosylated Muc1 tandem repeat peptide, which elicited modest antibody response (Sorensen et al., 2006). Lakshminarayanan et al. proposed a novel vaccine design containing an aberrantly glycosylated Muc1 peptide, a CD4+T cell epitope, and the immunoadjuvant Pam3CysSk4, a TLR2 agonist, all covalently linked. The resulting tripartite vaccine induced both humoral and cellular responses to glycosylated Muc1, and reduced tumor burden in a mouse breast tumor model (Lakshminarayanan et al., 2012).

As an alternative approach to the use of peptides or full-length proteins to target specific tumor antigens, transgenes coding for antigens can be inserted into recombinant plasmids to generate plasmid DNA/viral vector based vaccines. The presence of bacterial/viral proteins has the advantage of triggering innate cell activation and recruitment to vaccination site, thus potentiating immune activation. Poxviruses, a group of DNA-containing viruses that is used in the smallpox vaccine, have been explored as viral vectors for this purpose (Arlen, Gulley, Madan, Hodge, & Schlom, 2007). In an early study, vaccination using a recombinant vaccinia virus encoding prostate-specific antigen (PSA) was shown to induce T cell responses against PSA, but only after the first vaccination, as host immune responses to vaccinia proteins limited efficacy of subsequent doses and overall clinical activity of the vaccine (Eder et al., 2000). Enhanced clinical responses were seen in advanced prostate cancer patients using a prime/boost approach with vaccinia and fowlpox viruses, both expressing PSA (Kaufman et al., 2004). Applying the prime/boost vaccinia/fowlpox approach, with the recombinant viral vectors encoding PSA and three immunostimulatory molecules (LFA-3, ICAM1, and CD80), Kantoff and colleagues reported an 8-month improvement in overall survival in a phase 2 study for men with metastatic castration-resistant prostate cancer (Kantoff, Schuetz, et al., 2010). A corresponding phase 3 study for this vaccine approach (PROSTVAC-TRICOM) has recently completed accrual (Singh, Pal, Alex, & Agarwal, 2015). Recently, early humoral responses to a carbohydrate on the poxvirus were correlated with patient survival, suggesting viral glycans can modulate vaccine responses (Campbell et al., 2014). Attenuated *Listeria monocytogenes* (Lm), a Gram-positive bacterium that can be taken up by macrophages and phagocytic cells, has also been explored as a vector for cancer vaccine (Brockstedt et al., 2004; Wallecha, Maciag, Rivera, Paterson, & Shahabi, 2009). A live attenuated Lm strain engineered to express human mesothelin, an antigen expressed by a number of tumors, was shown to induce mesothelin-specific T cells in 37% of patients, although induction of anti-tumor immunity was not associated with clinical benefits (Le et al., 2012).

In vitro-transcribed (IVT) RNA encoding cancer antigens has also been shown to elicit antitumor immunity. Conry and colleagues first demonstrated that vaccination with mRNA transcripts encoding human carcinoembryonic antigen, administrated intramuscularly, elicited protective immunity against tumor challenge in a prophylactic tumor model (Conry et al., 1995). Mice vaccinated with a self-tumor antigen-encoding mRNA encapsulated in cationic liposomes were shown to elicit tumor-specific CTL responses, although the vaccine was insufficient to overcome tolerance in tumor-bearing mice (Hess, Boczkowski, Nair, Snyder, & Gilboa, 2006). The feasibility of this method in human was demonstrated by Probst and colleagues, who reported in vivo translation of mRNA that was injected into human dermis (Probst et al., 2007). In a phase 1/2 trial, vaccination with a protamine-stabilized RNA vaccine encoding multiple melanoma antigens plus systemic administration of GM-CSF was shown to reduce Treg and MDSC frequencies in melanoma patients following vaccination, although therapeutic efficacy was not reported (Weide et al., 2008, 2009). In another phase 1/2 nonrandomized study, intradermal administration of IVT-naked RNA was reported to induce both cellular and humoral responses to a number of tumor antigens in patients (Rittig et al., 2011) Single-stranded RNA (ssRNA) is recognized by TLR-7 and TLR-8 expressed on nonoverlapping innate cell populations as well as B cells and can stimulate potent immune responses (Diebold et al., 2004; Heil et al., 2004). Kreiter et al. showed that intranodal injection of a tumor antigen-encoding RNA vaccine resulted in induction of antitumor immunity in mice, a response which could be further enhanced by systemic administration of Flt3-ligand (Flt-3L) (Kreiter et al., 2011, 2010). Administration of the mTOR inhibitor rapamycin during the effector-to-memory transition phase postvaccination was also shown to enhance the therapeutic efficacy of the RNA vaccine in both early and advanced tumor models (Diken et al., 2013). An IVT-naked RNA vaccine is currently at the phase 3 stage of clinical testing in Europe (Sahin, Kariko, & Tureci, 2014). The immunostimulatory property of ssRNA has also been explored as a potential immunoadjuvant for peptide vaccines. A noncoding, long chain RNA was shown to be superior to poly (I:C) in its adjuvant activity for a peptide vaccine for HPV-16 E7 (Heidenreich et al., 2015).

Each vaccine approach discussed thus far has incorporated a single or limited number of target antigens. An alternative strategy is to exploit a much larger repertoire of targets. One personalized approach utilizes the high peptide binding properties of the gp96 family of heat-shock proteins, which can be purified from patient tumors and employed as an autologous vaccine (Wood & Mulders, 2009). Peptide–gp96 complexes can be internalized by antigen-presenting cells and processed to bind MHC class I molecules for presentation, thereby stimulating potent CTL responses (Tamura, Peng, Liu, Daou, & Srivastava, 1997). The gp96 vaccine (Vitespen) takes advantage of the strong immunogenicity of peptide–gp96 complexes harvested from a resected autologous tumor and hence is a fully individualized peptide vaccine. However, Vitespen did not improve overall survival of stage IV melanoma patients in a randomized phase 3 trial (Testori et al., 2008).

Together, these data highlight several major challenges for cancer vaccines based on peptides and proteins regardless of design and tumor type: (a) antigen/epitope selection for optimal binding to MHC molecules; (b) suitable adjuvant for each vaccine based on antigen property and desired immunological outcome; (c) presence of regulatory cells that suppress immune activation; and (d) effector cell exhaustion. These data provide important insights into the in vivo mechanisms that likely act in concert to hamper clinical efficacy of cancer vaccines and help to explain the modest clinical effects of many vaccines to date.

3.2 Whole Tumor Cell Vaccines

For whole cell cancer vaccines, retroviral or adenoviral transduction of tumor cells to express molecules relevant for immune activation has been explored as another strategy to improve tumor immunogenicity. Of many immunostimulatory mediators evaluated in the context of gene-modified tumor cell vaccines, GM-CSF emerged as the most potent in generating protective antitumor immunity (Antonia et al., 2002; Dranoff, 2002; Dranoff et al., 1993; Soiffer et al., 1998). Vaccination using irradiated, GM-CSF producing tumor cells (GVAX) consistently induced antitumor immunity across a wide range of experimental tumor models (Dranoff, 2002). While both Flt-3L and GM-CSF producing B16 tumor cells induced DC expansion in vivo, GM-CSF was more potent in stimulating protective tumor immunity in a direct comparison (Mach et al., 2000). The activity of GM-CSF involves the recruitment and maturation of DCs at vaccination sites, which express high levels of costimulatory molecules such as CD80 as well as CD1d, the minor nonclassical MHC protein, and induce both cellular and humoral responses (Dranoff et al., 1993; Mach et al., 2000; Zarei et al., 2009). Phase 1/2 clinical trials of GVAX have been conducted for a number of tumors (Jaffee et al., 2001; Salgia et al., 2003; Simons et al., 1997, 1999; Small et al., 2007; Soiffer et al., 2003). In most GVAX studies in solid tumors, enhanced tumor infiltration of lymphocytes was seen in a majority of the patients studied following vaccination, with objective clinical responses observed in a small proportion. In CLL patients who received allogeneic hematopoietic stem cell transplantation, early posttransplant vaccination by a variation of GVAX—a mixture of irradiated autologous tumor cells and GM-CSF producing bystander cells—induced significant expansion of tumor-reactive T cells (Burkhardt et al., 2013). As an alternative approach, the use of allogeneic tumor cell lines to replace autologous tumor cells, thus by-passing the need to harvest autologous tumors, had also been studied in a number of tumor types (Michael et al., 2005). However, a phase 3 clinical trial of this approach in advanced prostate carcinoma patients failed to demonstrate improved clinical outcome with an allogeneic prostate carcinoma cells secreting GM-CSF combined with docetaxel vs docetaxel alone (Madan, Aragon-Ching, Gulley, & Dahut, 2011). One potential limitation of this approach is that the antigenic repertoire of the allogeneic tumor cell lines may not reflect sufficiently the antigenic profile of individual patient tumors, particularly in a study with a large patient population (Gerlinger et al., 2012).

The modest clinical efficacy of GVAX can be attributed to a number of factors, some of which are common challenges to all cancer vaccines: the presence of suppressor cells such as Tregs and MDSCs either systemically or in the tumor microenvironment, the expression of inhibitory receptors by tumor and immune cells alike, and the infiltration of dysfunctional immune cells. Nonetheless, factors inherent to GM-CSF biology also likely play a role. Though generally considered as a pro-inflammatory cytokine, GM-CSF is now known to exert dual controls on immunity in a context dependent manner, as exemplified by development of inflammatory disease in GM-CSF-deficient mice (Enzler et al., 2003). Indeed,

GM-CSF-deficient mice also show a decrease in the frequency of CD4+ CD25+Fox3+ regulatory T cells, indicating that GM-CSF contributes to regulatory T cell homeostasis; under certain contexts, GM-CSF can also stimulate Treg expansion in vivo (Jinushi et al., 2007; Kared et al., 2008). This function of GM-CSF is in part related to the induction of milk-fat globule epidermal growth factor-8 (MFG-E8) in DCs and macrophages, which mediates uptake of apoptotic cells and downstream antigen degradation pathways that result in Treg induction and tolerance (Jinushi et al., 2007). In the B16 tumor model, combinatorial treatment with GVAX and a dominant-negative form of MFG-E8 attenuated Treg induction and potentiated GM-CSF efficacy, providing strong evidence for Treg targeting as a mean to increase the therapeutic potential of GVAX (Jinushi et al., 2007). A number of approaches to modulate Treg cells as well as inhibitory receptors on T cells have now shown efficacy both in animal models and in the clinic, paving the way for combinatorial immunotherapy that incorporates active immunization and immune modulators—an approach that can apply for all vaccine types.

One of the first combinatorial approaches in the context of GVAX that showed therapeutic synergy was blockade of the inhibitory receptor CTLA-4. In mice, a combination of GVAX and CTLA-4 blockade eradicated established B16 tumors that were unaffected by either treatment alone (van Elsas, Hurwitz, & Allison, 1999). Subsequent studies in murine models demonstrated that CTLA-4 blockade altered the intratumoral effector CD8+T cells to Foxp3+Tregs ratio, favoring infiltration of IFNγ+ CD8+T cells while dampening Tregs, thus enhancing GVAX-induced antitumor immunity (Quezada, Peggs, Curran, & Allison, 2006). Interestingly, though CTLA-4 can be expressed on both effector cells and Treg cells, the antitumor efficacy of ipilimumab was found to require antibody-mediated Treg depletion in murine models (Bulliard et al., 2013; Selby et al., 2013; Simpson et al., 2013).

The possible clinical efficacy of this combination was first reported by Hodi et al., who observed extensive tumor necrosis and immune cell infiltration following CTLA-4 blockade in patients who had previously received autologous GVAX (Hodi et al., 2003). In a follow-up study with shorter intervals between vaccination and ipilimumab administration, a high ratio of CD8+T cell/FoxP3+ Tregs in tumor biopsies after therapy was associated with immune-mediated tumor destruction; moreover, prolonged stable disease and/or regressions were observed in 8 of 11 metastatic melanoma patients and a subgroup of patients with ovarian carcinoma

(Hodi et al., 2008). An allogeneic GVAX + ipilimumab combination has also been explored in patients with metastatic castration-resistant prostate cancer, with meaningful clinical responses observed in a subset of patients (van den Eertwegh et al., 2012). In this study, ipilimumab and vaccination were administered concurrently, and some inflammatory toxicities were observed; improved survival was associated with the induction of antibody responses to prostate carcinoma-associated antigens. In a small comparison of GVAX + ipilimumab vs ipilimumab alone, pancreatic cancer patients who received the combination treatment showed superior overall survival, increased frequencies of tumor-specific T cells, and enhanced T cell repertoire (Le et al., 2013).

Blockade of PD-1 (programmed death-1) or its ligand PD-L1, another inhibitory pathway that is often harnessed by tumor cells for immune escape, has also been shown to augment GVAX-induced antitumor immunity in mouse tumor models (Li et al., 2009). The relevance of the PD-1/PD-L1 pathway in modulating GVAX efficacy was further supported by recent data demonstrating GVAX-mediated upregulation of PD-L1 on tumor cells in a pancreatic ductal carcinoma model; importantly, PD-1 blockade in this model augmented GVAX-induced tumor-specific IFNγ production by CD8+T cells (Soares et al., 2015). In patients, PD-1 expression was found on postvaccination tumor-infiltrating CD8+T cells in a clinical study of autologous GVAX for alveolar soft-part sarcoma (Goldberg et al., 2015). Dual blockade of CTLA-4 and PD-1 appears to further enhance protective antitumor immunity generated by either GVAX or the Flt-3L-based whole-cell vaccine in mouse tumor models (Curran, Montalvo, Yagita, & Allison, 2010; Duraiswamy, Kaluza, Freeman, & Coukos, 2013). In addition, depletion of Tregs using cyclophosphamide or by targeting the folate receptor 4, expressed at high levels on naturally occurring Tregs as well as Tregs induced by TGF-β, have also been explored (Liang, Moskalenko, Van Roey, & Jooss, 2013; Wada et al., 2009). In a mouse pancreatic tumor model, low-dose cyclophosphamide in combination with GVAX-induced intratumoral tertiary lymphoid structures that are indicative of immune activation in an otherwise nonimmunogenic tumor (Lutz et al., 2014). Recently, the regimen of cyclophosphamide following GVAX was shown to be well tolerated in patients (Lipson et al., 2015).

Cancer-targeted therapies, typically small-molecule inhibitors developed to target cancer-specific pathways or mutations, also have the potential to augment cancer vaccine responses (Vanneman & Dranoff, 2012). For example, the IAP inhibitor LCL161 appears to enhance costimulation through the activation of NFκB2 signaling, thereby potentiating GVAX activity (Dougan et al., 2010; Infante et al., 2014). Inhibitors to BRAF, developed for the $BRAF^{V600E}$ mutation frequently seen in melanoma patients, have also been shown to contribute to antitumor immunity by inducing tumor antigen expression and altering the profiles of tumor-produced chemokines, resulting in an increased intratumoral CD8+T cell:Treg ratio as well as NK cell infiltration, although exhaustion markers such as PD-1 and TIM-3 were also upregulated (Frederick et al., 2013; Knight et al., 2013). In addition to Treg cells, a high-dose GM-CSF-producing vaccine-induced MDSCs, which attenuated the antitumor response (Serafini et al., 2004). MDSC-mediated T cell suppression involves two major mechanisms: induction of the inducible nitric oxide synthetase (iNOS), leading to nitric oxide production and/or the activation of arginase (ARG) activity and L-arginine catabolism (Iwasaki-Arai, Iwasaki, Miyamoto, Watanabe, & Akashi, 2003). Moreover, consistent with observations from peptide vaccines, a high pretreatment level of MDSCs was also associated with unresponsiveness to GVAX + ipilimumab in castration-resistant prostate cancer patients (Santegoets et al., 2014). Hence, myeloid cell suppression, whether preexisting or induced under certain conditions by GM-CSF, presents yet another challenge for vaccine efficacy.

A number of approaches are currently being investigated for myeloid cell targeting (Talmadge & Gabrilovich, 2013). A pharmacologic inhibitor of CSF-1R signaling, a key pathway contributing to the proliferation and survival of monocytes/macrophages, attenuated MDSC tumor infiltration (Priceman et al., 2010). Blockade of stem-cell factor 1 (SCF1) may reduce MDSC in vivo expansion as well (Pan et al., 2008). A peptide that targets S100 family proteins expressed by MDSCs might also be developed as a therapeutic (Qin et al., 2014). The two main mediators of MDSC suppression, ARG-1 and iNOS, have also been explored as therapeutic targets (Noonan, Ghosh, Rudraraju, Bui, & Borrello, 2014; Serafini et al., 2006). In a recent study, treatment with tadalafil, an inhibitor of phosphodiesterase 5 that downregulates synthesis of ARG-1 and iNOS in MDSCs, was shown to reduce the frequencies of MDSCs in the tumor and circulation in patients with head and neck cancer (Weed et al., 2015). Moreover, T cells isolated from patients treated with tadalafil also showed augmented responses to tumor lysates (Califano et al., 2015). The efficacy of tadalafil in combination with a vaccine against telomerase (GV1001), recombinant GM-CSF, gemcitabine, and radiotherapy for patients with pancreatic cancer is currently underway (Crocenzi et al., 2013).

Efforts to improve GVAX efficacy have also focused on ways to facilitate DC activation and cross-presentation with adjuvants. The adjuvant activity of CpG ODN in this context was demonstrated for neuroblastoma-GVAX. A formulation of GVAX coupled with LPS also induced superior protective immunity against tumor challenge that was associated with enhanced DC activation and increased T cell tumor infiltration (Davis et al., 2011; Sandler et al., 2003). Recently, Narusawa et al. identified plasmacytoid DCs (pDCs), a DC subpopulation known for the capacity to produce high levels of IFNα upon stimulation, as an important mediator of GM-CSF induced immunity (Narusawa et al., 2014). The formulation of a TLR-7 agonist with GVAX further activated pDCs, which in turn potentiated antitumor responses. The stimulator of interferon (IFN) genes (STING) pathway, a TLR-independent pathway that is integral to innate defense against cytosolic double stranded DNA, can also be harnessed to enhance immune activation. Notably, STING activation triggers type 1 IFN responses, which are potent stimulators of DCs function. GVAX formulated with a STING agonist (STINGVAX) generated enhanced antitumor immunity compared to GVAX alone, and PD-1 blockade further potentiated tumor control (Fu et al., 2015). A prime/boost regimen of allogeneic GVAX (prime) and *L. monocytogenes*-expressing mesothelin plus low-dose cyclophosphamide also might improve overall survival for patients with metastatic pancreatic cancer in comparison to allogeneic GVAX + cyclophosphamide alone (Koks et al., 2015).

As data from both clinical and preclinical studies continue to shed light on the mechanisms influencing vaccine efficacy, research efforts are underway to understand in more detail the mechanisms of protective tumor immunity. The analysis of serum samples from patients responding to GVAX has identified antibody targets that likely contribute to immune-mediated tumor control. Therapy-induced antibodies against major histocompatibility chain-related protein-A (MICA), a ligand for the activating NK cell receptor NKG2D, were shown to overcome immune suppression mediated through soluble MICA and promote cytotoxicity (Jinushi, Hodi, & Dranoff, 2006). The induction of antibodies against multiple proangiogenic factors was correlated with improved outcomes in some vaccinated solid and hematologic cancer patients (Piesche et al., 2015; Schoenfeld et al., 2010). Some of the targets of CTL responses in patients treated with GVAX have also been defined. These include the melanoma inhibitor of apoptosis protein in some melanoma patients and mesothelin in some pancreatic cancer patients (Schmollinger et al., 2003; Thomas et al., 2004). Since whole-cell vaccines contain a multitude of peptides, much more work is required to understand more fully the range of CTL targets elicited with GVAX. In recent years, significant advances seen in deep sequencing techniques and powerful bioinformatics tools have opened the possibility for more thorough interrogations of the potential role of cancer neoantigens.

3.3 Neoantigens

Cancer immunoediting is a process through which T cell recognition of tumor antigens shapes tumor cell immunogenicity—T cell-mediated immune attacks eliminate highly immunogenic tumor cell clones, leaving behind subsets of tumor cells that can evade immunosurveillance (DuPage, Mazumdar, Schmidt, Cheung, & Jacks, 2012; Matsushita et al., 2012; Rooney, Shukla, Wu, Getz, & Hacohen, 2015). Immunoediting is also highly dynamic, with new mutations iteratively driving immune selection. Tumors with high mutational burdens such as melanoma are more likely to generate specific T cell responses against neoepitopes. Coulie et al. first reported an endogenous T cell response against a melanoma-specific neoepitope (Coulie et al., 1995). Using a method that involves whole-tumor exome sequencing and in silico screening for high-affinity candidate T cell epitopes, Robbins et al. subsequently identified neoantigen-specific effector CD8+T cells in melanoma patients who responded to adoptive-cell therapy (Dudley et al., 2002; Robbins et al., 2013). Substantial evidence for immunity against tumor neoantigens has since been generated (Lu et al., 2013; Rajasagi et al., 2014;van Rooij et al., 2013; Wick et al., 2014). Recognition of neoepitopes by CD4+T cells has also been reported in patients with melanoma and is a relatively frequent event (Linnemann et al., 2015). Tran et al. documented the clonal expansion of neoantigen-specific CD4+T cells in the setting of efficacious adoptive T cell transfer in a patient with metastatic cholangiocarcinoma (Tran et al., 2014). Retreatment of the patient with highly purified neoantigen reactive T cells upon disease progression effectuated a second period of disease control. CD4+T cell responses to neoepitopes are also characteristic of a number of transplantable murine tumor models (Kreiter et al., 2015).

The antitumor efficacy of immune checkpoint blockade similarly involves responses to neoepitopes. In a murine carcinogen-induced tumor model with a high mutation burden, cancer neoantigens were shown to constitute a major class of T cell rejection antigens post checkpoint blockade therapy with either anti-CTLA-4 or anti-PD-1 antibodies (Gubin et al., 2014). These neoepitopes also conferred protective immunity when used as cancer vaccines in the same models. In patients, neoepitopes were found to be targeted with antibodies to CTLA-4 or anti-PD-1, and there is a positive association between nonsynonymous mutation burden and response to therapy (Champiat, Ferte, Lebel-Binay, Eggermont, & Soria, 2014; Rizvi et al., 2015; Wolchok et al., 2013).

Neoantigens are increasingly being recognized as attractive targets for cancer vaccines. Schumacher et al. showed that epitopes derived from mutated isocitrate dehydrogenase type 1 (IDH1), an oncogenic lesion critical to the pathogenesis of low grade gliomas can be targeted with a vaccine, engendering specific CD4+T cells that contribute to tumor protection (Schumacher et al., 2014; Yan et al., 2009). In contrast to this recurrent lesion, most somatic mutations in cancer are heterogeneous, requiring the development of personalized vaccines. Novel methods for the discovery of immunogenic neoepitopes are being developed and applied for this purpose (Duan et al., 2014; Yadav et al., 2014). Indeed, a messenger RNA vaccine encoding a series of predicted MHC class II-binding epitopes induced cytotoxic T cells and protective tumor immunity in murine models (Kreiter et al., 2015). These experimental data are rapidly being translated into human trials, and an initial report of this approach (discussed further below) shows encouraging early results (Carreno et al., 2015).

3.4 Ex Vivo Dendritic Cell Vaccines

Mature DCs mediate three critical functions central for the clinical success of a cancer vaccine: (a) the capture of extracellular tumor-derived peptides or full-length proteins and subsequent processing to produce MHC-peptide complexes; (b) migration into lymph node for presentation of peptide–MHC complexes to T cells; (c) expression of costimulatory molecules, such as CD80, CD86, and CD40L, and production of cytokines such as IL-12 that are required for effector T cell activation and differentiation (Mellman & Steinman, 2001). Signals from PRRs as well as exposure to type I IFNs are also important in driving DC maturation. Immature DCs, found commonly in tumors, may also capture and present antigens to T cells, but in the absence of costimulatory molecules and pro-inflammatory cytokines, instead trigger Tregs and anergic T cells (Steinman, Hawiger, & Nussenzweig, 2003). The tumor microenvironment is known to produce a number of factors that suppress DC maturation and function (Gajewski,

Schreiber, & Fu, 2013). In patients with late stage disease, who are typically enrolled in various vaccine studies, systemic immunosuppression is common and could further hinder immune activation. Based on these observations, a therapeutic vaccination strategy in which autologous DCs are expanded and activated with cytokine combinations and then loaded with antigens was developed. In early studies, two major sources of DCs were used: circulating monocytes that were cultured in the presence of GM-CSF and IL-4, or CD34+ hematopoietic progenitors that were matured in a cocktail of GM-CSF, Flt-3L, and TNF (Banchereau et al., 2001; Nestle et al., 1998). DCs generated from CD34+ progenitors appeared to constitute two distinct mature DC subsets, one of which was similar to dermal Langerhans cells (LCs). Both ex vivo DC approaches were capable of generating antigen-specific effector responses. Importantly, DC-based delivery of gp100 was superior to a gp100 peptide vaccine in murine models (Yang et al., 1999).

Since these early studies, numerous phase 1/2 trials have explored DC vaccines for a wide range of cancer types, and this approach has proven to be safe (Ueno et al., 2010). Sipuleucel T, an autologous vaccine consisting of peripheral blood mononuclear cells pulsed with a fusion protein composed of full-length prostatic acid phosphatase and GM-CSF (to enhance antigen-presenting cell function) was approved by the FDA in 2010 as the first cell-based cancer immunotherapy, based on an overall survival benefit of 4 months in metastatic prostate cancer patients (Kantoff, Higano, et al., 2010). Data from the pivotal phase 3 clinical trial of 514 patients showed that the greatest therapeutic efficacy was associated with lower baseline PSA levels, suggesting that patients with a lesser disease burden might show better therapeutic benefits (Schellhammer et al., 2013). Two other autologous DC vaccines are currently in on-going phase 3 studies, including AGS-003, a vaccine comprised of monocyte-derived DCs electroporated with tumor-derived RNA and synthetic CD40L-encoding RNA for patients with advanced renal cell carcinoma. An earlier phase 2 study of AGS-003 suggested that improved clinical outcomes might be associated with the development of effector CTLs (Amin et al., 2015).

To improve the potency of these first generation DC vaccines, investigators have focused on optimizing ex vivo DC generation protocols and overcoming immunosuppressive mechanisms within the tumor microenvironment. One approach to improve the maturation status of DCs involves the addition of prostaglandin E2 (PGE-2) plus a cocktail of inflammatory cytokines including IL-1β, TNF-α, and IL-6. This manipulation endows

DCs with superior immunogenicity and migratory capacity compared to DCs matured with the conventional protocol that uses GM-CSF and IL-4 alone (Jonuleit et al., 1997; Luft et al., 2002). DCs matured in PGE-2 plus inflammatory cytokines expressed costimulatory molecules, but produced little IL-12p70, a cytokine essential for CD8+ T cell effector and memory responses (Pearce & Shen, 2007; Starbeck-Miller, Xue, & Harty, 2014). CD40L and IFN-γ induce transcription of the two IL-12 subunits in DCs and might enhance maturation further (Hayes, Wang, & Norcross, 1995; Mosca et al., 2000). Other studies have reported that DCs can be polarized into type 1 or type 2 functionalities based on additional microenvironment cues (Mailliard et al., 2004). TLR signaling, type I IFNs, and IFNγ appear critical for type 1 polarization (Del Corno et al., 2009; Mailliard et al., 2004). Notably, type 1-polarized DCs show enhanced production of IL-12 and increased migratory capacity. Vaccination with type 1-polarized, glioma-associated antigen (GAA)-pulsed DCs administrated in poly-ICLC adjuvant was well tolerated and induced GAA-specific responses that appeared associated with improved survival (Okada et al., 2011).

In another small clinical study, stage IV melanoma patients were vaccinated with DCs matured with CD40L and IFNγ and pulsed with gp100 peptides (Carreno et al., 2013). Six of the seven patients in the study developed immunity against gp100, and three showed clinical responses; moreover, progression-free survival was correlated with DC vaccine-derived IL-12 levels. This approach was subsequently adopted for a personalized neoantigen-based DC vaccine (Carreno et al., 2015). In this later study, tumor samples from stage III melanoma patients were analyzed for tumor missense mutations. Peptide sequences spanning the mutation were selected in silico for MHC binding and then pulsed onto autologous IL-12 producing DCs for vaccination. Peptides with the highest MHC binding affinity exhibited the strongest immunogenicity. Neoantigen-specific T cells isolated postvaccination exhibited CTL activity against cognate peptides at low concentration, suggesting high-avidity antigen recognition. A comparison of pre- and postvaccination responses revealed an increase in the breadth and diversity of TCR repertoire. Collectively, these observations provide strong evidence that DC vaccination has the potential to expand neoantigen-specific T cells.

The proteasomal-mediated processing of tumor antigens to peptides is another important consideration for DC vaccines. When normal cells are exposed to inflammatory mediators, the constitutive proteasome (cP), expressed on most cells, is turned into an inflammatory proteasome (iP). The iP shows different cleavage preferences than cP, altering the pool of peptides presented on the cell surface (Kloetzel, 2004). Cancer cells may block the switch from cP to iP, perhaps as a mechanism of immune escape. As such, a discrepancy in peptide targets may occur in the setting of DC vaccination. Through ex vivo activation and maturation, DCs switch from cP to iP to present iP-processed peptides to CD8+T cells, whereas tumor cells present cP-processed antigens on the cell surface—a mismatch of antigens that could hinder immune recognition by DC-activated CD8+T cells. To improve DC vaccine efficacy, Dannull et al. developed a protocol in which autologous monocytes were first matured ex vivo, transduced with RNA encoding melanoma antigens, and then again transduced with a siRNA sequence that inhibits conversion to iP, thus allowing the antigen transduced DCs to retain cP and antigen-processing machinery similar to tumor cells (Dannull et al., 2013). In a phase 1 clinical trial of this approach, enhanced antigen-specific CTL responses were observed with the double-transfected DCs.

Optimization of DC functionality ex vivo may still be limited by inhibitory mechanisms in vivo. To generate a microenvironment more permissive for DC function, Mitchell et al. explored a novel approach to precondition vaccine-sites using a potent recall antigen (Mitchell et al., 2015). This group employed tetanus toxoid as the preconditioning agent for DCs that were pulsed with RNA encoding cytomegalovirus proteins, which are expressed in most glioblastomas; a small phase 1 study showed that the preconditioning regimen significantly enhanced DC migration to lymph nodes, which might be associated with improved patient outcomes. Additional investigations in mice demonstrated that CCL3, a chemokine produced by CD4+T cells elicited with the recall response, promoted the migration of DCs to the lymph node and the induction of antitumor cytotoxic response.

Notwithstanding these important advances for ex vivo DC vaccines, substantive challenges that remain are the complexity and cost of generating clinical-grade DCs for individual patients. Sipuleucel T, though FDA approved and available as a therapeutic option, has not been widely applied as a standard prostate cancer treatment, perhaps due to the high treatment cost and modest impact on overall survival. An alternative and more cost efficient approach to harness DC functionality is the direct targeting of DCs in situ.

4. IN SITU VACCINATION

4.1 In Situ DC Targeting

There are several subsets of DCs that differ in PRR expression profiles and abilities to stimulate cellular and humoral responses. Three subsets can be found in human blood: pDCs, which specialize in the production of type I IFNs and antiviral responses; and two conventional DC subsets: CD1c (BDCA-1)+DCs and CD141(BCDA-3)+ DCs. CD141+ DCs are characterized by high TLR-3 expression and a potent capacity to cross-present soluble and necrotic cell antigens, particularly in the presence of TLR-3 agonists (Jongbloed et al., 2010). CD1c+ DCs can differentiate into skin LCs and prime CD103+CD8+ cytotoxic T cells through a mechanism that involves TGF-β production (Milne, Bigley, Gunawan, Haniffa, & Collin, 2015; Nizzoli et al., 2013; Yu et al., 2013). In the skin, LCs specialize in CD8+T cell priming, while CD14+ DCs promote humoral immunity (Mildner & Jung, 2014). Lymph node resident DCs, but not circulating DCs are responsible for the induction of protective immunity against protein vaccines (Anandasabapathy et al., 2014). Further, circulating CD141+DCs and CD1c+DCs are poor inducers of Th2 responses in comparison to their counterparts in the lymph node (Segura et al., 2012). Thus, targeting resident DCs in situ might harness the potent stimulatory functions of DC subsets that cannot otherwise be accessed for ex vivo DC vaccines.

One approach for DC targeting involves the conjugation of a candidate antigen to a monoclonal antibody directed against a specific DC receptor target. C-type lectin receptors such as DC-SIGN and CD23 mediate the internalization of receptor-bound antigens and thus might be exploited for cancer vaccination (Figdor, van Kooyk, & Adema, 2002). The lectin CLEC9A, expressed on CD141+DCs, recognizes actin that is found on necrotic cells, facilitating their internalization and cross-presentation (Sancho et al., 2009). Targeting of CLEC9A with HIV gag-24 peptide conjugated to CLECA mAb (CLEC9A/gag-24) induced Th1 and CD8+T cells responses against gag-24 in mouse models (Idoyaga et al., 2011). The combination of TLR7/8 agonists and a conjugate of MART-1/DCIR, the tyrosine-based inhibitory motif-containing DC immunoreceptor that is expressed on many subsets of DCs and other antigen-presenting cells, also induced specific CD8+T cell responses (Klechevsky et al., 2010).

The lectin receptor target that has progressed the furthest in development is DEC-205, expressed on lymph node CD141+DCs (Meixlsperger et al.,

2013). In the mouse, targeting antigens through DEC-205 (CD205) in the absence of activation signals result in Treg expansion and tolerance induction, whereas antigen internalization through the same receptor on activated DCs enhances antigen presentation (Hawiger et al., 2001; Kretschmer et al., 2005; Mahnke et al., 2000). A DEC-205/HIV gag-24 plus poly-ICLC vaccine elicited potent anti-gag-24 Th1 and CD8+T cell responses in non-human primates (Flynn et al., 2011). Minimal responses were noted without poly-ICLC, supporting the notion that activation signals for DCs are critical. Harnessing the property of DEC-205, a recent phase 1 study of patients with advanced malignancies investigated a conjugate of full-length NY-ESO-1 and DEC-205 mAb plus resiquimod and poly-ICLC (Dhodapkar et al., 2014). The NY-ESO-1/DEC-205 mAb conjugate induced detectable cellular and humoral immunity, which in some patients was associated with disease stabilization.

Other studies have investigated in situ DC targeting through TLRs. DC subsets express different TLRs, and thus might be targeted with specific TLR agonists. An ovalbumin/TLR7/8 conjugate vaccine activated migratory DCs and lymph node resident DCs to induce robust anti-OVA immunity (Kastenmuller et al., 2011). Vaccination with conjugates of long tumor antigen-derived peptides and TLR-2 agonists generated protective immunity against transplantable murine melanoma and lymphoma lines (Zom et al., 2014). TLR targeting is not limited to delivering peptide targets, as a TLR/siRNA approach, which allows for delivery of siRNA silencers, has also been used to inhibit Stat3 expression in tumor-associated myeloid cells (Kortylewski et al., 2009). Other potential receptor targets for in situ DC targeting include the Fcγ receptors (FcγRs), which are responsible for internalization of opsonized materials and immune complexes (Guilliams, Bruhns, Saeys, Hammad, & Lambrecht, 2014). High-affinity engagement of immune complexes with activating FcγRs is known to induce DC maturation and cross-presentation (Kalergis & Ravetch, 2002). Indeed, passive administration of rituximab, the therapeutic monoclonal antibody against CD20, which mediates direct cytotoxic killing of CD20+ leukemic cells through mechanisms that are dependent on FcγR engagement (ADCC), induces long-lasting antitumor immune responses with a single course of treatment (Abes, Gelize, Fridman, & Teillaud, 2010). Recently, DiLillo et al. showed that this vaccine-like effect involves engagement of FcγRIIA expressed on DCs, leading to DC activation and cross-presentation (DiLillo & Ravetch, 2015).

4.2 Immunogenic Cell Death

Tumor necrosis or apoptosis results in the production of a number of factors that can be recognized by DCs, leading to distinct downstream functional consequences. Damage-associated molecular patterns (DAMPs) that are released upon "immunogenic cell death" may activate DCs to promote antitumor immunity, providing an opportunity for in situ vaccination and DC targeting. Immunogenic cell death may be caused by certain chemotherapeutic agents or radiotherapy that results in activation of protective immunity against cancer (Green, Ferguson, Zitvogel, & Kroemer, 2009). The extracellular release of high-mobility group protein B1 (HMGB1) and ATP by dying cells, which binds PRRs such as TLR4 and P2X7 purinergic receptors on DCs, respectively, activates downstream signals that result in DC maturation and production of inflammatory cytokines (Krysko et al., 2012). Obeid et al. reported that cytotoxic agents that cause ER stress, such as anthracyclines, induce the cell surface translocation of calrecticulin, which enhances DC-mediated phagocytosis and cross-presentation (Obeid et al., 2007). Recently, Chao and colleagues showed that the dominant pro-phagocytic signal by calrecticulin can be countered by expression of CD47 on tumor cells (Chao et al., 2010). CD47 binds signal regulatory protein alpha (SIRPα) expressed on phagocytic cells including DCs and macrophages and inhibits phagocytosis (Chao, Weissman, & Majeti, 2012). Antibody-mediated CD47 blockade enhanced phagocytosis and augmented T cell priming, resulting in induction of antitumor T cell responses (Tseng et al., 2013).

The abscopal effect—regression of tumors at sites distal to local radiotherapy—is one example of immunogenic cell death triggering protective antitumor immunity. Radiation induces release of DAMPs by dying cells. Moreover, exposure of tumor DNA from dying cells can also stimulate type 1 IFN production by DCs via the STING pathway (Deng et al., 2014). Radiation-induced immune activation by these mechanisms, however, can be weakened by immunoregulatory factors from surviving, radiation-resistant tumor cells and stromal cells. Recently, several reports showed that TGF-β, expression of PD-1 on T cells, and its ligand PD-L1 on tumor and myeloid cells were all factors that could diminish abscopal effects in vivo (Park et al., 2015; Vanpouille-Box, Diamond, et al., 2015; Vanpouille-Box, Pilones, Wennerberg, Formenti, & Demaria, 2015).

Given the frequent expression of inhibitory receptors by immune cells in the tumor microenvironment, a potential synergy between the radiotherapy-induced abscopal effect and checkpoint blockade has been explored (Golden, Demaria, Schiff, Chachoua, & Formenti, 2013;

Postow et al., 2012). A detailed case report by Postow et al. of a melanoma patient treated with ipilimumab and radiation therapy highlighted the therapeutic potential of this combination. Immune analysis revealed that ipilimumab treatment increased the frequencies of activated CD4+T cells, whereas subsequent radiotherapy further induced NY-ESO-1 specific, IFNγ-producing CD4+T cells. CD14+ monocytes also showed an increase in HLA-DR expression postradiotherapy. The immunological changes were associated with a clinical response, suggesting that radiation therapy may have functioned as an in situ vaccine in a patient previously primed with CTLA-4 antibody blockade. Further, in a follow-up retrospective analysis, patients who received radiotherapy during maintenance phase of ipilimumab had a median overall survival of 39 months in comparison to 9 months seen in patients who received radiotherapy during induction phase of ipilimumab, patient selection bias notwithstanding (Barker et al., 2013). This suggests that correct timing of radiotherapy (in situ vaccine) and ipilimumab is likely important for optimal therapeutic benefits (Vanpouille-Box, Pilones, et al., 2015). However, when explored in a phase 1 study involving 18 patients with metastatic melanoma, the combination of radiotherapy and ipilimumab failed to improve clinical outcome in all patients (Twyman-Saint Victor et al., 2015). In corresponding mouse studies, the lack of response was found to be associated with PD-L1 upregulation on melanoma cells and T cell exhaustion, suggesting that tumor cells exploit the PD-L1/PD-1 pathway as an escape mechanism. Importantly, blockade of either PD-1 or PD-L1 appeared to be sufficient to reverse radiotherapy-induced T cell exhaustion in mouse models (Park et al., 2015; Twyman-Saint Victor et al., 2015). Together, these data provide strong rationale for dual CTLA-4 and PD-1/PD-L1 blockade to enhance abscopal responses to radiotherapy. To induce abscopal responses further by promoting DCs maturation and recruitment, Golden et al. recently proposed the use of systemic GM-CSF concurrently with radiotherapy (Golden et al., 2015). In the proof-of-principle study, abscopal responses were observed in a subset of patients under this treatment regimen. Thus, combinatorial approaches that incorporate strategies to overcome immunosuppressive pathways and potentiate immune activation warrant further investigation.

Another approach to enhance immune activation by immunogenic cell death, as is in the case of traditional cancer vaccines, is to provide additional immune stimuli in the form of adjuvants. The combination of local radiotherapy and intratumoral injection of TLR-9 agonists to the same site induced tumor-reactive memory CD8+T cells and objective clinical responses in a subset of lymphoma patients (Brody et al., 2010). Interestingly, in the same study, tumor cells harvested posttreatment from patients who failed to respond were shown to induce regulatory T cells rather than tumor-reactive T cells in vitro, suggesting involvement of other inhibitory mechanisms in these patients. In a mouse breast carcinoma model, the TLR-7 agonist imiquimod in combination with radiotherapy, but not either treatment alone, induced regression of the primary tumor and inhibited growth of secondary tumors at a distal site; moreover, this effect of imiquimod was CD8+T cell dependent, indicating immune activation (Dewan et al., 2012). The addition of cyclophosphamide as a Treg-depleting agent prior to radiotherapy further improved efficacy. In another case report, an objective clinical response and activation of antitumor immunity was observed in a patient with advanced facial embryonal rhabdomyosarcoma who received sequential intratumoral and intramuscular injections of poly-ICLC (Salazar, Erlich, Mark, Bhardwaj, & Herberman, 2014). In this setting, poly-ICLC promoted tumor cell death through innate mechanisms and DC activation through TLR signaling.

In the case of chemotherapy, a tumor cell autonomous contribution to immune activation may be an important determinant of chemotherapy efficacy. Anthracyclines induce TLR-3 on tumor cells leading to production of type I IFNs, which are otherwise typically produced upon viral infection (Sistigu et al., 2014). Type I IFNs act in both autocrine and paracrine fashions to stimulate release of the chemokine CXCL10, which likely plays a role in T cell recruitment. In patients, a type I IFN-related gene signature is predictive of clinical responsiveness to chemotherapy. The stimulation of type I IFN release is akin to an antiviral response, supporting the idea that a viral-infection-like inflammatory state in the tumor microenvironment is conducive for immune activation and generation of protective antitumor immunity. In a recent report, a single high dose of cyclic diguanylate (c-di-GMP), a ligand for STING, which killed tumor cells directly by induction of caspase 3, followed by repeated administrations of low-dose c-di-GMP, showed equal therapeutic efficacy as a combination of Lm-based vaccine and c-di-GMP (Chandra et al., 2014).

4.3 Oncolytic Viruses

The idea of using viruses to combat cancer has been studied for over a century, but oncolytic viral therapy has only recently emerged as a viable cancer treatment option, reflecting advances in virology and genetic engineering.

Oncolytic viruses can mediate tumor killing via three complementary mechanisms: (a) selective infection of tumor cells leading to their direct lysis (also known as viral oncolysis); (b) virus-induced disruption of the tumor vasculature; and (c) induction of antitumor immunity (Lichty, Breitbach, Stojdl, & Bell, 2014). Since the first two mechanisms have been reviewed in a number of recent works, the discussion here focuses on immune-mediated mechanisms of oncolytic viral therapy efficacy.

Early studies on oncolytic viruses largely focused on viral oncolysis as the main effector function, which, in turn, depends on efficiency of viral replication in infected cells (Chiocca, 2002). However, in spite of a relatively modest ability to replicate in infected cancer cells, oncolytic viruses such as adenovirus displayed some clinical activity, particularly when given in combination with other therapies (Khuri et al., 2000). Some agents, such as cyclophosphamide prolong viral replication in infected cells and thus enhance oncolysis; however, other chemotherapies, as discussed earlier, induce immunogenic cell death (Fulci et al., 2006; Green et al., 2009). The combination of a herpes simplex virus-1 (HSV-1)-derived oncolytic virus and mitoxantrone, a chemotherapeutic drug that induces immunogenic cell death, enhanced in vivo efficacy in a T cell-dependent manner (Workenhe, Pol, Lichty, Cummings, & Mossman, 2013). In another report, the combination of an oncolytic Parvovirus H-1 (H-1PV) with cisplatin or vincristine enhanced DC phagocytosis of infected tumor cells and maturation (Moehler et al., 2011).

Indeed, oncolytic viruses-induced cell deaths are mostly immunogenic. The HSV-1 virus induced expression of heat-shock proteins, a class of DAMPs, on the surface of infected tumor cells, resulting in their phagocytosis by DCs (Benencia, Courreges, Fraser, & Coukos, 2008; Toda, Martuza, & Rabkin, 2000; Todo et al., 1999). DAMPs together with viral-induced type I IFNs, abundant in infected tumors, also stimulate DC maturation, cross-presentation, and production of chemokines for recruitment of T cells into the tumor microenvironment (Benencia et al., 2005). In the highly aggressive 4T1 breast tumor model, HSV-1-infected primary tumors showed infiltration of CD4+T cells and CD8+T cells; control of metastatic spread was observed in immunocompetent animals, but not in SCID mice, which lack both T and B cells (Thomas & Fraser, 2003). Newcastle disease virus (NDV), an avian virus that is not pathogenic in humans, also triggered the release of HMGB1 by glioblastoma cells and improved DC antigen presentation, which likely contributed to the CD8+T cell-dependent prolonged survival of tumor-bearing mice (Koks et al., 2015). The antitumor effects of NDV, particularly in controlling uninfected metastatic tumors at distal sites, can be further potentiated by CTLA-4 blockade (Zamarin et al., 2014).

Oncolytic viruses have also been shown to activate DCs directly. Reovirus, a naturally occurring oncolytic virus that selectively replicates in cells carrying Ras mutation or constitutively activated EGF receptors, induced DC maturation and the production of pro-inflammatory cytokines through triggering NFκB and protein kinase receptor (PKR), resulting in cross-presentation of tumor antigens to CD8+T cells (Errington et al., 2008; Gujar, Pan, Marcato, Garant, & Lee, 2011). These effects translated to a marked therapeutic impact in vivo, with decreased primary tumor growth and metastatic burden that reflected an adaptive antitumor response (Prestwich et al., 2008, 2009). In a phase 1 clinical trial for patients with head and neck cancer, reovirus type 3 dearing (RT3D) was well tolerated with some levels of clinical response observed in a majority of patients (Karapanagiotou et al., 2012). The efficacy of oncolytic viruses can also be enhanced by immune mechanisms independent of adaptive immunity. For example, H-1PV infection upregulates the expression of NK cell ligands on infected tumor cells, leading to augmentation of NK cell activity and killing (Bhat, Dempe, Dinsart, & Rommelaere, 2011; Bhat & Rommelaere, 2013).

The strong association between immune activation and oncolytic virus efficacy has prompted engineering of viruses to encode immunostimulatory molecules. An adenovirus designed to express CD40L induces expression of DAMPs on infected cells, which enhances antigen presentation and the activation of Th1 response in vivo (Diaconu et al., 2012). GM-CSF has also been introduced into several oncolytic viruses. The intratumoral injection of HSV-1 engineered to express GM-CSF (Oncovex$^{\text{GM-CSF}}$) leads to the production of GM-CSF in the tumor microenvironment, with a consequent DC recruitment and in situ maturation (Toda et al., 2000). In a phase 2 study of oncovex$^{\text{GM-CSF}}$ in metastatic melanoma, patients whose tumor regressed in response to treatment showed induction of intratumoral MART-1-specific T cells and a corresponding reduction in suppressor cell populations (Kaufman et al., 2010). This approach has progressed to a phase 3 testing in advanced melanoma patients, with an improvement in the durability of regressions of noninjected lesions compared to recombinant GM-CSF alone. The adenovirus expressing GM-CSF (CGTG-102) also showed antitumor activity in patients with advanced disease, with a correlation between the frequencies of antiviral T cells and antitumor T cells (Kanerva et al., 2013). In a phase 1 study, a poxvirus JX-594 engineered to encode GM-CSF that was delivered systemically through intravenous injection was well tolerated and effectuated complete responses in some patients (Breitbach et al., 2011). The major mechanism associated with clinical response was the induction of antibody-mediated complement-dependent cancer cell cytotoxicity (CDC) (Kim et al., 2013). In a dose-escalation study of JX-594 for liver cancer, similar CDC responses were observed and the efficacy of the in situ vaccine appeared to be dose-dependent (Heo et al., 2013). JX-594 also appears to infect tumor-associated endothelial cells, disrupting the tumor vasculature while sparing normal vessels (Breitbach et al., 2013).

Though promising with clear clinical benefits seen in early trials, the anti-tumor efficacy of in situ vaccination strategies remains to be established in large randomized studies. As with other cancer vaccine strategies, in situ vaccination has to overcome an immunosuppressive tumor microenvironment where a multitude of tumor escape mechanisms may dampen immune activation (Fig. 1).

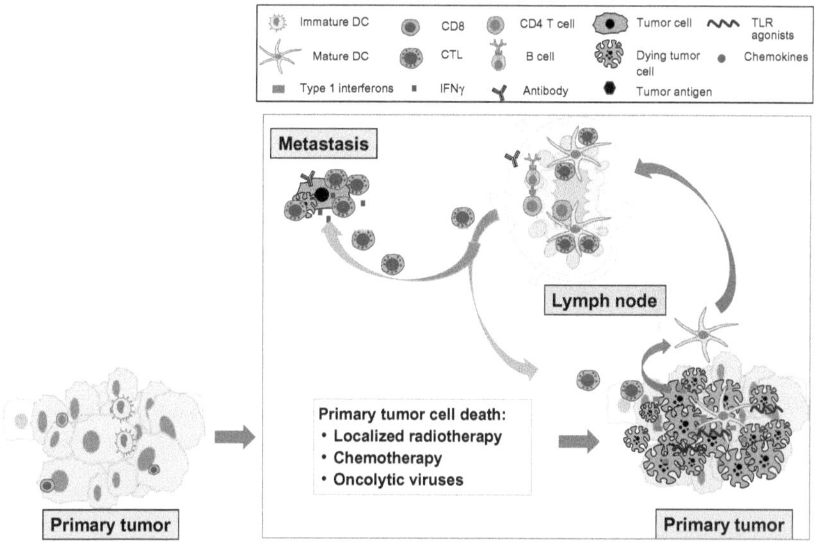

Figure 1 In situ vaccination: tumors are normally infiltrated by immature dendritic cells, which induce regulatory T cells and other suppressor populations, rendering the tumor microenvironment immunosuppressive. Upon cell death by chemotoxic agents, radiotherapy, oncolytic viruses and the release of PAMPs and/or type I interferons, or upon intratumoral injection of immunostimulants such as STING agonists or TLR agonists, intratumoral DCs mature, phagocytose and possess dying tumor cells, and traffic to the lymph node, where they encounter and present tumor antigens to CD8+T cells and CD4+T cells, inducing antitumor CTL responses and humoral responses. Induction of antitumor immunity in the local tumor can provide systemic protection for metastatic disease. Activated tumor-specific CD8+T cells produce IFNγ. (See the color plate.)

5. ENGINEERING APPROACHES

5.1 Materials Engineering and Cancer Immunotherapy

The field of materials engineering has gone beyond its traditional application in medical device engineering, and biomaterials are now the basis of many drug and cell delivery, drug targeting, and tissue regeneration therapies (Huebsch & Mooney, 2009). It is now feasible to engineer sophisticated and multifaceted technologies for cellular and molecular targeting and delivery, and these technologies are likely to integrate synergistically with current cancer immunotherapies. This section will discuss the potential advantages of material engineering and recent key advances in material engineering to develop novel vaccines for cancer immunotherapy (Fig. 2).

Materials-based technologies have a number of attractive features that are likely to augment current cancer vaccine strategies (Kim & Mooney, 2011; Li & Mooney, 2013; Mehta, Moynihan, & Irvine, 2015). These technologies can be multifunctional and are designed to package multiple immune-modulating agents into a single material system; doing so allows one to efficiently deliver multiple moieties into a cell and gain understanding of how different agents modulate immune cells. Materials can promote cell–cell crosstalk and enhance both innate and adaptive immunity. They can target specific organs and deliver payloads to cells within these organs, potentially increasing the potency of the payloads while reducing systemic toxicity. Lastly, they can prolong the in vivo bioactivity of immune-modulating agents by protecting them against degradation. Recent advances can be broadly divided into using synthetic scaffolds to host and direct immune cell function, or micro- and nanoparticles to directly target immune organs and cells.

5.2 Synthetic Scaffold Vaccines

Material scaffolds that incorporate key physical and chemical cues to create synthetic microenvironments may be widely useful to program immune cells directly in vivo. This approach can harness the potent stimulatory function of specific subsets of immune cells and stimulate both innate and adaptive responses that cannot otherwise be achieved by ex vivo vaccines. Macroporous 3D scaffolds that allow cell infiltration and subsequent emigration from the scaffold provide control over cell–cell and cell–material interactions (Bencherif et al., 2015; Capes, Ando, & Cameron, 2005;

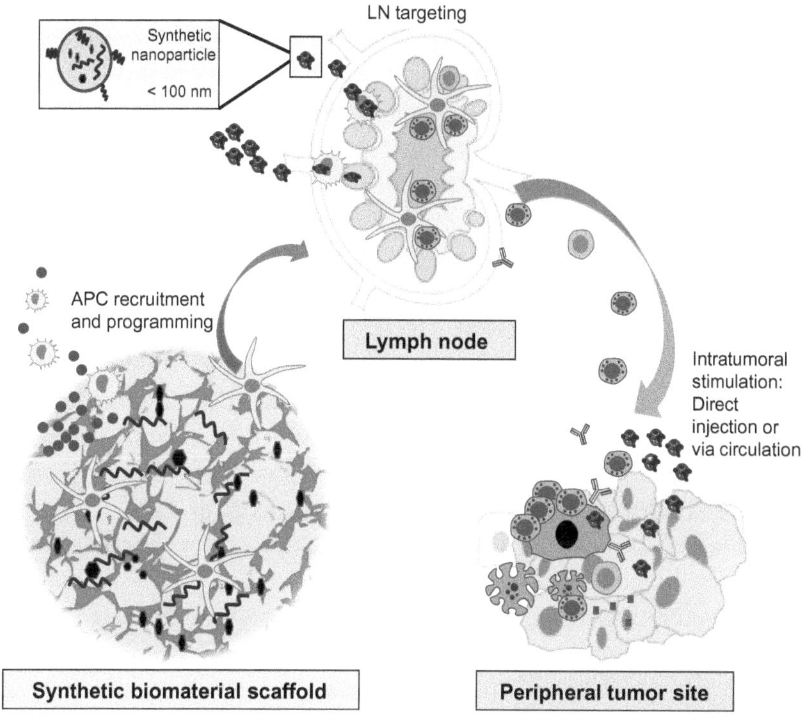

Figure 2 Three major engineering approaches to cancer vaccine are depicted: scaffold vaccines incorporated with chemo-attractants, immunostimulants, and tumor antigens serve as an in situ reservoir for DC maturation and uptake of tumor antigens. DCs matured in the 3D structure of the scaffold traffic to the lymph node and initiate antitumor immune responses. Alternatively, micro- or nanoparticles carrying tumor antigens or immunostimulants target lymph node resident DCs, which contain populations with potent cross-priming capabilities, thus triggering DC cross-presentation of tumor antigens and priming of tumor-specific T cells. These particles can also be targeted to the tumor directly, to stimulate maturation of intratumoral DCs. Trafficking of matured DCs to the lymph node results in antigen presentation and induction of antitumor immunity. (See the color plate.)

Henderson et al., 2013; Kim, Bencherif, Li, & Mooney, 2014). These scaffolds can serve as a site to expose immune cells with biochemical and physical cues to direct cell function (Hori, Winans, Huang, Horrigan, & Irvine, 2008). One of the first demonstrations of in vivo immune cell modulation with a biomaterial used a nonporous polymeric rod implant (Kumamoto et al., 2002). These centimeter-long rods were loaded with CCL19 and tumor-associated antigens, and implanted subcutaneously. Dermal and migratory APCs such as LCs were attracted to the rods, and the devices led to tumor prevention and regression. Subsequently, a poly(lactide-co-glycolide) (PLGA)-based scaffold system that contained more than 90% open and interconnected macropores was demonstrated to enhance immune cell infiltration into the scaffold (Ali, Emerich, Dranoff, & Mooney, 2009; Ali, Huebsch, Cao, Dranoff, & Mooney, 2009). PLGA is an FDA approved and biodegradable material that has been used widely in suture materials. GM-CSF was released from the scaffold in a sustained manner over a period of more than 30 days. Immature myeloid DCs migrated toward the gradient of GM-CSF and populated the scaffold. A single implant of the complete vaccine, composed of GM-CSF, CpG oligonucleotide condensed with polyethyleneimine and melanoma tumor lysate, led to >80% melanoma protection. More impressively, two implantations of this vaccine led to complete melanoma regression in more than 40% of the mice (Ali, Emerich, et al., 2009; Ali et al., 2014). Tumor efficacy in this vaccine system is correlated with an increased CD8+ T effector cell to CD4+Foxp3+ T regulatory cell ratio in the scaffold site and increased production of pro-inflammatory type 1 cytokines in the tissue surrounding the scaffold. Based on these results, this vaccine has moved to a first-in-human phase 1 clinical trial in patients with metastatic melanoma (Hodi, 2012).

A number of recent technologies have sought to generate injectable scaffolds and antigen depots in situ to reduce surgical burden and widen the application pool of scaffold-based vaccines (Bencherif et al., 2012; Verbeke & Mooney, 2015). For example, one vaccine system incorporates immune-modulatory agents into chitosan and hydroxyapatite, which are coinjected with a cross-linking solution containing tripolyphosphate (Chua et al., 2015). When the two components are mixed in situ, a hydrogel is formed that can slowly release the immune-modulatory agents, leading to a more prolonged antibody response. Since it is likely that macropores are important in inducing potent antitumor immunity due to the ability of immune cells to reside within these material while being manipulated, significant effort has been put into engineering injectable scaffolds with interconnected macropores. For instance, the hydrogel polymer alginate was polymerized in a partially frozen state, allowing the ice crystals to form throughout the hydrogel. Once the hydrogel is completely polymerized, the ice crystals were thawed, leaving behind macropores throughout the polymer (Bencherif et al., 2012). The shape-memory properties of this hydrogel make it injectable via a needle and recruit APCs in vivo. Using irradiated whole tumor cells as the antigen source, this vaccine system showed superior prophylactic and therapeutic efficacy compared to a bolus irradiated tumor cell vaccine (Bencherif et al., 2015). The degradation profile in vivo of these types of materials can be readily manipulated by altering the base polymer forming the hydrogel (Koshy, Ferrante, Lewin, & Mooney, 2014); a long-term presence of the scaffold may not be desirable in some applications. Strikingly, it has recently been demonstrated that rod-shaped microparticles can spontaneously assemble into a 3D scaffold after subcutaneous injection; the low packing density of the high aspect ratio particles resulted in macropores between the particles that allowed for cell infiltration (Kim et al., 2015). This technology demonstrates a facile strategy to formulate and inject a scaffold vaccine. When formulated with GM-CSF, CpG, and a model antigen ovalbumin, this vaccine showed significantly enhanced DC trafficking to the draining lymph nodes (dLNs), high and durable antigen-specific antibody responses and delayed onset of tumor growth as compared with the bolus formulation without the material scaffold. The high loading volume and facile surface chemistry of this scaffold will likely enable the incorporation of various types of immune-modulatory agents such as small-molecule drugs, oligonucleotides, and proteins.

5.3 Micro- and Nanoparticle Vaccines

Engineered nanoparticles that directly target immune cells in various tissues such as dLNs, mucosal sites, and the tumor provide an alternative approach to recruiting and programming immune cells with a biomaterial. At the site of interest, nanoparticles can target specific cell types and subcellular compartments. Nanoparticles can be synthesized using a wide range of materials, and immune-modulatory agents can be incorporated into the particles prior to delivery by exploiting the chemical and physical properties of the specific material.

LN targeting nanoparticle systems for cancer vaccination have shown significant promise in preclinical models. DC subsets exist in high concentrations in LNs, and targeting the LN can enhance antigen processing and DC–T cell crosstalk. The efficiency of nanoparticle localization to the dLN is controlled mostly by their size, as a number of papers have demonstrated that nanoparticles 20–50 nm in diameter transport more efficiently into lymphatic capillaries and their dLNs (~60%) compared to 100 nm or larger nanoparticles (<10%) (Fifis et al., 2004; Kourtis et al., 2013; Manolova et al., 2008; Reddy, Rehor, Schmoekel, Hubbell, & Swartz, 2006; Reddy et al., 2007). Once in the dLN, nanoparticles can release their payloads to enhance multiple immune processes such as antigen cross-presentation (Hanson et al., 2015; Jewell, Lopez, & Irvine, 2011; Kourtis et al., 2013; St John, Chan, Staats, Leong, & Abraham, 2012; Thomas, Vokali, Lund, Hubbell, & Swartz, 2014), and release and presentation of the payload can be controlled by the incorporation method (Hirosue, Kourtis, van der Vlies, Hubbell, & Swartz, 2010; Hu et al., 2009; Scott et al., 2012; Selbo et al., 2010; Shen et al., 2006; Stano, Scott, Dane, Swartz, & Hubbell, 2013). Nanoparticles that release encapsulated IL-12 and TNF into the dLN in a controlled and sustained manner have been shown to prolong a polarized systemic TH1 response. This is likely a result of prolonging immune cell action by protecting the cytokines from degradation until release (St John et al., 2012). Interbilayer-cross-linked multilamellar vesicle (ICMV) nanoparticles encapsulating both ovalbumin and MPLA elicited greater antigen-specific CD8+ T cell and sera antibody response than soluble forms of ovalbumin mixed with MPLA. More strikingly, this work demonstrated that incorporating the MPLA within the layers of the nanoparticle increased the OVA-specific CD8+ response by over threefold compared to incorporating the MPLA simply on the outside of the nanoparticles (Moon, Suh, Li, et al., 2012; Moon, Suh, Polhemus, et al., 2012). Finally, conjugation of CpG and peptides to hydrocarbon lipid tails that bind to the fatty acid-binding pocket of serum albumin prolonged CpG and antigen retention at the dLN more than 10-fold compared to the unconjugated forms. These lipid-conjugated vaccines significantly slowed tumor growth at doses that were ineffective for unconjugated vaccines (Liu et al., 2014). Altogether, these data suggest that not only can nanoparticles efficiently deliver immune-modulatory payloads to APCs in the dLNs but also indicate the design of payload incorporation may significantly alter the resulting immune response.

Complementary to delivering antigens and adjuvants to target cells in the LNs, the materials used to fabricate nanoparticle carriers can possess inherent adjuvant effects, acting as "danger signals." Synthetic materials such as PLGA, silica, poly(propylene sulfide), and poly (ethylene glycol) have been shown to activate the complement pathway and pattern recognition receptors, and their adjuvant effects can be harnessed in a vaccine system (Demento et al., 2009; Hamad et al., 2008; Morishige et al., 2010). For example, nanoparticles synthesized from a-Al_2O_3, in contrast to Alum, were shown to induce autophagy and effective cross-presentation of the loaded antigen (Li, Li, Jiao, & Hu, 2011). Mice vaccinated with a-Al_2O_3, nanoparticles conjugated to autophagosomes derived from tumor cells, without additional adjuvants, showed drastically enhanced percentage of

IFN secreting CD8+ cells and decreased lung metastasis in a 3LL lung tumor model.

In addition to dLN targeting, nanoparticles can be used to target mucosal tissues, where many primary tumors originate. Mucosal tissue targeting for cancer immunotherapy has focused on respiratory tissues (Li et al., 2013; Nembrini et al., 2011; Nochi et al., 2010; Stylianou et al., 2014), but there is an increasing body of work targeting the gastrointestinal (GI) track and vaginal mucus (Cu, Booth, & Saltzman, 2011; Klavinskis, Barnfield, Gao, & Parker, 1999; Zhu et al., 2012). One key consideration for pulmonary vaccination is the biocompatibility of the material. Studies have shown that having a targeting motif built into the nanoparticles significantly reduces local pulmonary inflammation, and that materials with hydrophilic properties are better tolerated than hydrophobic materials. Intranasal vaccination using a number of nanoparticle systems, including cholesterol-bearing pullulan polysaccharide nanogels, ICMV nanoparticles and poly(propylene sulfide) nanoparticles have resulted in significantly enhanced antigen-specific CD8+ T cell and an overall Th1-biased response in the lungs (Li et al., 2013; Nembrini et al., 2011; Nochi et al., 2010; Stanos, Bruckenthal, & Barkin, 2012). One vaccine using amphiphilic poly (γ-glutamic acid) nanoparticles encapsulating ovalbumin as a model antigen significantly reduced the number of lung metastasis of B16-OVA cells, compared to bolus ovalbumin vaccination (Matsuo et al., 2011).

Targeting the tumor directly as a vaccine strategy has two advantages. First, it may bypass the need for predicting and discovering novel tumor-associated antigens, as it uses the tumor directly as the antigen depot. Immune tissue targeting nanoparticles, in contrast, require a source of antigens, but it may be difficult to successfully predict the array of defined protein or peptide antigens that are needed for a potent cancer vaccine as a tumor mutates (Joyce & Fearon, 2015). Second, it could break tumor-associated immune suppression, as pro-inflammatory cytokines and adjuvants can be targeted to immune and stromal cells in the tumor microenvironment that mediates suppression (Hori, Winans, & Irvine, 2009). Tumor-targeting materials can be introduced systemically and locally. Systemically introduced nanoparticles can accumulate in the leaky regions of a tumor because of the enhanced permeation and retention effect. The enhanced permeability is a result of the deregulation of angiogenesis during tumor growth, during which regions of leaky vasculature and discontinuous endothelial layer are formed. The fenestrations between endothelial cells can be between 300 and 4700 nm in size in these vessels, allowing circulating nanoparticles to accumulate in these tissues (Hashizume et al., 2000; Nehoff, Parayath, Domanovitch, Taurin, & Greish, 2014). Liposome nanogels carrying IL-2 and a TGF-β inhibitor, when delivered systemically, led to increased retention of both the cytokine and small-molecule drug in the tumor (Park et al., 2011, 2012). When evaluated in a B16 lung metastasis model, the nanogel formulation significantly enhanced survival and decreased the number of metastases compared to the bolus formulations. The prolonged presentation and release of immune-stimulating agents also enhanced CD8 + T cell and NK cell activity in the tumor (Park et al., 2012). Peri- and intratumoral injection of immune-modulatory agents carrying nanoparticles could also reduce the toxicity elicited from the agents, as the nanoparticle will diminish diffusion of the agent from the injection site (Dominguez & Lustgarten, 2010; Egilmez et al., 2000; Sano, Nakajima, Choyke, & Kobayashi, 2013; Velluto, Thomas, Simeoni, Swartz, & Hubbell, 2011). For example, intratumoral injection of bolus anti-CD137 and IL-12, or anti-CD40 and CpG led to significant systemic toxicity as demonstrated by increased serum IL-6, TNF-α, enlarged spleens, and weight loss; this is likely due to the leaking of these agents from the tumor into systemic circulation (Kwong, Gai, Elkhader, Wittrup, & Irvine, 2013; Kwong, Liu, & Irvine, 2011). However, tethering these agents onto liposome nanoparticles prevented their leakage from the tumor microenvironment, thereby reducing systemic toxicity while retaining their bioactivity.

Biomaterial-based immune-modulatory approaches have demonstrated strong potential in preclinical rodent and primate models, and one PLGA scaffold-based melanoma vaccine is currently under phase 1 clinical testing. Future studies will likely explore the impact of varying the delivery timing and kinetics of various immune modulators, and new materials with precise structure control over receptor–ligand interaction. These are expected to yield a deeper understanding of the interactions between materials and the immune system. Altogether, we anticipate these will lead to more biomaterial approaches in cancer vaccine design moving into clinical testing over the next several years.

6. CONCLUDING REMARKS

Notwithstanding several decades of concentrated efforts from scientists and clinicians alike, cancer vaccines have not yet matured as a major component of cancer treatment. An improved understanding of negative immune checkpoints and immunosuppressive pathways operative within the tumor microenvironment has helped clarify the substantive obstacles to efficacious cancer vaccination. These insights together with advances in DC biology, germline-encoded pattern recognition receptors, and the regulatory networks that link innate and adaptive immunity provide new opportunities for intensifying cancer vaccines. Genomic technologies that permit facile characterization of patient neoantigens and the cross-disciplinary efforts of material science engineers and cancer immunologists are generating new tools to augment immune priming. Much in the same way that basic science has informed the design of more potent vaccines against infectious agents, the next few years are likely to witness significant advances in the crafting of improved therapeutic cancer vaccines.

REFERENCES

Abes, R., Gelize, E., Fridman, W. H., & Teillaud, J. L. (2010). Long-lasting antitumor protection by anti-CD20 antibody through cellular immune response. *Blood*, *116*, 926.

Agrawal, B., Krantz, M. J., Reddish, M. A., & Longenecker, B. M. (1998). Rapid induction of primary human CD4+ and CD8+ T cell responses against cancer-associated MUC1 peptide epitopes. *International Immunology*, *10*, 1907.

Ali, O. A., Emerich, D., Dranoff, G., & Mooney, D. J. (2009). In situ regulation of DC subsets and T cells mediates tumor regression in mice. *Science Translational Medicine*, *1*, 8ra19.

Ali, O. A., Huebsch, N., Cao, L., Dranoff, G., & Mooney, D. J. (2009). Infection-mimicking materials to program dendritic cells in situ. *Nature Materials*, *8*, 151.

Ali, O. A., Verbeke, C., Johnson, C., Sands, R. W., Lewin, S. A., White, D., et al. (2014). Identification of immune factors regulating antitumor immunity using polymeric vaccines with multiple adjuvants. *Cancer Research*, *74*, 1670.

Amin, A., Dudek, A. Z., Logan, T. F., Lance, R. S., Holzbeierlein, J. M., Knox, J. J., et al. (2015). Survival with AGS-003, an autologous dendritic cell-based immunotherapy, in combination with sunitinib in unfavorable risk patients with advanced renal cell carcinoma (RCC): Phase 2 study results. *Journal for Immunotherapy of Cancer*, *3*, 14.

Anandasabapathy, N., Feder, R., Mollah, S., Tse, S. W., Longhi, M. P., & Mehandru, S. (2014). Classical Flt3L-dependent dendritic cells control immunity to protein vaccine. *The Journal of Experimental Medicine*, *211*, 1875.

Antonia, S. J., Seigne, J., Diaz, J., Muro-Cacho, C., Extermann, M., & Farmelo, M. J. (2002). Phase I trial of a B7-1 (CD80) gene modified autologous tumor cell vaccine in combination with systemic interleukin-2 in patients with metastatic renal cell carcinoma. *The Journal of Urology*, *167*, 1995.

Arlen, P. M., Gulley, J. L., Madan, R. A., Hodge, J. W., & Schlom, J. (2007). Preclinical and clinical studies of recombinant poxvirus vaccines for carcinoma therapy. *Critical Reviews in Immunology*, *27*, 451.

Atanackovic, D., Altorki, N. K., Cao, Y., Ritter, E., Ferrara, C. A., & Ritter, G. (2008). Booster vaccination of cancer patients with MAGE-A3 protein reveals long-term immunological memory or tolerance depending on priming. *Proceedings of the National Academy of Sciences of the United States of America*, *105*, 1650.

Ayyoub, M., Rimoldi, D., Guillaume, P., Romero, P., Cerottini, J. C., Valmori, D., et al. (2003). Tumor-reactive, SSX-2-specific CD8+ T cells are selectively expanded during immune responses to antigen-expressing tumors in melanoma patients. *Cancer Research*, *63*, 5601.

Baitsch, L., Baumgaertner, P., Devevre, E., Raghav, S. K., Legat, A., Barba, L., et al. (2011). Exhaustion of tumor-specific CD8(+) T cells in metastases from melanoma patients. *The Journal of Clinical Investigation, 121*, 2350.

Banchereau, J., Palucka, A. K., Dhodapkar, M., Burkeholder, S., Taquet, N., Rolland, A., et al. (2001). Immune and clinical responses in patients with metastatic melanoma to CD34(+) progenitor-derived dendritic cell vaccine. *Cancer Research, 61*, 6451.

Barker, C. A., Postow, M. A., Khan, S. A., Beal, K., Parhar, P. K., Yamada, Y., et al. (2013). Concurrent radiotherapy and ipilimumab immunotherapy for patients with melanoma. *Cancer Immunology Research, 1*, 92.

Bencherif, S. A., Sands, R. W., Bhatta, D., Arany, P., Verbeke, C. S., Edwards, D. A., et al. (2012). Injectable preformed scaffolds with shape-memory properties. *Proceedings of the National Academy of Sciences of the United States of America, 109*, 19590.

Bencherif, S. A., Warren Sands, R., Ali, O. A., Li, W. A., Lewin, S. A., Braschler, T. M., et al. (2015). Injectable cryogel-based whole-cell cancer vaccines. *Nature Communications, 6*, 7556.

Benencia, F., Courreges, M. C., Conejo-Garcia, J. R., Mohamed-Hadley, A., Zhang, L., Buckanovich, R. J., et al. (2005). HSV oncolytic therapy upregulates interferon-inducible chemokines and recruits immune effector cells in ovarian cancer. *Molecular Therapy, 12*, 789.

Benencia, F., Courreges, M. C., Fraser, N. W., & Coukos, G. (2008). Herpes virus oncolytic therapy reverses tumor immune dysfunction and facilitates tumor antigen presentation. *Cancer Biology & Therapy*, 7, 1194.

Bhat, R., Dempe, S., Dinsart, C., & Rommelaere, J. (2011). Enhancement of NK cell antitumor responses using an oncolytic parvovirus. *International Journal of Cancer, 128*, 908.

Bhat, R., & Rommelaere, J. (2013). NK-cell-dependent killing of colon carcinoma cells is mediated by natural cytotoxicity receptors (NCRs) and stimulated by parvovirus infection of target cells. *BMC Cancer, 13*, 367.

Billiau, A., & Matthys, P. (2001). Modes of action of Freund's adjuvants in experimental models of autoimmune diseases. *Journal of Leukocyte Biology, 70*, 849.

Bowne, W. B., Wolchok, J. D., Hawkins, W. G., Srinivasan, R., Gregor, P., Blachere, N. E., et al. (1999). Injection of DNA encoding granulocyte-macrophage colony-stimulating factor recruits dendritic cells for immune adjuvant effects. *Cytokines, Cellular & Molecular Therapy, 5*, 217.

Breitbach, C. J., Arulanandam, R., De Silva, N., Thorne, S. H., Patt, R., Daneshmand, M., et al. (2013). Oncolytic vaccinia virus disrupts tumor-associated vasculature in humans. *Cancer Research, 73*, 1265.

Breitbach, C. J., Burke, J., Jonker, D., Stephenson, J., Haas, A. R., Chow, L. Q., et al. (2011). Intravenous delivery of a multi-mechanistic cancer-targeted oncolytic poxvirus in humans. *Nature, 477*, 99.

Brockstedt, D. G., Giedlin, M. A., Leong, M. L., Bahjat, K. S., Gao, Y., Luckett, W., et al. (2004). Listeria-based cancer vaccines that segregate immunogenicity from toxicity. *Proceedings of the National Academy of Sciences of the United States of America, 101*, 13832.

Brody, J. D., Ai, W. Z., Czerwinski, D. K., Torchia, J. A., Levy, M., Advani, R. H., et al. (2010). In situ vaccination with a TLR9 agonist induces systemic lymphoma regression: A phase I/II study. *Journal of Clinical Oncology, 28*, 4324.

Bulliard, Y., Jolicoeur, R., Windman, M., Rue, S. M., Ettenberg, S., Knee, D. A., et al. (2013). Activating Fc gamma receptors contribute to the antitumor activities of immunoregulatory receptor-targeting antibodies. *The Journal of Experimental Medicine, 210*, 1685.

Burkhardt, U. E., Hainz, U., Stevenson, K., Goldstein, N. R., Pasek, M., Naito, M., et al. (2013). Autologous CLL cell vaccination early after transplant induces leukemia-specific T cells. *The Journal of Clinical Investigation, 123*, 3756.

Butts, C., Murray, N., Maksymiuk, A., Goss, G., Marshall, E., Soulieres, D., et al. (2005). Randomized phase IIB trial of BLP25 liposome vaccine in stage IIIB and IV non-small-cell lung cancer. *Journal of Clinical Oncology, 23*, 6674.

Butts, C., Socinski, M. A., Mitchell, P. L., Thatcher, N., Havel, L., Krzakowski, M., et al. (2014). Tecemotide (L-BLP25) versus placebo after chemoradiotherapy for stage III non-small-cell lung cancer (START): A randomised, double-blind, phase 3 trial. *The Lancet Oncology, 15*, 59.

Califano, J. A., Khan, Z., Noonan, K. A., Rudraraju, L., Zhang, Z., Wang, H., et al. (2015). Tadalafil augments tumor specific immunity in patients with head and neck squamous cell carcinoma. *Clinical Cancer Research, 21*, 30.

Campbell, C. T., Gulley, J. L., Oyelaran, O., Hodge, J. W., Schlom, J., & Gildersleeve, J. C. (2014). Humoral response to a viral glycan correlates with survival on PROSTVAC-VF. *Proceedings of the National Academy of Sciences of the United States of America, 111*, E1749.

Capes, J. S., Ando, H. Y., & Cameron, R. E. (2005). Fabrication of polymeric scaffolds with a controlled distribution of pores. *Journal of Materials Science. Materials in Medicine, 16*, 1069.

Carreno, B. M., Anderson, R. W., Coligan, J. E., & Biddison, W. E. (1990). LA-B37 and HLA-A2.1 molecules bind largely nonoverlapping sets of peptides. *Proceedings of the National Academy of Sciences of the United States of America, 87*, 3420.

Carreno, B. M., Becker-Hapak, M., Huang, A., Chan, M., Alyasiry, A., Lie, W. R., et al. (2013). IL-12p70-producing patient DC vaccine elicits Tc1-polarized immunity. *The Journal of Clinical Investigation, 123*, 3383.

Carreno, B. M., Magrini, V., Becker-Hapak, M., Kaabinejadian, S., Hundal, J., Petti, A. A., et al. (2015). Cancer immunotherapy. A dendritic cell vaccine increases the breadth and diversity of melanoma neoantigen-specific T cells. *Science, 348*, 803.

Casey, K. A., & Mescher, M. F. (2007). IL-21 promotes differentiation of naive CD8 T cells to a unique effector phenotype. *Journal of Immunology, 178*, 7640.

Champiat, S., Ferte, C., Lebel-Binay, S., Eggermont, A., & Soria, J. C. (2014). Exomics and immunogenics: Bridging mutational load and immune checkpoints efficacy. *Oncoimmunology, 3*, e27817.

Chandra, D., Quispe-Tintaya, W., Jahangir, A., Asafu-Adjei, D., Ramos, I., Sintim, H. O., et al. (2014). STING ligand c-di-GMP improves cancer vaccination against metastatic breast cancer. *Cancer Immunology Research, 2*, 901.

Chao, M. P., Jaiswal, S., Weissman-Tsukamoto, R., Alizadeh, A. A., Gentles, A. J., Volkmer, J., et al. (2010). Calreticulin is the dominant pro-phagocytic signal on multiple human cancers and is counterbalanced by CD47. *Science Translational Medicine, 2*, 63ra94.

Chao, M. P., Weissman, I. L., & Majeti, R. (2012). The CD47-SIRPalpha pathway in cancer immune evasion and potential therapeutic implications. *Current Opinion in Immunology, 24*, 225.

Chen, Y. T. (2000). Cancer vaccine: Identification of human tumor antigens by SEREX. *The Cancer Journal, 6*(Suppl. 3), S208.

Chen, Q., Jackson, H., Parente, P., Luke, T., Rizkalla, M., Tai, T. Y., et al. (2004). Immunodominant CD4+ responses identified in a patient vaccinated with full-length NY-ESO-1 formulated with ISCOMATRIX adjuvant. *Proceedings of the National Academy of Sciences of the United States of America, 101*, 9363.

Chen, Y. T., Scanlan, M. J., Sahin, U., Tureci, O., Gure, A. O., Tsang, S., et al. (1997). A testicular antigen aberrantly expressed in human cancers detected by autologous antibody screening. *Proceedings of the National Academy of Sciences of the United States of America, 94*, 1914.

Chiocca, E. A. (2002). Oncolytic viruses. *Nature Reviews. Cancer, 2*, 938.

Chua, B. Y., Sekiya, T., Al Kobaisi, M., Short, K. R., Mainwaring, D. E., & Jackson, D. C. (2015). A single dose biodegradable vaccine depot that induces persistently high levels of antibody over a year. *Biomaterials, 53*, 50.

Clancy-Thompson, E., King, L. K., Nunnley, L. D., Mullins, I. M., Slingluff, C. L., Jr., & Mullins, D. W. (2013). Peptide vaccination in Montanide adjuvant induces and GM-CSF increases CXCR3 and cutaneous lymphocyte antigen expression by tumor antigen-specific CD8 T cells. *Cancer Immunology Research*, *1*, 332.

Colombo, M. P., Ferrari, G., Stoppacciaro, A., Parenza, M., Rodolfo, M., & Mavilio, F. (1991). Granulocyte colony-stimulating factor gene transfer suppresses tumorigenicity of a murine adenocarcinoma in vivo. *The Journal of Experimental Medicine*, *173*, 889.

Conry, R. M., LoBuglio, A. F., Wright, M., Sumerel, L., Pike, M. J., Johanning, F., et al. (1995). Characterization of a messenger RNA polynucleotide vaccine vector. *Cancer Research*, *55*, 1397.

Coulie, P. G., Brichard, V., Van Pel, A., Wolfel, T., Schneider, J., Traversari, C., et al. (1994). A new gene coding for a differentiation antigen recognized by autologous cytolytic T lymphocytes on HLA-A2 melanomas. *The Journal of Experimental Medicine*, *180*, 35.

Coulie, P. G., Lehmann, F., Lethe, B., Herman, J., Lurquin, C., & Andrawiss, M. (1995). A mutated intron sequence codes for an antigenic peptide recognized by cytolytic T lymphocytes on a human melanoma. *Proceedings of the National Academy of Sciences of the United States of America*, *92*, 7976.

Coulie, P. G., Van den Eynde, B. J., van der Bruggen, P., & Boon, T. (2014). Tumour antigens recognized by T lymphocytes: At the core of cancer immunotherapy. *Nature Reviews. Cancer*, *14*, 135.

Cox, A. L., Skipper, J., Chen, Y., Henderson, R. A., Darrow, T. L., Shabanowitz, J., et al. (1994). Identification of a peptide recognized by five melanoma-specific human cytotoxic T cell lines. *Science*, *264*, 716.

Crocenzi, T. S., Crittenden, M., Wolf, R., Hammill, C., Newell, P., Savoy, A., et al. (2013). An exploratory phase I trial of immunochemoradiotherapy in locally advanced and borderline resectable (LA/BR) pancreatic adenocarcinoma (PC). *Journal of Clinical Oncology*, *31*(abstr 174).

Cu, Y., Booth, C. J., & Saltzman, W. M. (2011). In vivo distribution of surface-modified PLGA nanoparticles following intravaginal delivery. *Journal of Controlled Release*, *156*, 258.

Curran, M. A., Montalvo, W., Yagita, H., & Allison, J. P. (2010). PD-1 and CTLA-4 combination blockade expands infiltrating T cells and reduces regulatory T and myeloid cells within B16 melanoma tumors. *Proceedings of the National Academy of Sciences of the United States of America*, *107*, 4275.

Dannull, J., Haley, N. R., Archer, G., Nair, S., Boczkowski, D., Harper, M., et al. (2013). Melanoma immunotherapy using mature DCs expressing the constitutive proteasome. *The Journal of Clinical Investigation*, *123*, 3135.

Davenport, M. P., Smith, K. J., Barouch, D., Reid, S. W., Bodnar, W. M., Willis, A. C., et al. (1997). HLA class I binding motifs derived from random peptide libraries differ at the COOH terminus from those of eluted peptides. *The Journal of Experimental Medicine*, *185*, 367.

Davis, I. D., Chen, W., Jackson, H., Parente, P., Shackleton, M., Hopkins, W., et al. (2004). Recombinant NY-ESO-1 protein with ISCOMATRIX adjuvant induces broad integrated antibody and CD4(+) and CD8(+) T cell responses in humans. *Proceedings of the National Academy of Sciences of the United States of America*, *101*, 10697.

Davis, M. B., Vasquez-Dunddel, D., Fu, J., Albesiano, E., Pardoll, D., & Kim, Y. J. (2011). Intratumoral administration of TLR4 agonist absorbed into a cellular vector improves antitumor responses. *Clinical Cancer Research*, *17*, 3984.

Del Corno, M., Michienzi, A., Masotti, A., Da Sacco, L., Bottazzo, G. F., Belardelli, F., et al. (2009). CC chemokine ligand 2 down-modulation by selected Toll-like receptor agonist combinations contributes to T helper 1 polarization in human dendritic cells. *Blood*, *114*, 796.

Demento, S. L., Eisenbarth, S. C., Foellmer, H. G., Platt, C., Caplan, M. J., Mark Saltzman, W., et al. (2009). Inflammasome-activating nanoparticles as modular systems for optimizing vaccine efficacy. *Vaccine, 27*, 3013.

Deng, L., Liang, H., Xu, M., Yang, X., Burnette, B., Arina, A., et al. (2014). STING-dependent cytosolic DNA sensing promotes radiation-induced type I interferon-dependent antitumor immunity in immunogenic tumors. *Immunity, 41*, 843.

Derre, L., Rivals, J. P., Jandus, C., Pastor, S., Rimoldi, D., Romero, P., et al. (2010). BTLA mediates inhibition of human tumor-specific CD8+ T cells that can be partially reversed by vaccination. *The Journal of Clinical Investigation, 120*, 157.

Dewan, M. Z., Vanpouille-Box, C., Kawashima, N., DiNapoli, S., Babb, J. S., Formenti, S. C., et al. (2012). Synergy of topical toll-like receptor 7 agonist with radiation and low-dose cyclophosphamide in a mouse model of cutaneous breast cancer. *Clinical Cancer Research, 18*, 6668.

Dhodapkar, M. V., Sznol, M., Zhao, B., Wang, D., Carvajal, R. D., Keohan, M. L., et al. (2014). Induction of antigen-specific immunity with a vaccine targeting NY-ESO-1 to the dendritic cell receptor DEC-205. *Science Translational Medicine, 6*, 232ra51.

Diaconu, I., Cerullo, V., Hirvinen, M. L., Escutenaire, S., Ugolini, M., Pesonen, S. K., et al. (2012). Immune response is an important aspect of the antitumor effect produced by a CD40L-encoding oncolytic adenovirus. *Cancer Research, 72*, 2327.

Diebold, S. S., Kaisho, T., Hemmi, H., Akira, S., & Reis e Sousa, C. (2004). Innate antiviral responses by means of TLR7-mediated recognition of single-stranded RNA. *Science, 303*, 1529.

Diken, M., Kreiter, S., Vascotto, F., Selmi, A., Attig, S., Diekmann, J., et al. (2013). mTOR inhibition improves antitumor effects of vaccination with antigen-encoding RNA. *Cancer Immunology Research, 1*, 386.

DiLillo, D. J., & Ravetch, J. V. (2015). Differential Fc-receptor engagement drives an antitumor vaccinal effect. *Cell, 161*, 1035.

Disis, M. L., Bernhard, H., Shiota, F. M., Hand, S. L., Gralow, J. R., Huseby, E. S., et al. (1996). Granulocyte-macrophage colony-stimulating factor: An effective adjuvant for protein and peptide-based vaccines. *Blood, 88*, 202.

Dominguez, A. L., & Lustgarten, J. (2010). Targeting the tumor microenvironment with anti-neu/anti-CD40 conjugated nanoparticles for the induction of antitumor immune responses. *Vaccine, 28*, 1383.

Dougan, M., Dougan, S., Slisz, J., Firestone, B., Vanneman, M., Draganov, D., et al. (2010). IAP inhibitors enhance co-stimulation to promote tumor immunity. *The Journal of Experimental Medicine, 207*, 2195.

Dranoff, G. (2002). GM-CSF-based cancer vaccines. *Immunological Reviews, 188*, 147.

Dranoff, G., Jaffee, E., Lazenby, A., Golumbek, P., Levitsky, H., Brose, K., et al. (1993). Vaccination with irradiated tumor cells engineered to secrete murine granulocyte-macrophage colony-stimulating factor stimulates potent, specific, and long-lasting anti-tumor immunity. *Proceedings of the National Academy of Sciences of the United States of America, 90*, 3539.

Duan, F., Duitama, J., Al Seesi, S., Ayres, C. M., Corcelli, S. A., Pawashe, A. P., et al. (2014). Genomic and bioinformatic profiling of mutational neoepitopes reveals new rules to predict anticancer immunogenicity. *The Journal of Experimental Medicine, 211*, 2231.

Dubensky, T. W., Jr., & Reed, S. G. (2010). Adjuvants for cancer vaccines. *Seminars in Immunology, 22*, 155.

Dudley, M. E., Wunderlich, J. R., Robbins, P. F., Yang, J. C., Hwu, P., Schwartzentruber, D. J., et al. (2002). Cancer regression and autoimmunity in patients after clonal repopulation with antitumor lymphocytes. *Science, 298*, 850.

DuPage, M., Mazumdar, C., Schmidt, L. M., Cheung, A. F., & Jacks, T. (2012). Expression of tumour-specific antigens underlies cancer immunoediting. *Nature, 482*, 405.

Duraiswamy, J., Kaluza, K. M., Freeman, G. J., & Coukos, G. (2013). Dual blockade of PD-1 and CTLA-4 combined with tumor vaccine effectively restores T-cell rejection function in tumors. *Cancer Research, 73*, 3591.

Ebert, L. M., MacRaild, S. E., Zanker, D., Davis, I. D., Cebon, J., & Chen, W. (2012). A cancer vaccine induces expansion of NY-ESO-1-specific regulatory T cells in patients with advanced melanoma. *PloS One*, 7, e48424.

Eder, J. P., Kantoff, P. W., Roper, K., Xu, G. X., Bubley, G. J., Boyden, J., et al. (2000). A phase I trial of a recombinant vaccinia virus expressing prostate-specific antigen in advanced prostate cancer. *Clinical Cancer Research, 6*, 1632.

Egilmez, N. K., Jong, Y. S., Hess, S. D., Jacob, J. S., Mathiowitz, E., & Bankert, R. B. (2000). Cytokines delivered by biodegradable microspheres promote effective suppression of human tumors by human peripheral blood lymphocytes in the SCID-Winn model. *Journal of Immunotherapy, 23*, 190.

Enzler, T., Gillessen, S., Manis, J. P., Ferguson, D., Fleming, J., Alt, F. W., et al. (2003). Deficiencies of GM-CSF and interferon gamma link inflammation and cancer. *The Journal of Experimental Medicine, 197*, 1213.

Errington, F., Steele, L., Prestwich, R., Harrington, K. J., Pandha, H. S., Vidal, L., et al. (2008). Reovirus activates human dendritic cells to promote innate antitumor immunity. *Journal of Immunology, 180*, 6018.

Falk, K., Rotzschke, O., Stevanovic, S., Jung, G., & Rammensee, H. G. (1991). Allele-specific motifs revealed by sequencing of self-peptides eluted from MHC molecules. *Nature, 351*, 290.

Fifis, T., Gamvrellis, A., Crimeen-Irwin, B., Pietersz, G. A., Li, J., Mottram, P. L., et al. (2004). Size-dependent immunogenicity: Therapeutic and protective properties of nano-vaccines against tumors. *Journal of Immunology, 173*, 3148.

Figdor, C. G., van Kooyk, Y., & Adema, G. J. (2002). C-type lectin receptors on dendritic cells and Langerhans cells. *Nature Reviews. Immunology, 2*, 77.

Flynn, B. J., Kastenmuller, K., Wille-Reece, U., Tomaras, G. D., Alam, M., Lindsay, R. W., et al. (2011). Immunization with HIV Gag targeted to dendritic cells followed by recombinant New York vaccinia virus induces robust T-cell immunity in nonhuman primates. *Proceedings of the National Academy of Sciences of the United States of America, 108*, 7131.

Francois, V., Ottaviani, S., Renkvist, N., Stockis, J., Schuler, G., Thielemans, K., et al. (2009). The CD4(+) T-cell response of melanoma patients to a MAGE-A3 peptide vaccine involves potential regulatory T cells. *Cancer Research, 69*, 4335.

Frederick, D. T., Piris, A., Cogdill, A. P., Cooper, Z. A., Lezcano, C., Ferrone, C. R., et al. (2013). BRAF inhibition is associated with enhanced melanoma antigen expression and a more favorable tumor microenvironment in patients with metastatic melanoma. *Clinical Cancer Research, 19*, 1225.

Frohlich, A., Kisielow, J., Schmitz, I., Freigang, S., Shamshiev, A. T., Weber, J., et al. (2009). IL-21R on T cells is critical for sustained functionality and control of chronic viral infection. *Science, 324*, 1576.

Fu, J., Kanne, D. B., Leong, M., Glickman, L. H., McWhirter, S. M., Lemmens, E., et al. (2015). STING agonist formulated cancer vaccines can cure established tumors resistant to PD-1 blockade. *Science Translational Medicine*, 7, 283ra52.

Fulci, G., Breymann, L., Gianni, D., Kurozomi, K., Rhee, S. S., Yu, J., et al. (2006). Cyclophosphamide enhances glioma virotherapy by inhibiting innate immune responses. *Proceedings of the National Academy of Sciences of the United States of America, 103*, 12873.

Gajewski, T. F., Schreiber, H., & Fu, Y. X. (2013). Innate and adaptive immune cells in the tumor microenvironment. *Nature Immunology, 14*, 1014.

Galon, J., Angell, H. K., Bedognetti, D., & Marincola, F. M. (2013). The continuum of cancer immunosurveillance: Prognostic, predictive, and mechanistic signatures. *Immunity*, *39*, 11.

Gerlinger, M., Rowan, A. J., Horswell, S., Larkin, J., Endesfelder, D., Gronroos, E., et al. (2012). Intratumor heterogeneity and branched evolution revealed by multiregion sequencing. *The New England Journal of Medicine*, *366*, 883.

Goldberg, J. M., Fisher, D. E., Demetri, G. D., Neuberg, D., Allsop, S. A., Fonseca, C., et al. (2015). Biologic activity of autologous, granulocyte-macrophage colony-stimulating factor secreting alveolar soft-part sarcoma and clear cell sarcoma vaccines. *Clinical Cancer Research*, *21*, 3178.

Golden, E. B., Chhabra, A., Chachoua, A., Adams, S., Donach, M., Fenton-Kerimian, M., et al. (2015). Local radiotherapy and granulocyte-macrophage colony-stimulating factor to generate abscopal responses in patients with metastatic solid tumours: A proof-of-principle trial. *The Lancet Oncology*, *16*, 795.

Golden, E. B., Demaria, S., Schiff, P. B., Chachoua, A., & Formenti, S. C. (2013). An abscopal response to radiation and ipilimumab in a patient with metastatic non-small cell lung cancer. *Cancer Immunology Research*, *1*, 365.

Green, D. R., Ferguson, T., Zitvogel, L., & Kroemer, G. (2009). Immunogenic and tolerogenic cell death. *Nature Reviews. Immunology*, *9*, 353.

Gubin, M. M., Zhang, X., Schuster, H., Caron, E., Ward, J. P., Noguchi, T., et al. (2014). Checkpoint blockade cancer immunotherapy targets tumour-specific mutant antigens. *Nature*, *515*, 577.

Guilliams, M., Bruhns, P., Saeys, Y., Hammad, H., & Lambrecht, B. N. (2014). The function of Fcgamma receptors in dendritic cells and macrophages. *Nature Reviews. Immunology*, *14*, 94.

Gujar, S. A., Pan, D. A., Marcato, P., Garant, K. A., & Lee, P. W. (2011). Oncolytic virus-initiated protective immunity against prostate cancer. *Molecular Therapy*, *19*, 797.

Hailemichael, Y., Dai, Z., Jaffarzad, N., Ye, Y., Medina, M. A., Huang, X. F., et al. (2013). Persistent antigen at vaccination sites induces tumor-specific CD8(+) T cell sequestration, dysfunction and deletion. *Nature Medicine*, *19*, 465.

Hamad, I., Christy Hunter, A., Rutt, K. J., Liu, Z., Dai, H., & Moein Moghimi, S. (2008). Complement activation by PEGylated single-walled carbon nanotubes is independent of C1q and alternative pathway turnover. *Molecular Immunology*, *45*, 3797.

Hansen, T. H., & Bouvier, M. (2009). MHC class I antigen presentation: Learning from viral evasion strategies. *Nature Reviews Immunology*, *9*, 503.

Hanson, M. C., Crespo, M. P., Abraham, W., Moynihan, K. D., Szeto, G. L., Chen, S. H., et al. (2015). Nanoparticulate STING agonists are potent lymph node-targeted vaccine adjuvants. *The Journal of Clinical Investigation*, *125*, 2532.

Hashizume, H., Baluk, P., Morikawa, S., McLean, J. W., Thurston, G., Roberge, S., et al. (2000). Openings between defective endothelial cells explain tumor vessel leakiness. *The American Journal of Pathology*, *156*, 1363.

Hawiger, D., Inaba, K., Dorsett, Y., Guo, M., Mahnke, K., Rivera, M., et al. (2001). Dendritic cells induce peripheral T cell unresponsiveness under steady state conditions in vivo. *The Journal of Experimental Medicine*, *194*, 769.

Hayes, M. P., Wang, J., & Norcross, M. A. (1995). Regulation of interleukin-12 expression in human monocytes: Selective priming by interferon-gamma of lipopolysaccharide-inducible p35 and p40 genes. *Blood*, *86*, 646.

Heidenreich, R., Jasny, E., Kowalczyk, A., Lutz, J., Probst, J., Baumhof, P., et al. (2015). A novel RNA-based adjuvant combines strong immunostimulatory capacities with a favorable safety profile. *International Journal of Cancer*, *137*, 372.

Heil, F., Hemmi, H., Hochrein, H., Ampenberger, F., Kirschning, C., Akira, S., et al. (2004). Species-specific recognition of single-stranded RNA via toll-like receptor 7 and 8. *Science*, *303*, 1526.

Henderson, T. M., Ladewig, K., Haylock, D. N., McLeanb, K. M., & O'Connor, A. J. (2013). Cryogels for biomedical applications. *Journal of Materials Chemistry B, 1*, 2682.

Heo, J., Reid, T., Ruo, L., Breitbach, C. J., Rose, S., Bloomston, M., et al. (2013). Randomized dose-finding clinical trial of oncolytic immunotherapeutic vaccinia JX-594 in liver cancer. *Nature Medicine, 19*, 329.

Hess, P. R., Boczkowski, D., Nair, S. K., Snyder, D., & Gilboa, E. (2006). Vaccination with mRNAs encoding tumor-associated antigens and granulocyte-macrophage colony-stimulating factor efficiently primes CTL responses, but is insufficient to overcome tolerance to a model tumor/self antigen. *Cancer Immunology, Immunotherapy, 55*, 672.

Hirosue, S., Kourtis, I. C., van der Vlies, A. J., Hubbell, J. A., & Swartz, M. A. (2010). Antigen delivery to dendritic cells by poly(propylene sulfide) nanoparticles with disulfide conjugated peptides: Cross-presentation and T cell activation. *Vaccine, 28*, 7897.

Hodi, F. S. (2012). *Dendritic cell activating scaffold in melanoma.* (Clinicaltrials.gov).

Hodi, F. S., Butler, M., Oble, D. A., Seiden, M. V., Haluska, F. G., Kruse, A., et al. (2008). Immunologic and clinical effects of antibody blockade of cytotoxic T lymphocyte-associated antigen 4 in previously vaccinated cancer patients. *Proceedings of the National Academy of Sciences of the United States of America, 105*, 3005.

Hodi, F. S., Mihm, M. C., Soiffer, R. J., Haluska, F. G., Butler, M., Seiden, M. V., et al. (2003). Biologic activity of cytotoxic T lymphocyte-associated antigen 4 antibody blockade in previously vaccinated metastatic melanoma and ovarian carcinoma patients. *Proceedings of the National Academy of Sciences of the United States of America, 100*, 4712.

Hodi, F. S., O'Day, S. J., McDermott, D. F., Weber, R. W., Sosman, J. A., Haanen, J. B., et al. (2010). Improved survival with ipilimumab in patients with metastatic melanoma. *The New England Journal of Medicine, 363*, 711.

Hori, Y., Winans, A. M., Huang, C. C., Horrigan, E. M., & Irvine, D. J. (2008). Injectable dendritic cell-carrying alginate gels for immunization and immunotherapy. *Biomaterials, 29*, 3671.

Hori, Y., Winans, A. M., & Irvine, D. J. (2009). Modular injectable matrices based on alginate solution/microsphere mixtures that gel in situ and co-deliver immunomodulatory factors. *Acta Biomaterialia, 5*, 969.

Houghton, A. N., & Guevara-Patino, J. A. (2004). Immune recognition of self in immunity against cancer. *The Journal of Clinical Investigation, 114*, 468.

Hu, Y., Atukorale, P. U., Lu, J. J., Moon, J. J., Um, S. H., Cho, E. C., et al. (2009). Cytosolic delivery mediated via electrostatic surface binding of protein, virus, or siRNA cargos to pH-responsive core-shell gel particles. *Biomacromolecules, 10*, 756.

Huebsch, N., & Mooney, D. J. (2009). Inspiration and application in the evolution of biomaterials. *Nature, 462*, 426.

Idoyaga, J., Lubkin, A., Fiorese, C., Lahoud, M. H., Caminschi, I., Huang, Y., et al. (2011). Comparable T helper 1 (Th1) and CD8 T-cell immunity by targeting HIV gag p24 to CD8 dendritic cells within antibodies to Langerin, DEC205, and Clec9A. *Proceedings of the National Academy of Sciences of the United States of America, 108*, 2384.

Infante, J. R., Dees, E. C., Olszanski, A. J., Dhuria, S. V., Sen, S., Cameron, S., et al. (2014). Phase I dose-escalation study of LCL161, an oral inhibitor of apoptosis proteins inhibitor, in patients with advanced solid tumors. *Journal of Clinical Oncology, 32*, 3103.

Iwasaki-Arai, J., Iwasaki, H., Miyamoto, T., Watanabe, S., & Akashi, K. (2003). Enforced granulocyte/macrophage colony-stimulating factor signals do not support lymphopoiesis, but instruct lymphoid to myelomonocytic lineage conversion. *The Journal of Experimental Medicine, 197*, 1311.

Jaffee, E. M., Hruban, R. H., Biedrzycki, B., Laheru, D., Schepers, K., Sauter, P. R., et al. (2001). Novel allogeneic granulocyte-macrophage colony-stimulating factor-secreting tumor vaccine for pancreatic cancer: A phase I trial of safety and immune activation. *Journal of Clinical Oncology, 19*, 145.

Jandus, C., Bioley, G., Dojcinovic, D., Derre, L., Baitsch, L., Wieckowski, S., et al. (2009). Tumor antigen-specific FOXP3+ CD4 T cells identified in human metastatic melanoma: Peptide vaccination results in selective expansion of Th1-like counterparts. *Cancer Research, 69*, 8085.

Jewell, C. M., Lopez, S. C., & Irvine, D. J. (2011). In situ engineering of the lymph node microenvironment via intranodal injection of adjuvant-releasing polymer particles. *Proceedings of the National Academy of Sciences of the United States of America, 108*, 15745.

Jinushi, M., Hodi, F. S., & Dranoff, G. (2006). Therapy-induced antibodies to MHC class I chain-related protein A antagonize immune suppression and stimulate antitumor cytotoxicity. *Proceedings of the National Academy of Sciences of the United States of America, 103*, 9190.

Jinushi, M., Nakazaki, Y., Dougan, M., Carrasco, D. R., Mihm, M., & Dranoff, G. (2007). MFG-E8-mediated uptake of apoptotic cells by APCs links the pro- and antiinflammatory activities of GM-CSF. *The Journal of Clinical Investigation, 117*, 1902.

Jongbloed, S. L., Kassianos, A. J., McDonald, K. J., Clark, G. J., Ju, X., Angel, C. E., et al. (2010). Human CD141+ (BDCA-3)+ dendritic cells (DCs) represent a unique myeloid DC subset that cross-presents necrotic cell antigens. *The Journal of Experimental Medicine, 207*, 1247.

Jonuleit, H., Kuhn, U., Muller, G., Steinbrink, K., Paragnik, L., Schmitt, E., et al. (1997). Pro-inflammatory cytokines and prostaglandins induce maturation of potent immunostimulatory dendritic cells under fetal calf serum-free conditions. *European Journal of Immunology, 27*, 3135.

Joyce, J. A., & Fearon, D. T. (2015). T cell exclusion, immune privilege, and the tumor microenvironment. *Science, 348*, 74.

Kalergis, A. M., & Ravetch, J. V. (2002). Inducing tumor immunity through the selective engagement of activating Fcgamma receptors on dendritic cells. *The Journal of Experimental Medicine, 195*, 1653.

Kanerva, A., Nokisalmi, P., Diaconu, I., Koski, A., Cerullo, V., Liikanen, I., et al. (2013). Antiviral and antitumor T-cell immunity in patients treated with GM-CSF-coding oncolytic adenovirus. *Clinical Cancer Research, 19*, 2734.

Kantoff, P. W., Higano, C. S., Shore, N. D., Berger, E. R., Small, E. J., Penson, D. F., et al. (2010). Sipuleucel-T immunotherapy for castration-resistant prostate cancer. *The New England Journal of Medicine, 363*, 411.

Kantoff, P. W., Schuetz, T. J., Blumenstein, B. A., Glode, L. M., Bilhartz, D. L., Wyand, M., et al. (2010). Overall survival analysis of a phase II randomized controlled trial of a Poxviral-based PSA-targeted immunotherapy in metastatic castration-resistant prostate cancer. *Journal of Clinical Oncology, 28*, 1099.

Karapanagiotou, E. M., Roulstone, V., Twigger, K., Ball, M., Tanay, M., Nutting, C., et al. (2012). Phase I/II trial of carboplatin and paclitaxel chemotherapy in combination with intravenous oncolytic reovirus in patients with advanced malignancies. *Clinical Cancer Research, 18*, 2080.

Karbach, J., Gnjatic, S., Bender, A., Neumann, A., Weidmann, E., Yuan, J., et al. (2010). Tumor-reactive CD8+ T-cell responses after vaccination with NY-ESO-1 peptide, CpG 7909 and Montanide ISA-51: Association with survival. *International Journal of Cancer, 126*, 909.

Kared, H., Leforban, B., Montandon, R., Renand, A., Layseca Espinosa, E., Chatenoud, L., et al. (2008). Role of GM-CSF in tolerance induction by mobilized hematopoietic progenitors. *Blood, 112*, 2575.

Kastenmuller, K., Wille-Reece, U., Lindsay, R. W., Trager, L. R., Darrah, P. A., Flynn, B. J., et al. (2011). Protective T cell immunity in mice following protein-TLR7/8 agonist-conjugate immunization requires aggregation, type I IFN, and multiple DC subsets. *The Journal of Clinical Investigation, 121*, 1782.

Kaufman, H. L., Kim, D. W., DeRaffele, G., Mitcham, J., Coffin, R. S., & Kim-Schulze, S. (2010). Local and distant immunity induced by intralesional vaccination with an oncolytic herpes virus encoding GM-CSF in patients with stage IIIc and IV melanoma. *Annals of Surgical Oncology*, *17*, 718.

Kaufman, H. L., Wang, W., Manola, J., DiPaola, R. S., Ko, Y. J., Sweeney, C., et al. (2004). Phase II randomized study of vaccine treatment of advanced prostate cancer (E7897): A trial of the Eastern Cooperative Oncology Group. *Journal of Clinical Oncology*, *22*, 2122.

Kawakami, Y., Eliyahu, S., Sakaguchi, K., Robbins, P. F., Rivoltini, L., Yannelli, J. R., et al. (1994). Identification of the immunodominant peptides of the MART-1 human melanoma antigen recognized by the majority of HLA-A2-restricted tumor infiltrating lymphocytes. *The Journal of Experimental Medicine*, *180*, 347.

Kenter, G. G., Welters, M. J., Valentijn, A. R., Lowik, M. J., Berends-van der Meer, D. M., Vloon, A. P., et al. (2009). Vaccination against HPV-16 oncoproteins for vulvar intraepithelial neoplasia. *The New England Journal of Medicine*, *361*, 1838.

Khuri, F. R., Nemunaitis, J., Ganly, I., Arseneau, J., Tannock, I. F., Romel, L., et al. (2000). A controlled trial of intratumoral ONYX-015, a selectively-replicating adenovirus, in combination with cisplatin and 5-fluorouracil in patients with recurrent head and neck cancer. *Nature Medicine*, *6*, 879.

Kim, J., Bencherif, S. A., Li, W. A., & Mooney, D. J. (2014). Cell-friendly inverse opal-like hydrogels for a spatially separated co-culture system. *Macromolecular Rapid Communications*, *35*, 1578.

Kim, M. K., Breitbach, C. J., Moon, A., Heo, J., Lee, Y. K., Cho, M., et al. (2013). Oncolytic and immunotherapeutic vaccinia induces antibody-mediated complement-dependent cancer cell lysis in humans. *Science Translational Medicine*, *5*, 185ra63.

Kim, J., Li, W. A., Choi, Y., Lewin, S. A., Verbeke, C. S., Dranoff, G., et al. (2015). Injectable, spontaneously assembling, inorganic scaffolds modulate immune cells in vivo and increase vaccine efficacy. *Nature Biotechnology*, *33*, 64.

Kim, J., & Mooney, D. J. (2011). In vivo modulation of dendritic cells by engineered materials: Towards new cancer vaccines. *Nano Today*, *6*, 466.

Klavinskis, L. S., Barnfield, C., Gao, L., & Parker, S. (1999). Intranasal immunization with plasmid DNA-lipid complexes elicits mucosal immunity in the female genital and rectal tracts. *Journal of Immunology*, *162*, 254.

Klechevsky, E., Flamar, A. L., Cao, Y., Blanck, J. P., Liu, M., O'Bar, A., et al. (2010). Cross-priming CD8+ T cells by targeting antigens to human dendritic cells through DCIR. *Blood*, *116*, 1685.

Klein, O., Davis, I. D., McArthur, G. A., Chen, L., Haydon, A., Parente, P., et al. (2015). Low-dose cyclophosphamide enhances antigen-specific CD4(+) T cell responses to NY-ESO-1/ISCOMATRIX vaccine in patients with advanced melanoma. *Cancer Immunology, Immunotherapy*, *64*, 507.

Kloetzel, P. M. (2004). Generation of major histocompatibility complex class I antigens: Functional interplay between proteasomes and TPPII. *Nature Immunology*, *5*, 661.

Knight, D. A., Ngiow, S. F., Li, M., Parmenter, T., Mok, S., Cass, A., et al. (2013). Host immunity contributes to the anti-melanoma activity of BRAF inhibitors. *The Journal of Clinical Investigation*, *123*, 1371.

Koks, C. A., Garg, A. D., Ehrhardt, M., Riva, M., Vandenberk, L., Boon, L., et al. (2015). Newcastle disease virotherapy induces long-term survival and tumor-specific immune memory in orthotopic glioma through the induction of immunogenic cell death. *International Journal of Cancer*, *136*, E313.

Kortylewski, M., Swiderski, P., Herrmann, A., Wang, L., Kowolik, C., Kujawski, M., et al. (2009). In vivo delivery of siRNA to immune cells by conjugation to a TLR9 agonist enhances antitumor immune responses. *Nature Biotechnology*, *27*, 925.

Koshy, S. T., Ferrante, T. C., Lewin, S. A., & Mooney, D. J. (2014). Injectable, porous, and cell-responsive gelatin cryogels. *Biomaterials*, *35*, 2477.

Kourtis, I. C., Hirosue, S., de Titta, A., Kontos, S., Stegmann, T., & Hubbell, J. A. (2013). Peripherally administered nanoparticles target monocytic myeloid cells, secondary lymphoid organs and tumors in mice. *PloS One*, *8*, e61646.

Kreiter, S., Diken, M., Selmi, A., Diekmann, J., Attig, S., Husemann, Y., et al. (2011). FLT3 ligand enhances the cancer therapeutic potency of naked RNA vaccines. *Cancer Research*, *71*, 6132.

Kreiter, S., Selmi, A., Diken, M., Koslowski, M., Britten, C. M., Huber, C., et al. (2010). Intranodal vaccination with naked antigen-encoding RNA elicits potent prophylactic and therapeutic antitumoral immunity. *Cancer Research*, *70*, 9031.

Kreiter, S., Vormehr, M., van de Roemer, N., Diken, M., Lower, M., Diekmann, J., et al. (2015). Mutant MHC class II epitopes drive therapeutic immune responses to cancer. *Nature*, *520*, 692.

Kretschmer, K., Apostolou, I., Hawiger, D., Khazaie, K., Nussenzweig, M. C., & von Boehmer, H. (2005). Inducing and expanding regulatory T cell populations by foreign antigen. *Nature Immunology*, *6*, 1219.

Kruit, W. H., Suciu, S., Dreno, B., Mortier, L., Robert, C., Chiarion-Sileni, V., et al. (2013). Selection of immunostimulant AS15 for active immunization with MAGE-A3 protein: Results of a randomized phase II study of the European Organisation for Research and Treatment of Cancer Melanoma Group in Metastatic Melanoma. *Journal of Clinical Oncology*, *31*, 2413.

Krysko, D. V., Garg, A. D., Kaczmarek, A., Krysko, O., Agostinis, P., & Vandenabeele, P. (2012). Immunogenic cell death and DAMPs in cancer therapy. *Nature Reviews. Cancer*, *12*, 860.

Kuksa, P. P., Min, M. R., Dugar, R., & Gerstein, M. (2015). High-order neural networks and kernel methods for peptide-MHC binding prediction. *Bioinformatics*, *31*(22), 3600–3607.

Kumamoto, T., Huang, E. K., Paek, H. J., Morita, A., Matsue, H., Valentini, R. F., et al. (2002). Induction of tumor-specific protective immunity by in situ Langerhans cell vaccine. *Nature Biotechnology*, *20*, 64.

Kwong, B., Gai, S. A., Elkhader, J., Wittrup, K. D., & Irvine, D. J. (2013). Localized immunotherapy via liposome-anchored Anti-CD137 + IL-2 prevents lethal toxicity and elicits local and systemic antitumor immunity. *Cancer Research*, *73*, 1547.

Kwong, B., Liu, H., & Irvine, D. J. (2011). Induction of potent anti-tumor responses while eliminating systemic side effects via liposome-anchored combinatorial immunotherapy. *Biomaterials*, *32*, 5134.

Lakshminarayanan, V., Thompson, P., Wolfert, M. A., Buskas, T., Bradley, J. M., Pathangey, L. B., et al. (2012). Immune recognition of tumor-associated mucin MUC1 is achieved by a fully synthetic aberrantly glycosylated MUC1 tripartite vaccine. *Proceedings of the National Academy of Sciences of the United States of America*, *109*, 261.

Le, D. T., Brockstedt, D. G., Nir-Paz, R., Hampl, J., Mathur, S., Nemunaitis, J., et al. (2012). A live-attenuated Listeria vaccine (ANZ-100) and a live-attenuated Listeria vaccine expressing mesothelin (CRS-207) for advanced cancers: Phase I studies of safety and immune induction. *Clinical Cancer Research*, *18*, 858.

Le, D. T., Lutz, E., Uram, J. N., Sugar, E. A., Onners, B., Solt, S., et al. (2013). Evaluation of ipilimumab in combination with allogeneic pancreatic tumor cells transfected with a GM-CSF gene in previously treated pancreatic cancer. *Journal of Immunotherapy*, *36*, 382.

Leone, P., Shin, E. C., Perosa, F., Vacca, A., Dammacco, F., & Racanelli, V. (2013). MHC class I antigen processing and presenting machinery: Organization, function, and defects in tumor cells. *Journal of the National Cancer Institute*, *105*, 1172.

Li, H., Li, Y., Jiao, J., & Hu, H. M. (2011). Alpha-alumina nanoparticles induce efficient autophagy-dependent cross-presentation and potent antitumour response. *Nature Nanotechnology*, *6*, 645.

Li, A. V., Moon, J. J., Abraham, W., Suh, H., Elkhader, J., Seidman, M. A., et al. (2013). Generation of effector memory T cell-based mucosal and systemic immunity with pulmonary nanoparticle vaccination. *Science Translational Medicine*, *5*, 204ra130.

Li, W. A., & Mooney, D. J. (2013). Materials based tumor immunotherapy vaccines. *Current Opinion in Immunology*, *25*, 238.

Li, B., VanRoey, M., Wang, C., Chen, T. H., Korman, A., & Jooss, K. (2009). Anti-programmed death-1 synergizes with granulocyte macrophage colony-stimulating factor–secreting tumor cell immunotherapy providing therapeutic benefit to mice with established tumors. *Clinical Cancer Research*, *15*, 1623.

Liang, S. C., Moskalenko, M., Van Roey, M., & Jooss, K. (2013). Depletion of regulatory T cells by targeting folate receptor 4 enhances the potency of a GM-CSF-secreting tumor cell immunotherapy. *Clinical Immunology*, *148*, 287.

Lichty, B. D., Breitbach, C. J., Stojdl, D. F., & Bell, J. C. (2014). Going viral with cancer immunotherapy. *Nature Reviews. Cancer*, *14*, 559.

Linnemann, C., van Buuren, M. M., Bies, L., Verdegaal, E. M., Schotte, R., Calis, J. J., et al. (2015). High-throughput epitope discovery reveals frequent recognition of neo-antigens by CD4+ T cells in human melanoma. *Nature Medicine*, *21*, 81.

Lipson, E. J., Sharfman, W. H., Chen, S., McMiller, T. L., Pritchard, T. S., Salas, J. T., et al. (2015). Safety and immunologic correlates of Melanoma GVAX, a GM-CSF secreting allogeneic melanoma cell vaccine administered in the adjuvant setting. *Journal of Translational Medicine*, *13*, 214.

Liu, H., Moynihan, K. D., Zheng, Y., Szeto, G. L., Li, A. V., Huang, B., et al. (2014). Structure-based programming of lymph-node targeting in molecular vaccines. *Nature*, *507*, 519.

Lu, Y. C., Yao, X., Li, Y. F., El-Gamil, M., Dudley, M. E., Yang, J. C., et al. (2013). Mutated PPP1R3B is recognized by T cells used to treat a melanoma patient who experienced a durable complete tumor regression. *Journal of Immunology*, *190*, 6034.

Luft, T., Jefford, M., Luetjens, P., Toy, T., Hochrein, H., Masterman, K. A., et al. (2002). Functionally distinct dendritic cell (DC) populations induced by physiologic stimuli: Prostaglandin E(2) regulates the migratory capacity of specific DC subsets. *Blood*, *100*, 1362.

Lutz, E. R., Wu, A. A., Bigelow, E., Sharma, R., Mo, G., Soares, K., et al. (2014). Immunotherapy converts nonimmunogenic pancreatic tumors into immunogenic foci of immune regulation. *Cancer Immunology Research*, *2*, 616.

Mach, N., Gillessen, S., Wilson, S. B., Sheehan, C., Mihm, M., & Dranoff, G. (2000). Differences in dendritic cells stimulated in vivo by tumors engineered to secrete granulocyte-macrophage colony-stimulating factor or Flt3-ligand. *Cancer Research*, *60*, 3239.

Madan, R. A., Aragon-Ching, J. B., Gulley, J. L., & Dahut, W. L. (2011). From clinical trials to clinical practice: Therapeutic cancer vaccines for the treatment of prostate cancer. *Expert Review of Vaccines*, *10*, 743.

Mahnke, K., Guo, M., Lee, S., Sepulveda, H., Swain, S. L., Nussenzweig, M., et al. (2000). The dendritic cell receptor for endocytosis, DEC-205, can recycle and enhance antigen presentation via major histocompatibility complex class II-positive lysosomal compartments. *The Journal of Cell Biology*, *151*, 673.

Mailliard, R. B., Wankowicz-Kalinska, A., Cai, Q., Wesa, A., Hilkens, C. M., Kapsenberg, M. L., et al. (2004). alpha-type-1 polarized dendritic cells: A novel immunization tool with optimized CTL-inducing activity. *Cancer Research*, *64*, 5934.

Manolova, V., Flace, A., Bauer, M., Schwarz, K., Saudan, P., & Bachmann, M. F. (2008). Nanoparticles target distinct dendritic cell populations according to their size. *European Journal of Immunology*, *38*, 1404.

Maryanski, J. L., Romero, P., Van Pel, A., Boon, T., Salemme, F. R., Cerottini, J. C., et al. (1991). The identification of tyrosine as a common key residue in unrelated H-2Kd restricted antigenic peptides. *International Immunology*, *3*, 1035.

Matsuo, K., Koizumi, H., Akashi, M., Nakagawa, S., Fujita, T., Yamamoto, A., et al. (2011). Intranasal immunization with poly(gamma-glutamic acid) nanoparticles entrapping antigenic proteins can induce potent tumor immunity. *Journal of Controlled Release*, *152*, 310.

Matsushita, H., Vesely, M. D., Koboldt, D. C., Rickert, C. G., Uppaluri, R., Magrini, V. J., et al. (2012). Cancer exome analysis reveals a T-cell-dependent mechanism of cancer immunoediting. *Nature*, *482*, 400.

Mehta, N. K., Moynihan, K. D., & Irvine, D. J. (2015). Engineering new approaches to cancer vaccines. *Cancer Immunology Research*, *3*, 836.

Meixlsperger, S., Leung, C. S., Ramer, P. C., Pack, M., Vanoaica, L. D., Breton, G., et al. (2013). CD141+ dendritic cells produce prominent amounts of IFN-alpha after dsRNA recognition and can be targeted via DEC-205 in humanized mice. *Blood*, *121*, 5034.

Mellman, I., & Steinman, R. M. (2001). Dendritic cells: Specialized and regulated antigen processing machines. *Cell*, *106*, 255.

Michael, A., Ball, G., Quatan, N., Wushishi, F., Russell, N., Whelan, J., et al. (2005). Delayed disease progression after allogeneic cell vaccination in hormone-resistant prostate cancer and correlation with immunologic variables. *Clinical Cancer Research*, *11*, 4469.

Mildner, A., & Jung, S. (2014). Development and function of dendritic cell subsets. *Immunity*, *40*, 642.

Milne, P., Bigley, V., Gunawan, M., Haniffa, M., & Collin, M. (2015). CD1c+ blood dendritic cells have Langerhans cell potential. *Blood*, *125*, 470.

Mitchell, D. A., Batich, K. A., Gunn, M. D., Huang, M. N., Sanchez-Perez, L., Nair, S. K., et al. (2015). Tetanus toxoid and CCL3 improve dendritic cell vaccines in mice and glioblastoma patients. *Nature*, *519*, 366.

Moehler, M., Sieben, M., Roth, S., Springsguth, F., Leuchs, B., Zeidler, M., et al. (2011). Activation of the human immune system by chemotherapeutic or targeted agents combined with the oncolytic parvovirus H-1. *BMC Cancer*, *11*, 464.

Moon, J. J., Suh, H., Li, A. V., Ockenhouse, C. F., Yadava, A., & Irvine, D. J. (2012). Enhancing humoral responses to a malaria antigen with nanoparticle vaccines that expand Tfh cells and promote germinal center induction. *Proceedings of the National Academy of Sciences of the United States of America*, *109*, 1080.

Moon, J. J., Suh, H., Polhemus, M. E., Ockenhouse, C. F., Yadava, A., & Irvine, D. J. (2012). Antigen-displaying lipid-enveloped PLGA nanoparticles as delivery agents for a Plasmodium vivax malaria vaccine. *PloS One*, *7*, e31472.

Morishige, T., Yoshioka, Y., Inakura, H., Tanabe, A., Yao, X., Narimatsu, S., et al. (2010). The effect of surface modification of amorphous silica particles on NLRP3 inflammasome mediated IL-1beta production, ROS production and endosomal rupture. *Biomaterials*, *31*, 6833.

Mosca, P. J., Hobeika, A. C., Clay, T. M., Nair, S. K., Thomas, E. K., & Morse, M. A. (2000). A subset of human monocyte-derived dendritic cells expresses high levels of interleukin-12 in response to combined CD40 ligand and interferon-gamma treatment. *Blood*, *96*, 3499.

Narusawa, M., Inoue, H., Sakamoto, C., Matsumura, Y., Takahashi, A., Inoue, T., et al. (2014). TLR7 ligand augments GM-CSF-initiated antitumor immunity through activation of plasmacytoid dendritic cells. *Cancer Immunology Research*, *2*, 568.

Nehoff, H., Parayath, N. N., Domanovitch, L., Taurin, S., & Greish, K. (2014). Nanomedicine for drug targeting: Strategies beyond the enhanced permeability and retention effect. *International Journal of Nanomedicine*, *9*, 2539.

Nembrini, C., Stano, A., Dane, K. Y., Ballester, M., van der Vlies, A. J., Marsland, B. J., et al. (2011). Nanoparticle conjugation of antigen enhances cytotoxic T-cell responses in pulmonary vaccination. *Proceedings of the National Academy of Sciences of the United States of America*, *108*, E989.

Nestle, F. O., Alijagic, S., Gilliet, M., Sun, Y., Grabbe, S., Dummer, R., et al. (1998). Vaccination of melanoma patients with peptide- or tumor lysate-pulsed dendritic cells. *Nature Medicine*, *4*, 328.

Nicholaou, T., Ebert, L. M., Davis, I. D., McArthur, G. A., Jackson, H., Dimopoulos, N., et al. (2009). Regulatory T-cell-mediated attenuation of T-cell responses to the NY-ESO-1 ISCOMATRIX vaccine in patients with advanced malignant melanoma. *Clinical Cancer Research*, *15*, 2166.

Nizzoli, G., Krietsch, J., Weick, A., Steinfelder, S., Facciotti, F., Gruarin, P., et al. (2013). Human CD1c+ dendritic cells secrete high levels of IL-12 and potently prime cytotoxic T-cell responses. *Blood*, *122*, 932.

Nochi, T., Yuki, Y., Takahashi, H., Sawada, S., Mejima, M., Kohda, T., et al. (2010). Nanogel antigenic protein-delivery system for adjuvant-free intranasal vaccines. *Nature Materials*, *9*, 572.

Noonan, K. A., Ghosh, N., Rudraraju, L., Bui, M., & Borrello, I. (2014). Targeting immune suppression with PDE5 inhibition in end-stage multiple myeloma. *Cancer Immunology Research*, *2*, 725.

Obeid, M., Tesniere, A., Ghiringhelli, F., Fimia, G. M., Apetoh, L., Perfettini, J. L., et al. (2007). Calreticulin exposure dictates the immunogenicity of cancer cell death. *Nature Medicine*, *13*, 54.

Odunsi, K., Qian, F., Matsuzaki, J., Mhawech-Fauceglia, P., Andrews, C., Hoffman, E. W., et al. (2007). Vaccination with an NY-ESO-1 peptide of HLA class I/II specificities induces integrated humoral and T cell responses in ovarian cancer. *Proceedings of the National Academy of Sciences of the United States of America*, *104*, 12837.

Okada, H., Kalinski, P., Ueda, R., Hoji, A., Kohanbash, G., Donegan, T. E., et al. (2011). Induction of CD8+ T-cell responses against novel glioma-associated antigen peptides and clinical activity by vaccinations with {alpha}-type 1 polarized dendritic cells and polyinosinic-polycytidylic acid stabilized by lysine and carboxymethylcellulose in patients with recurrent malignant glioma. *Journal of Clinical Oncology*, *29*, 330.

Pan, P. Y., Wang, G. X., Yin, B., Ozao, J., Ku, T., Divino, C. M., et al. (2008). Reversion of immune tolerance in advanced malignancy: Modulation of myeloid-derived suppressor cell development by blockade of stem-cell factor function. *Blood*, *111*, 219.

Park, S. S., Dong, H., Liu, X., Harrington, S. M., Krco, C. J., Grams, M. P., et al. (2015). PD-1 restrains radiotherapy-induced abscopal effect. *Cancer Immunology Research*, *3*, 610.

Park, J., Gao, W., Whiston, R., Strom, T. B., Metcalfe, S., & Fahmy, T. M. (2011). Modulation of CD4+ T lymphocyte lineage outcomes with targeted, nanoparticle-mediated cytokine delivery. *Molecular Pharmaceutics*, *8*, 143.

Park, J., Wrzesinski, S. H., Stern, E., Look, M., Criscione, J., Ragheb, R., et al. (2012). Combination delivery of TGF-beta inhibitor and IL-2 by nanoscale liposomal polymeric gels enhances tumour immunotherapy. *Nature Materials*, *11*, 895.

Pearce, E. L., & Shen, H. (2007). Generation of CD8 T cell memory is regulated by IL-12. *Journal of Immunology*, *179*, 2074.

Peters, B., Tong, W., Sidney, J., Sette, A., & Weng, Z. (2003). Examining the independent binding assumption for binding of peptide epitopes to MHC-I molecules. *Bioinformatics*, *19*, 1765.

Piesche, M., Ho, V. T., Kim, H., Nakazaki, Y., Nehil, M., Yaghi, N. K., et al. (2015). Angiogenic cytokines are antibody targets during graft-versus-leukemia reactions. *Clinical Cancer Research*, *21*, 1010.

Postow, M. A., Callahan, M. K., Barker, C. A., Yamada, Y., Yuan, J., Kitano, S., et al. (2012). Immunologic correlates of the abscopal effect in a patient with melanoma. *The New England Journal of Medicine*, *366*, 925.

Prestwich, R. J., Errington, F., Ilett, E. J., Morgan, R. S., Scott, K. J., Kottke, T., et al. (2008). Tumor infection by oncolytic reovirus primes adaptive antitumor immunity. *Clinical Cancer Research*, *14*, 7358.

Prestwich, R. J., Ilett, E. J., Errington, F., Diaz, R. M., Steele, L. P., & Kottke, T. (2009). Immune-mediated antitumor activity of reovirus is required for therapy and is independent of direct viral oncolysis and replication. *Clinical Cancer Research*, *15*, 4374.

Priceman, S. J., Sung, J. L., Shaposhnik, Z., Burton, J. B., Torres-Collado, A. X., Moughon, D. L., et al. (2010). Targeting distinct tumor-infiltrating myeloid cells by inhibiting CSF-1 receptor: Combating tumor evasion of antiangiogenic therapy. *Blood*, *115*, 1461.

Probst, J., Weide, B., Scheel, B., Pichler, B. J., Hoerr, I., Rammensee, H. G., et al. (2007). Spontaneous cellular uptake of exogenous messenger RNA in vivo is nucleic acid-specific, saturable and ion dependent. *Gene Therapy*, *14*, 1175.

Qin, H., Lerman, B., Sakamaki, I., Wei, G., Cha, S. C., Rao, S. S., et al. (2014). Generation of a new therapeutic peptide that depletes myeloid-derived suppressor cells in tumor-bearing mice. *Nature Medicine*, *20*, 676.

Quezada, S. A., Peggs, K. S., Curran, M. A., & Allison, J. P. (2006). CTLA4 blockade and GM-CSF combination immunotherapy alters the intratumor balance of effector and regulatory T cells. *The Journal of Clinical Investigation*, *116*, 1935.

Rajasagi, M., Shukla, S. A., Fritsch, E. F., Keskin, D. B., DeLuca, D., Carmona, E., et al. (2014). Systematic identification of personal tumor-specific neoantigens in chronic lymphocytic leukemia. *Blood*, *124*, 453.

Rammensee, H., Bachmann, J., Emmerich, N. P., Bachor, O. A., & Stevanovic, S. (1999). SYFPEITHI: Database for MHC ligands and peptide motifs. *Immunogenetics*, *50*, 213.

Reddy, S. T., Rehor, A., Schmoekel, H. G., Hubbell, J. A., & Swartz, M. A. (2006). In vivo targeting of dendritic cells in lymph nodes with poly(propylene sulfide) nanoparticles. *Journal of Controlled Release*, *112*, 26.

Reddy, S. T., van der Vlies, A. J., Simeoni, E., Angeli, V., Randolph, G. J., O'Neil, C. P., et al. (2007). Exploiting lymphatic transport and complement activation in nanoparticle vaccines. *Nature Biotechnology*, *25*, 1159.

Rittig, S. M., Haentschel, M., Weimer, K. J., Heine, A., Muller, M. R., Brugger, W., et al. (2011). Intradermal vaccinations with RNA coding for TAA generate CD8+ and CD4+ immune responses and induce clinical benefit in vaccinated patients. *Molecular Therapy*, *19*, 990.

Rizvi, N. A., Hellmann, M. D., Snyder, A., Kvistborg, P., Makarov, V., Havel, J. J., et al. (2015). Cancer immunology. Mutational landscape determines sensitivity to PD-1 blockade in non-small cell lung cancer. *Science*, *348*, 124.

Robbins, P. F., Lu, Y. C., El-Gamil, M., Li, Y. F., Gross, C., Gartner, J., et al. (2013). Mining exomic sequencing data to identify mutated antigens recognized by adoptively transferred tumor-reactive T cells. *Nature Medicine*, *19*, 747.

Rooney, M. S., Shukla, S. A., Wu, C. J., Getz, G., & Hacohen, N. (2015). Molecular and genetic properties of tumors associated with local immune cytolytic activity. *Cell*, *160*, 48.

Rosenberg, S. A., Packard, B. S., Aebersold, P. M., Solomon, D., Topalian, S. L., & Toy, S. T. (1988). Use of tumor-infiltrating lymphocytes and interleukin-2 in the

immunotherapy of patients with metastatic melanoma. A preliminary report. *The New England Journal of Medicine*, *319*, 1676.
Rosenberg, S. A., Yang, J. C., Schwartzentruber, D. J., Hwu, P., Marincola, F. M., Topalian, S. L., et al. (1998). Immunologic and therapeutic evaluation of a synthetic peptide vaccine for the treatment of patients with metastatic melanoma. *Nature Medicine*, *4*, 321.
Sabado, R. L., Pavlick, A., Gnjatic, S., Cruz, C. M., Vengco, I., Hasan, F., et al. (2015). Resiquimod as an immunologic adjuvant for NY-ESO-1 protein vaccination in patients with high-risk melanoma. *Cancer Immunology Research*, *3*, 278.
Sahin, U., Kariko, K., & Tureci, O. (2014). mRNA-based therapeutics–developing a new class of drugs. *Nature Reviews. Drug Discovery*, *13*, 759.
Sahin, U., Tureci, O., Schmitt, H., Cochlovius, B., Johannes, T., Schmits, R., et al. (1995). Human neoplasms elicit multiple specific immune responses in the autologous host. *Proceedings of the National Academy of Sciences of the United States of America*, *92*, 11810.
Salazar, A. M., Erlich, R. B., Mark, A., Bhardwaj, N., & Herberman, R. B. (2014). Therapeutic in situ autovaccination against solid cancers with intratumoral poly-ICLC: Case report, hypothesis, and clinical trial. *Cancer Immunology Research*, *2*, 720.
Salgaller, M. L., Marincola, F. M., Cormier, J. N., & Rosenberg, S. A. (1996). Immunization against epitopes in the human melanoma antigen gp100 following patient immunization with synthetic peptides. *Cancer Research*, *56*, 4749.
Salgia, R., Lynch, T., Skarin, A., Lucca, J., Lynch, C., Jung, K., et al. (2003). Vaccination with irradiated autologous tumor cells engineered to secrete granulocyte-macrophage colony-stimulating factor augments antitumor immunity in some patients with metastatic non-small-cell lung carcinoma. *Journal of Clinical Oncology*, *21*, 624.
Sancho, D., Joffre, O. P., Keller, A. M., Rogers, N. C., Martinez, D., Hernanz-Falcon, P., et al. (2009). Identification of a dendritic cell receptor that couples sensing of necrosis to immunity. *Nature*, *458*, 899.
Sandler, A. D., Chihara, H., Kobayashi, G., Zhu, X., Miller, M. A., Scott, D. L., et al. (2003). CpG oligonucleotides enhance the tumor antigen-specific immune response of a granulocyte macrophage colony-stimulating factor-based vaccine strategy in neuroblastoma. *Cancer Research*, *63*, 394.
Sano, K., Nakajima, T., Choyke, P. L., & Kobayashi, H. (2013). Markedly enhanced permeability and retention effects induced by photo-immunotherapy of tumors. *ACS Nano*, *7*, 717.
Santegoets, S. J., Stam, A. G., Lougheed, S. M., Gall, H., Jooss, K., Sacks, N., et al. (2014). Myeloid derived suppressor and dendritic cell subsets are related to clinical outcome in prostate cancer patients treated with prostate GVAX and ipilimumab. *Journal for Immunotherapy of Cancer*, *2*, 31.
Schellhammer, P. F., Chodak, G., Whitmore, J. B., Sims, R., Frohlich, M. W., & Kantoff, P. W. (2013). Lower baseline prostate-specific antigen is associated with a greater overall survival benefit from sipuleucel-T in the Immunotherapy for Prostate Adenocarcinoma Treatment (IMPACT) trial. *Urology*, *81*, 1297.
Schmollinger, J. C., Vonderheide, R. H., Hoar, K. M., Maecker, B., Schultze, J. L., Hodi, F. S., et al. (2003). Melanoma inhibitor of apoptosis protein (ML-IAP) is a target for immune-mediated tumor destruction. *Proceedings of the National Academy of Sciences of the United States of America*, *100*, 3398.
Schoenfeld, J., Jinushi, M., Nakazaki, Y., Wiener, D., Park, J., Soiffer, R., et al. (2010). Active immunotherapy induces antibody responses that target tumor angiogenesis. *Cancer Research*, *70*, 10150.
Schreiber, R. D., Old, L. J., & Smyth, M. J. (2011). Cancer immunoediting: Integrating immunity's roles in cancer suppression and promotion. *Science*, *331*, 1565.

Schumacher, T., Bunse, L., Pusch, S., Sahm, F., Wiestler, B., Quandt, J., et al. (2014). A vaccine targeting mutant IDH1 induces antitumour immunity. *Nature*, *512*, 324.

Schwartzentruber, D. J., Lawson, D. H., Richards, J. M., Conry, R. M., Miller, D. M., Treisman, J., et al. (2011). gp100 peptide vaccine and interleukin-2 in patients with advanced melanoma. *The New England Journal of Medicine*, *364*, 2119.

Scott, E. A., Stano, A., Gillard, M., Maio-Liu, A. C., Swartz, M. A., & Hubbell, J. A. (2012). Dendritic cell activation and T cell priming with adjuvant- and antigen-loaded oxidation-sensitive polymersomes. *Biomaterials*, *33*, 6211.

Segura, E., Valladeau-Guilemond, J., Donnadieu, M. H., Sastre-Garau, X., Soumelis, V., & Amigorena, S. (2012). Characterization of resident and migratory dendritic cells in human lymph nodes. *The Journal of Experimental Medicine*, *209*, 653.

Selbo, P. K., Weyergang, A., Hogset, A., Norum, O. J., Berstad, M. B., Vikdal, M., et al. (2010). Photochemical internalization provides time- and space-controlled endolysosomal escape of therapeutic molecules. *Journal of Controlled Release*, *148*, 2.

Selby, M. J., Engelhardt, J. J., Quigley, M., Henning, K. A., Chen, T., Srinivasan, M., et al. (2013). Anti-CTLA-4 antibodies of IgG2a isotype enhance antitumor activity through reduction of intratumoral regulatory T cells. *Cancer Immunology Research*, *1*, 32.

Serafini, P., Carbley, R., Noonan, K. A., Tan, G., Bronte, V., & Borrello, I. (2004). High-dose granulocyte-macrophage colony-stimulating factor-producing vaccines impair the immune response through the recruitment of myeloid suppressor cells. *Cancer Research*, *64*, 6337.

Serafini, P., Meckel, K., Kelso, M., Noonan, K., Califano, J., Koch, W., et al. (2006). Phosphodiesterase-5 inhibition augments endogenous antitumor immunity by reducing myeloid-derived suppressor cell function. *The Journal of Experimental Medicine*, *203*, 2691.

Shen, H., Ackerman, A. L., Cody, V., Giodini, A., Hinson, E. R., Cresswell, P., et al. (2006). Enhanced and prolonged cross-presentation following endosomal escape of exogenous antigens encapsulated in biodegradable nanoparticles. *Immunology*, *117*, 78.

Sikora, A. G., Jaffarzad, N., Hailemichael, Y., Gelbard, A., Stonier, S. W., Schluns, K. S., et al. (2009). IFN-alpha enhances peptide vaccine-induced CD8+ T cell numbers, effector function, and antitumor activity. *Journal of Immunology*, *182*, 7398.

Simons, J. W., Jaffee, E. M., Weber, C. E., Levitsky, H. I., Nelson, W. G., Carducci, M. A., et al. (1997). Bioactivity of autologous irradiated renal cell carcinoma vaccines generated by ex vivo granulocyte-macrophage colony-stimulating factor gene transfer. *Cancer Research*, *57*, 1537.

Simons, J. W., Mikhak, B., Chang, J. F., DeMarzo, A. M., Carducci, M. A., Lim, M., et al. (1999). Induction of immunity to prostate cancer antigens: Results of a clinical trial of vaccination with irradiated autologous prostate tumor cells engineered to secrete granulocyte-macrophage colony-stimulating factor using ex vivo gene transfer. *Cancer Research*, *59*, 5160.

Simpson, T. R., Li, F., Montalvo-Ortiz, W., Sepulveda, M. A., Bergerhoff, K., Arce, F., et al. (2013). Fc-dependent depletion of tumor-infiltrating regulatory T cells co-defines the efficacy of anti-CTLA-4 therapy against melanoma. *The Journal of Experimental Medicine*, *210*, 1695.

Singh, P., Pal, S. K., Alex, A., & Agarwal, N. (2015). Development of PROSTVAC immunotherapy in prostate cancer. *Future Oncology*, *11*, 2137.

Sistigu, A., Yamazaki, T., Vacchelli, E., Chaba, K., Enot, D. P., Adam, J., et al. (2014). Cancer cell-autonomous contribution of type I interferon signaling to the efficacy of chemotherapy. *Nature Medicine*, *20*, 1301.

Small, E. J., Sacks, N., Nemunaitis, J., Urba, W. J., Dula, E., Centeno, A. S., et al. (2007). Granulocyte macrophage colony-stimulating factor–secreting allogeneic cellular immunotherapy for hormone-refractory prostate cancer. *Clinical Cancer Research*, *13*, 3883.

Soares, K. C., Rucki, A. A., Wu, A. A., Olino, K., Xiao, Q., Chai, Y., et al. (2015). PD-1/PD-L1 blockade together with vaccine therapy facilitates effector T-cell infiltration into pancreatic tumors. *Journal of Immunotherapy*, *38*, 1.

Soiffer, R., Hodi, F. S., Haluska, F., Jung, K., Gillessen, S., Singer, S., et al. (2003). Vaccination with irradiated, autologous melanoma cells engineered to secrete granulocyte-macrophage colony-stimulating factor by adenoviral-mediated gene transfer augments antitumor immunity in patients with metastatic melanoma. *Journal of Clinical Oncology*, *21*, 3343.

Soiffer, R., Lynch, T., Mihm, M., Jung, K., Rhuda, C., Schmollinger, J. C., et al. (1998). Vaccination with irradiated autologous melanoma cells engineered to secrete human granulocyte-macrophage colony-stimulating factor generates potent antitumor immunity in patients with metastatic melanoma. *Proceedings of the National Academy of Sciences of the United States of America*, *95*, 13141.

Sorensen, A. L., Reis, C. A., Tarp, M. A., Mandel, U., Ramachandran, K., Sankaranarayanan, V., et al. (2006). Chemoenzymatically synthesized multimeric Tn/STn MUC1 glycopeptides elicit cancer-specific anti-MUC1 antibody responses and override tolerance. *Glycobiology*, *16*, 96.

Speiser, D. E., Lienard, D., Rufer, N., Rubio-Godoy, V., Rimoldi, D., Lejeune, F., et al. (2005). Rapid and strong human CD8+ T cell responses to vaccination with peptide, IFA, and CpG oligodeoxynucleotide 7909. *The Journal of Clinical Investigation*, *115*, 739.

St John, A. L., Chan, C. Y., Staats, H. F., Leong, K. W., & Abraham, S. N. (2012). Synthetic mast-cell granules as adjuvants to promote and polarize immunity in lymph nodes. *Nature Materials*, *11*, 250.

Stano, A., Scott, E. A., Dane, K. Y., Swartz, M. A., & Hubbell, J. A. (2013). Tunable T cell immunity towards a protein antigen using polymersomes vs. solid-core nanoparticles. *Biomaterials*, *34*, 4339.

Stanos, S. P., Bruckenthal, P., & Barkin, R. L. (2012). Strategies to reduce the tampering and subsequent abuse of long-acting opioids: Potential risks and benefits of formulations with physical or pharmacologic deterrents to tampering. *Mayo Clinic Proceedings*, *87*, 683.

Starbeck-Miller, G. R., Xue, H. H., & Harty, J. T. (2014). IL-12 and type I interferon prolong the division of activated CD8 T cells by maintaining high-affinity IL-2 signaling in vivo. *The Journal of Experimental Medicine*, *211*, 105.

Steinman, R. M., Hawiger, D., & Nussenzweig, M. C. (2003). Tolerogenic dendritic cells. *Annual Review of Immunology*, *21*, 685.

Stylianou, E., Diogo, G. R., Pepponi, I., van Dolleweerd, C., Arias, M. A., Locht, C., et al. (2014). Mucosal delivery of antigen-coated nanoparticles to lungs confers protective immunity against tuberculosis infection in mice. *European Journal of Immunology*, *44*, 440.

Sun, H. X., Xie, Y., & Ye, Y. P. (2009). ISCOMs and ISCOMATRIX. *Vaccine*, *27*, 4388.

Talmadge, J. E., & Gabrilovich, D. I. (2013). History of myeloid-derived suppressor cells. *Nature Reviews. Cancer*, *13*, 739.

Tamura, Y., Peng, P., Liu, K., Daou, M., & Srivastava, P. K. (1997). Immunotherapy of tumors with autologous tumor-derived heat shock protein preparations. *Science*, *278*, 117.

Testori, A., Richards, J., Whitman, E., Mann, G. B., Lutzky, J., Camacho, L., et al. (2008). Phase III comparison of vitespen, an autologous tumor-derived heat shock protein gp96 peptide complex vaccine, with physician's choice of treatment for stage IV melanoma: The C-100-21 Study Group. *Journal of Clinical Oncology*, *26*, 955.

Thomas, D. L., & Fraser, N. W. (2003). HSV-1 therapy of primary tumors reduces the number of metastases in an immune-competent model of metastatic breast cancer. *Molecular Therapy*, *8*, 543.

Thomas, A. M., Santarsiero, L. M., Lutz, E. R., Armstrong, T. D., Chen, Y. C., Huang, L. Q., et al. (2004). Mesothelin-specific CD8(+) T cell responses provide evidence of in vivo cross-priming by antigen-presenting cells in vaccinated pancreatic cancer patients. *The Journal of Experimental Medicine*, *200*, 297.

Thomas, S. N., Vokali, E., Lund, A. W., Hubbell, J. A., & Swartz, M. A. (2014). Targeting the tumor-draining lymph node with adjuvanted nanoparticles reshapes the anti-tumor immune response. *Biomaterials*, *35*, 814.

Toda, M., Martuza, R. L., & Rabkin, S. D. (2000). Tumor growth inhibition by intratumoral inoculation of defective herpes simplex virus vectors expressing granulocyte-macrophage colony-stimulating factor. *Molecular Therapy*, *2*, 324.

Todo, T., Rabkin, S. D., Sundaresan, P., Wu, A., Meehan, K. R., Herscowitz, H. B., et al. (1999). Systemic antitumor immunity in experimental brain tumor therapy using a multimutated, replication-competent herpes simplex virus. *Human Gene Therapy*, *10*, 2741.

Tran, E., Turcotte, S., Gros, A., Robbins, P. F., Lu, Y. C., Dudley, M. E., et al. (2014). Cancer immunotherapy based on mutation-specific CD4+ T cells in a patient with epithelial cancer. *Science*, *344*, 641.

Tseng, D., Volkmer, J. P., Willingham, S. B., Contreras-Trujillo, H., Fathman, J. W., Fernhoff, N. B., et al. (2013). Anti-CD47 antibody-mediated phagocytosis of cancer by macrophages primes an effective antitumor T-cell response. *Proceedings of the National Academy of Sciences of the United States of America*, *110*, 11103.

Tsuji, T., Sabbatini, P., Jungbluth, A. A., Ritter, E., Pan, L., Ritter, G., et al. (2013). Effect of Montanide and poly-ICLC adjuvant on human self/tumor antigen-specific CD4+ T cells in phase I overlapping long peptide vaccine trial. *Cancer Immunology Research*, *1*, 340.

Twyman-Saint Victor, C., Rech, A. J., Maity, A., Rengan, R., Pauken, K. E., Stelekati, E., et al. (2015). Radiation and dual checkpoint blockade activate non-redundant immune mechanisms in cancer. *Nature*, *520*, 373.

Ueno, H., Schmitt, N., Klechevsky, E., Pedroza-Gonzalez, A., Matsui, T., Zurawski, G., et al. (2010). Harnessing human dendritic cell subsets for medicine. *Immunological Reviews*, *234*, 199.

Ulloa-Montoya, F., Louahed, J., Dizier, B., Gruselle, O., Spiessens, B., Lehmann, F. F., et al. (2013). Predictive gene signature in MAGE-A3 antigen-specific cancer immunotherapy. *Journal of Clinical Oncology*, *31*, 2388.

Valmori, D., Souleimanian, N. E., Tosello, V., Bhardwaj, N., Adams, S., O'Neill, D., et al. (2007). Vaccination with NY-ESO-1 protein and CpG in Montanide induces integrated antibody/Th1 responses and CD8 T cells through cross-priming. *Proceedings of the National Academy of Sciences of the United States of America*, *104*, 8947.

van den Eertwegh, A. J., Versluis, J., van den Berg, H. P., Santegoets, S. J., van Moorselaar, R. J., van der Sluis, T. M., et al. (2012). Combined immunotherapy with granulocyte-macrophage colony-stimulating factor-transduced allogeneic prostate cancer cells and ipilimumab in patients with metastatic castration-resistant prostate cancer: A phase 1 dose-escalation trial. *The Lancet Oncology*, *13*, 509.

van der Bruggen, P., Traversari, C., Chomez, P., Lurquin, C., De Plaen, E., Van den Eynde, B., et al. (1991). A gene encoding an antigen recognized by cytolytic T lymphocytes on a human melanoma. *Science*, *254*, 1643.

van Elsas, A., Hurwitz, A. A., & Allison, J. P. (1999). Combination immunotherapy of B16 melanoma using anti-cytotoxic T lymphocyte-associated antigen 4 (CTLA-4) and granulocyte/macrophage colony-stimulating factor (GM-CSF)-producing vaccines induces rejection of subcutaneous and metastatic tumors accompanied by autoimmune depigmentation. *The Journal of Experimental Medicine*, *190*, 355.

van Rooij, N., van Buuren, M. M., Philips, D., Velds, A., Toebes, M., Heemskerk, B., et al. (2013). Tumor exome analysis reveals neoantigen-specific T-cell reactivity in an ipilimumab-responsive melanoma. *Journal of Clinical Oncology, 31*, e439.

Vanneman, M., & Dranoff, G. (2012). Combining immunotherapy and targeted therapies in cancer treatment. *Nature Reviews. Cancer, 12*, 237.

Vanpouille-Box, C., Diamond, J. M., Pilones, K. A., Zavadil, J., Babb, J. S., Formenti, S. C., et al. (2015). TGFbeta is a master regulator of radiation therapy-induced antitumor immunity. *Cancer Research, 75*, 2232.

Vanpouille-Box, C., Pilones, K. A., Wennerberg, E., Formenti, S. C., & Demaria, S. (2015). In situ vaccination by radiotherapy to improve responses to anti-CTLA-4 treatment. *Vaccine, 33*, 7415.

Vansteenkiste, J., Zielinski, M., Linder, A., Dahabreh, J., Gonzalez, E. E., Malinowski, W., et al. (2013). Adjuvant MAGE-A3 immunotherapy in resected non-small-cell lung cancer: Phase II randomized study results. *Journal of Clinical Oncology, 31*, 2396.

Vansteenkiste, J. F., Cho, B., Vanakesa, T., et al. (2014). MAGRIT, a double-blind, randomized, placebo-controlled Phase III study to assess the efficacy of the recMAGE-A3 + AS15 cancer immunotherapeutic as adjunct therapy in patients with resected MAGE-A3-positive non-small cell lung cancer (NSCLC). *Annals of Oncology, 25*(Suppl. 4), iv409–iv416. http://dx.doi.org/10.1093/annonc/mdu347.

Velluto, D., Thomas, S. N., Simeoni, E., Swartz, M. A., & Hubbell, J. A. (2011). PEG-b-PPS-b-PEI micelles and PEG-b-PPS/PEG-b-PPS-b-PEI mixed micelles as non-viral vectors for plasmid DNA: Tumor immunotoxicity in B16F10 melanoma. *Biomaterials, 32*, 9839.

Verbeke, C. S., & Mooney, D. J. (2015). Injectable, pore-forming hydrogels for in vivo enrichment of immature dendritic cells. *Advanced Healthcare Materials, 4*, 2677–2687.

Viatte, S., Alves, P. M., & Romero, P. (2006). Reverse immunology approach for the identification of CD8 T-cell-defined antigens: Advantages and hurdles. *Immunology and Cell Biology, 84*, 318.

Vose, B. M., & Bonnard, G. D. (1982a). Human tumour antigens defined by cytotoxicity and proliferative responses of cultured lymphoid cells. *Nature, 296*, 359.

Vose, B. M., & Bonnard, G. D. (1982b). Specific cytotoxicity against autologous tumour and proliferative responses of human lymphocytes grown in interleukin 2. *International Journal of Cancer, 29*, 33.

Wada, S., Yoshimura, K., Hipkiss, E. L., Harris, T. J., Yen, H. R., Goldberg, M. V., et al. (2009). Cyclophosphamide augments antitumor immunity: Studies in an autochthonous prostate cancer model. *Cancer Research, 69*, 4309.

Wallecha, A., Maciag, P. C., Rivera, S., Paterson, Y., & Shahabi, V. (2009). Construction and characterization of an attenuated Listeria monocytogenes strain for clinical use in cancer immunotherapy. *Clinical and Vaccine Immunology, 16*, 96.

Walter, S., Weinschenk, T., Stenzl, A., Zdrojowy, R., Pluzanska, A., Szczylik, C., et al. (2012). Multipeptide immune response to cancer vaccine IMA901 after single-dose cyclophosphamide associates with longer patient survival. *Nature Medicine, 18*, 1254.

Weed, D. T., Vella, J. L., Reis, I. M., De la Fuente, A. C., Gomez, C., Sargi, Z., et al. (2015). Tadalafil reduces myeloid-derived suppressor cells and regulatory T cells and promotes tumor immunity in patients with head and neck squamous cell carcinoma. *Clinical Cancer Research, 21*, 39.

Weide, B., Carralot, J. P., Reese, A., Scheel, B., Eigentler, T. K., Hoerr, I., et al. (2008). Results of the first phase I/II clinical vaccination trial with direct injection of mRNA. *Journal of Immunotherapy, 31*, 180.

Weide, B., Pascolo, S., Scheel, B., Derhovanessian, E., Pflugfelder, A., Eigentler, T. K., et al. (2009). Direct injection of protamine-protected mRNA: Results of a phase 1/2 vaccination trial in metastatic melanoma patients. *Journal of Immunotherapy, 32*, 498.

Weinschenk, T., Gouttefangeas, C., Schirle, M., Obermayr, F., Walter, S., Schoor, O., et al. (2002). Integrated functional genomics approach for the design of patient-individual antitumor vaccines. *Cancer Research, 62*, 5818.

Whitmire, J. K., Tan, J. T., & Whitton, J. L. (2005). Interferon-gamma acts directly on CD8 + T cells to increase their abundance during virus infection. *The Journal of Experimental Medicine, 201*, 1053.

Wick, D. A., Webb, J. R., Nielsen, J. S., Martin, S. D., Kroeger, D. R., Milne, K., et al. (2014). Surveillance of the tumor mutanome by T cells during progression from primary to recurrent ovarian cancer. *Clinical Cancer Research, 20*, 1125.

Wolchok, J. D., Kluger, H., Callahan, M. K., Postow, M. A., Rizvi, N. A., Lesokhin, A. M., et al. (2013). Nivolumab plus ipilimumab in advanced melanoma. *The New England Journal of Medicine, 369*, 122.

Wood, C. G., & Mulders, P. (2009). Vitespen: A preclinical and clinical review. *Future Oncology, 5*, 763.

Workenhe, S. T., Pol, J. G., Lichty, B. D., Cummings, D. T., & Mossman, K. L. (2013). Combining oncolytic HSV-1 with immunogenic cell death-inducing drug mitoxantrone breaks cancer immune tolerance and improves therapeutic efficacy. *Cancer Immunology Research, 1*, 309.

Yadav, M., Jhunjhunwala, S., Phung, Q. T., Lupardus, P., Tanguay, J., Bumbaca, S., et al. (2014). Predicting immunogenic tumour mutations by combining mass spectrometry and exome sequencing. *Nature, 515*, 572.

Yan, H., Parsons, D. W., Jin, G., McLendon, R., Rasheed, B. A., Yuan, W., et al. (2009). IDH1 and IDH2 mutations in gliomas. *The New England Journal of Medicine, 360*, 765.

Yang, S., Vervaert, C. E., Burch, J., Jr., Grichnik, J., Seigler, H. F., & Darrow, T. L. (1999). Murine dendritic cells transfected with human GP100 elicit both antigen-specific CD8 (+) and CD4(+) T-cell responses and are more effective than DNA vaccines at generating anti-tumor immunity. *International Journal of Cancer, 83*, 532.

Yu, C. I., Becker, C., Wang, Y., Marches, F., Helft, J., Leboeuf, M., et al. (2013). Human CD1c+ dendritic cells drive the differentiation of CD103+ CD8+ mucosal effector T cells via the cytokine TGF-beta. *Immunity, 38*, 818.

Zamarin, D., Holmgaard, R. B., Subudhi, S. K., Park, J. S., Mansour, M., Palese, P., et al. (2014). Localized oncolytic virotherapy overcomes systemic tumor resistance to immune checkpoint blockade immunotherapy. *Science Translational Medicine, 6*, 226ra32.

Zarei, S., Schwenter, F., Luy, P., Aurrand-Lions, M., Morel, P., Kopf, M., et al. (2009). Role of GM-CSF signaling in cell-based tumor immunization. *Blood, 113*, 6658.

Zeng, R., Spolski, R., Finkelstein, S. E., Oh, S., Kovanen, P. E., Hinrichs, C. S., et al. (2005). Synergy of IL-21 and IL-15 in regulating CD8+ T cell expansion and function. *The Journal of Experimental Medicine, 201*, 139.

Zhu, Q., Talton, J., Zhang, G., Cunningham, T., Wang, Z., Waters, R. C., et al. (2012). Large intestine-targeted, nanoparticle-releasing oral vaccine to control genitorectal viral infection. *Nature Medicine, 18*, 1291.

Zom, G. G., Khan, S., Britten, C. M., Sommandas, V., Camps, M. G., Loof, N. M., et al. (2014). Efficient induction of antitumor immunity by synthetic toll-like receptor ligand-peptide conjugates. *Cancer Immunology Research, 2*, 756.

CHAPTER SIX

Combinatorial Cancer Immunotherapies

Matthew D. Hellmann, Claire F. Friedman, Jedd D. Wolchok[1]
Memorial Sloan Kettering Cancer Center, New York, NY, United States
Weill Cornell Medical College, New York, NY, United States
[1]Corresponding author: e-mail address: wolchokj@mskcc.org

Contents

Abstract

T cell checkpoint blockade therapies are revolutionizing the treatment of patients with cancer. Highlighted by the recent success of PD-1 plus CTLA-4 blockade in patients with melanomas, synergistic immunotherapy combinations of modalities represent an important opportunity to improve responses and outcomes for patients. We review the rationale and experience with T cell checkpoint blockade in combination with targeting of other coinhibitory or costimulatory checkpoints, immunomodulatory molecules in the tumor microenvironment, and other anticancer modalities such as vaccines, chemotherapy, and radiation.

Advances in Immunology, Volume 130
ISSN 0065-2776
http://dx.doi.org/10.1016/bs.ai.2015.12.005

1. INTRODUCTION

Building upon initial hypotheses by Coley (1893), Ehrlich (1909; reviewed in Strebhardt & Ullrich, 2008), Burnet (1957a, 1957b), and Thomas (1959), among others, that the host immune system plays a critical role in shaping malignant tumors (Schreiber, Old, & Smyth, 2011); the recent successes of several immunotherapy strategies have confirmed the potential to harness the immune system for the treatment of patients with cancer (Andtbacka, Kaufman, Collichio, et al., 2015; Escudier, Pluzanska, Koralewski, et al., 2007; Le, Wang-Gillam, Picozzi, et al., 2015; Morales, Eidinger, & Bruce, 1976; Rosenberg, Yang, Sherry, et al., 2011, Rosenberg, Yang, Topalian, et al., 1994). In particular, T cell checkpoint inhibitors targeting either CTLA-4 (Hodi, O'Day, McDermott, et al., 2010; Robert, Thomas, Bondarenko, et al., 2011; Wolchok, Neyns, Linette, et al., 2010) or PD-1/PD-L1 (Ansell, Lesokhin, Borrello, et al., 2015; Brahmer, Drake, Wollner, et al., 2010, Brahmer, Reckamp, Baas, et al., 2015; Garon, Rizvi, Hui, et al., 2015; Hamid, Robert, Daud, et al., 2013; Herbst, Soria, Kowanetz, et al., 2014; Motzer, Rini, McDermott, et al., 2015; Powles, Eder, Fine, et al., 2014; Rizvi, Mazieres, Planchard, et al., 2015; Robert, Ribas, Wolchok, et al., 2014, Robert, Schachter, Long, et al., 2015; Topalian, Hodi, Brahmer, et al., 2012; Weber, D'Angelo, Minor, et al., 2015) have demonstrated tremendous efficacy in a diverse array of malignancies. As single agents, these therapies have response rates in the range of 10–35%, with responses characterized by a durability that is rarely, if ever, achieved with other types of anticancer therapy. Despite this remarkable progress, it remains only a minority of patients who respond to T cell checkpoint inhibitor monotherapy. One of the critical challenges at present is the development of alternative treatment strategies that improve the proportion of patients who may respond to immunotherapy. This chapter will focus on recent and ongoing efforts to use T cell checkpoint inhibitors in combination with other therapeutic approaches to address this challenge. The organizational framework for this chapter is structured around anti-CTLA-4 or anti-PD-1/PD-L1 therapies in combination with (a) other coinhibitory checkpoints, (b) costimulatory checkpoints, (c) other molecules in the tumor microenvironment, and (d) other anticancer treatment modalities. Initial results highlight the transformative potential of combination immunotherapy and collectively move the field closer to the ultimate goal of cure for patients with advanced cancers.

2. COMBINATIONS OF COINHIBITORY CHECKPOINTS

2.1 Targeting CTLA-4 + PD-1/PD-L1

Although both CTLA-4 and PD-1 function as inhibitory coreceptors expressed on T cells, their ligands and functions are distinct. When localized to the T cell surface after antigen-driven TCR-mediated T cell activation (Linsley et al., 1996), CTLA-4 binds to CD80 and CD86 (Linsley et al., 1991) and prevents effector T cell activation and proliferation (Krummel & Allison, 1995, 1996; Walunas, Lenschow, Bakker, et al., 1994) by competitively preventing binding of B7 ligands to the costimulatory receptor CD-28 (Freeman, Borriello, Hodes, et al., 1993; Greene, Leytze, Emswiler, et al., 1996; Linsley et al., 1994; Walunas, Bakker, & Bluestone, 1996) and inhibition of intracellular signaling pathways (Brunner et al., 1999; Carreno, Bennett, Chau, et al., 2000; Schneider, Downey, Smith, et al., 2006). PD-1 is similarly shuttled to the surface of effector T cell upon activation (Agata, Kawasaki, Nishimura, et al., 1996), where PD-1 can bind to ligands PD-L1 (Dong, Zhu, Tamada, & Chen, 1999; Freeman, Long, Iwai, et al., 2000) and PD-L2 (Latchman, Wood, Chernova, et al., 2001) and inhibit proliferation (Parry, Chemnitz, Frauwirth, et al., 2005), cytokine production (Bennett, Luxenberg, Ling, et al., 2003; Saunders, Hendrycks, Lidinsky, & Woods, 2005), and survival (Chemnitz, Parry, Nichols, June, & Riley, 2004; Ishida, Agata, Shibahara, & Honjo, 1992), which is characteristic of the state of T cell exhaustion (Barber, Wherry, Masopust, et al., 2006; Day, Kaufmann, Kiepiela, et al., 2006).

In this context, CTLA-4 and PD-1 can produce nonredundant effects on effector T cells, including inhibition of early activation and differentiation by CTLA-4 and modulation of effector function by PD-1 (Pardoll, 2012; Ribas, 2012). Preclinical observations that mice deficient for CTLA-4 (Tivol et al., 1995; Waterhouse, Penninger, Timms, et al., 1995) or PD-1 (Nishimura, Nose, Hiai, Minato, & Honjo, 1999, Nishimura, Okazaki, Tanaka, et al., 2001; Okazaki, Tanaka, Nishio, et al., 2003) had distinct patterns of toxicity further highlighted their differing properties and prompted efforts to examine the effects of targeting these pathways in combination. In a murine model of melanoma, the combination of CTLA-4 and PD-1 blockade significantly increased tumor rejection compared to either molecule alone (Curran, Montalvo, Yagita, & Allison, 2010). In humans, the observation that patients with melanoma treated with

PD-1 blockade had equivalent responses irrespective of prior exposure to CTLA-4 blockade (Hamid et al., 2013) has also demonstrated the non-redundant nature of these therapies.

The first study of combination of T cell checkpoint blockade was performed in patients with advanced melanomas treated with the PD-1 inhibitor nivolumab and the CTLA-4 inhibitor ipilimumab (Wolchok, Kluger, Callahan, et al., 2013). This phase I dose-escalation study treated a total 86 of patients. Thirty-three had previously received ipilimumab within 12 weeks and were then treated sequentially with nivolumab monotherapy (sequential regimen); 53 patients were ipilimumab naïve and received ipilimumab and nivolumab concurrently (concurrent regimen). The resulting clinical activity was unprecedented, particularly in the concurrent arm. In patients treated with the concurrent regimen, 40% had objective partial response and 65% derived clinical benefit. In patients treated with the sequential regimen, the objective response rate was 20%, and 43% derived clinical benefit. Remarkably, the majority of responses seen in the concurrent arm were rapid (most occurring by week 12 of therapy), deep (1/3 achieving ≥80% reduction in tumor burden), and durable (78% of patients alive at 2 years (Sznol, Kluger, Callahan, et al., 2014)). Notably, there were some important toxicities. In the concurrent regimen, treatment-related grade 3–4 elevated liver enzymes occurred in 15%, gastrointestinal toxicities occurred in 9%, rash in 4%, and pneumonitis and endocrinopathy occurred in 2% each. Still, most toxicity was manageable and nivolumab 1 mg/kg plus ipilimumab 3 mg/kg every 3 weeks for four doses followed by nivolumab 3 mg/kg every 2 weeks was determined to be the optimal dosing regimen for further development.

A subsequent double-blind, phase II, randomized study of nivolumab plus ipilimumab compared to ipilimumab in patients with advanced melanoma has confirmed the remarkable activity of this combination (Postow, Chesney, Pavlick, et al., 2015). The objective response rate to nivolumab plus ipilimumab was 59%, compared to 11% with ipilimumab alone. Among those treated with combination therapy, the median change in tumor volume was a 68% decrease and 22% of patients achieved a complete response.

Most recently, a double-blind, phase III, randomized study of nivolumab plus ipilimumab vs nivolumab vs ipilimumab was performed in patients with treatment-naïve advanced melanoma (Larkin, Chiarion-Sileni, Gonzalez, et al., 2015). Notably, and indicative of the interest from both patients and investigators, 945 patients were accrued and randomized in 9 months time. Again, the response rate with combination therapy was 57.6%, compared to 43.7% in the nivolumab monotherapy arm and 19% in the ipilimumab monotherapy group. Progression-free survival (PFS) also favored combination therapy, with hazard ratio of 0.57 (95% CI 0.43–0.76) compared to ipilimumab and 0.74 (95% CI 0.60–0.92) compared to nivolumab.

The impact of PD-L1 expression on tumor cells, which has been postulated to be a predictor of response to anti-PD-1 (PD-L1) therapy (Garon et al., 2015; Herbst et al., 2014; Taube, Klein, Brahmer, et al., 2014; Topalian et al., 2012; Tumeh, Harview, Yearley, et al., 2014), was also assessed. Responses to both combination therapy and nivolumab monotherapy were enriched in PD-L1 positive patients (72.1% and 57.5%, respectively, compared to 54.8% and 41.3% in PD-L1 negative patients). Among PD-L1 positive patients, PFS was relatively similar in patients treated with either combination therapy or nivolumab monotherapy, but follow-up remains short and many patients remain on therapy. Further follow-up and ultimately overall survival analysis will be needed to determine whether PD-L1 may be a meaningful tool for determining optimal therapy (combination vs PD-1 blockade monotherapy) for patients. Regardless, with responses exceeding 50% of the population and median PFS reaching nearly a year, it is clear that combination therapy has reshaped the treatment landscape of advanced melanoma and profoundly improved the outcomes for these patients.

Building upon the remarkable activity seen in patients with melanoma, several studies have begun to explore this combination in other diseases as well. In a trial of renal cell carcinoma (Hammers, Plimack, Infante, et al., 2014), 44 patients were randomized to receive induction therapy with four doses every 3 weeks of either nivolumab 1 mg/kg plus ipilimumab 3 mg/kg (the melanoma regimen, Nivo1/Ipi3) or nivolumab 3 mg/kg plus ipilimumab 1 mg/kg (Nivo3/Ipi1), followed by nivolumab 3 mg/kg every 3 weeks thereafter in both arms. Response rates were again substantial, with ~45% of patients achieving a confirmed objective response and ~65% of patients being progression free at 24 weeks after beginning therapy. Although toxicities were modest in both arms, toxicity was higher in the Nivo1/Ipi3 regimen with treatment-related grade 3–4 liver toxicity occurring in 26.1% of patients and GI toxicity occurring in 17.4%, compared to 0% and 4.8%, respectively, in the Nivo3/Ipi1 regimen. Treatment-related events leading to treatment discontinuation occurred in 26.1% of patients treated with Nivo1/Ipi3, compared to 9.5% in the Nivo3/Ipi1 arm. As a result, and differing from melanoma, the Nivo3/Ipi1 schedule has been chosen for further development in patients with renal cell carcinoma. A Phase 3 study of nivolumab plus ipilimumab vs sunitinib in patients with untreated, metastatic renal cell carcinoma is currently ongoing (NCT02231749).

In non-small cell lung cancer (NSCLC), development of PD-1/PD-L1 in combination with CTLA-4 blockade has required thorough evaluation of the optimal treatment regimen. In an initial study of nivolumab and ipilimumab, patients with NSCLC were initially treated with the same dose and schedule that was evaluated in melanoma and renal cell carcinoma (Nivo1/Ipi3 or Nivo3/Ipi1 every 3 weeks for four doses, followed by Nivo3 every 2 weeks) (Antonia, Gettinger, Chow, et al., 2014). A total of 49 patients were treated and frequent toxicity was seen in both groups, with 49% patients experiencing treatment-related grade 3–4 toxicity and 35% of patients discontinuing therapy due to treatment-related events. Some responses were seen (16% overall), but efficacy was difficult to interpret in the setting of the rate of toxicity. Subsequently, the study was amended to examine several additional treatment regimens, including giving lower doses of both nivolumab and ipilimumab (1 mg/kg each) or increasing the interval between ipilimumab administration (every 6 or 12 weeks). Recently reported results (Rizvi, Gettinger, Goldman, et al., 2015) from these arms have demonstrated significantly improved tolerability with 28–35% treatment-related grade 3–4 toxicities and 5–13% of patients discontinuing treatment related to toxicity. Efficacy was also promising and responses were greatest in arms with nivolumab given at 3 mg/kg (Nivo 1 mg/kg every 2 weeks + Ipi 1 mg/kg every 3 or 6 weeks: ORR, 13–25%; Nivo 3 mg/kg every 2 weeks + Ipi 1 mg/kg every 6 or 12 weeks: ORR, 31–39%). Overall response was greatest in those who were PD-L1 positive (defined as >1% on tumor cells). A phase 3 study of nivolumab plus ipilimumab vs nivolumab monotherapy vs chemotherapy is currently ongoing (NCT02477826) and will stratify patients based on PD-L1 expression.

Results were also recently reported from the dose-escalation study of the anti-PD-L1 inhibitor durvalumab in combination with the CTLA-4 inhibitor tremelimumab in patients with NSCLC (Antonia, Goldberg, Balmanoukian, et al., 2015). One hundred two patients were treated across 10 dose-escalation cohorts. Durvalumab was given at 10 mg/kg every 2 weeks or 3, 10, 15, 20 mg/kg every 4 weeks, and tremelimumab was given at 1, 3, or 10 mg/kg every 4 weeks. Similarly, frequent toxicity was seen in some cohorts, particularly with higher doses of tremelimumab. In cohorts treated with tremelimumab 3 or 10 mg/kg, the rate of treatment-related grade 3–4 toxicity was 50–80% and 34–45% led to treatment discontinuation. Toxicity was more modest in cohorts of tremelimumab 1 mg/kg, with grade 3–4 toxicity occurring in 29% of patients, and only 7% resulting in discontinuation of therapy. In addition to manageable toxicity with lower doses of tremelimumab, initial efficacy was also promising with objective responses seen in 33% of patients. Fifty-two percent of patients were progression-free 16 weeks after beginning therapy. Perhaps most notably, in contrast to the ipilimumab + nivolumab results above as well as experience with anti-PD-1/PD-L1 monotherapy in NSCLC, there was no difference in response rate by PD-L1 expression in this study.

These results demonstrated that in combination with durvalumab, doses of tremelimumab 3 mg/kg or higher are likely intolerable but 1 mg/kg appears to be reasonable and manageable. The regimen of durvalumab 20 mg/kg plus tremelimumab 1 mg/kg given together every 4 weeks has been chosen for further development. Early efficacy is promising with response rates of ~30% that numerically exceeds (nearly double) what has been reported with anti-PD-1/PD-L1 monotherapy in NSCLC. Additionally, the equivalent response rates in PD-L1 positive and negative patients demonstrate that the combination appears permit a broader population of patients to response to immunotherapy compared to monotherapy. Phase 3 studies of durvalumab plus tremelimumab vs durvalumab vs chemotherapy in untreated, advanced NSCLC has recently begun (NCT02453282).

It remains an interesting yet unsolved question as to why patients with lung cancers have experienced more toxicity than was seen in melanomas or renal cell carcinomas (Table 1). Nonetheless, this experience across these

Table 1 Summary of Toxicity, Efficacy, and Dosing Schedule of CTLA-4 Plus PD-1(PD-L1) in Patients with Melanoma, Renal Cell Carcinoma, and Non-Small Cell Lung Cancer

	Melanoma	**RCC**	**Non-Small Cell Lung Cancer**	
Treatment	Nivolumab + ipilimumab	Nivolumab + ipilimumab	Nivolumab + ipilimumab	Durvalumab + tremelimumab
Optimal dose/ schedule in development	Nivo 1 mg/kg + Ipi 3 mg/kg every 3 weeks for four doses, followed by Nivo 3 mg/kg every 2 weeks	Nivo 3 mg/kg + Ipi 1 mg/kg every 3 weeks for four doses, followed by Nivo 3 mg/kg every 2 weeks	Nivo 3 mg/kg + Ipi 1 mg/kg every 6 weeks	Durvalumab 20 mg/kg and tremelimumab 1 mg/kg every 4 weeks
ORR	57.6%	43%	31%	33%
Treatment-related toxicity requiring treatment discontinuation	36.4%	9.5%	10%	7%

diseases has highlighted the importance of vigilant evaluation of potential toxicity in patients treated with combination immunotherapy as well as the virtue of close industry/academia collaboration to carefully identify an optimal dose and schedule of therapy. Studies examining the efficacy and safety of combination PD-1/PD-L1 plus CTLA-4 blockade are now ongoing in several other cancers, including small cell lung cancers (Antonia, Bendell, Taylor, et al., 2015), as well as gastric cancer and bladder cancer (NCT01928394). In sum, the combination of CTLA-4 plus PD-1/PD-L1 blockade is associated with unequivocally improved responses in patients with melanoma and has sparked development and promising early results across a wide range of malignancies, highlighting the power of T cell checkpoint blockade to transcend tumor histology.

2.2 Targeting LAG-3 + PD-1/PD-L1

Lymphocyte activation gene-3 (LAG-3) is a coinhibitory transmembrane receptor (Triebel, Jitsukawa, Baixeras, et al., 1990) that binds MHC class II (Baixeras, Huard, Miossec, et al., 1992), enhances Treg activity (Huang, Workman, Flies, et al., 2004), and regulates T cell proliferation, differentiation, and effector function (Grosso, Goldberg, Getnet, et al., 2009; Sierro, Romero, & Speiser, 2011; Workman & Vignali, 2003). Blockade of LAG-3 alone *in vitro* augments T cell proliferation and cytokine production (Huard, Tournier, Hercend, Triebel, & Faure, 1994). In mouse models, LAG-3 monotherapy blockade results in increased accumulation and effector function of tumor antigen-specific CD8+ T cells (Grosso, Kelleher, Harris, et al., 2007). Yet, as a monotherapy, LAG-3 blockade appears to have relatively modest effect on antitumor efficacy in preclinical models (Woo, Turnis, Goldberg, et al., 2012). Consistent with this somewhat subtle immune modulation, LAG-3 knockout mice have no evident autoimmunity (Miyazaki, Dierich, Benoist, & Mathis, 1996). However, in combination with PD-1 blockade, murine data are suggestive of significant synergistic potential. Dual knockout of LAG-3 and PD-1 results in substantial lymphoproliferation and autoimmune organ toxicity (Okazaki, Okazaki, Wang, et al., 2011). Additionally, dual blockade of PD-1 and LAG-3 produced substantially increased accumulation of CD8+ T cells and improved clearance of established tumors compared to targeting of either receptor individually (Woo et al., 2012). Given this promising preclinical work, monoclonal antibodies directed against LAG-3 are being tested in early phase clinical trials (eg, BMS-986016 and LAG525), particularly in combination with PD1 blockade (NCT01968109, NCT02460224), although clinical results have not yet been reported.

2.3 Targeting TIM-3 + PD-1/PD-L1

T cell immunoglobulin and mucin domain 3 (TIM-3) is an inhibitory receptor expressed by IFN-γ secreting CD4+ helper T cells and CD8+ cytotoxic T cells (Monney, Sabatos, Gaglia, et al., 2002). High TIM-3 expression has been reported on "exhausted" T cells, characterized by decreased T cell proliferation, decreased IFN-γ, TNF-α, and IL-2 secretion, and increased IL-10 secretion (Fourcade, Sun, Benallaoua, et al., 2010; Jin, Anderson, Tan, et al., 2010; McMahan, Golden-Mason, Nishimura, et al., 2010; Sakuishi et al., 2010). In preclinical models, blockade of TIM-3 can increase cytokine-producing, tumor-specific T cells, and increased antitumor activity was also observed when combination with PD-L1 blockade (Ngiow et al., 2011; Sakuishi et al., 2010). Although preclinical development is ongoing, there are no trials currently enrolling patients for treatment with anti-TIM-3 antibodies.

3. COMBINATIONS WITH COSTIMULATORY CHECKPOINTS

3.1 Targeting OX40 + CTLA-4 or PD-1/PD-L1

Tumor necrosis factor receptor superfamily (TNFRSF), member 4 (OX40) is a member of the TNFR family of receptors. It is a secondary costimulatory immune checkpoint molecule expressed by activated T cells (al-Shamkhani et al., 1996; Bansal-Pakala, Halteman, Cheng, & Croft, 2004; Mallett, Fossum, & Barclay, 1990). Upon engaging its ligand, which is expressed primarily on antigen presenting cells, OX40 enhances T cell expansion (Baum, Gayle, Ramsdell, et al., 1994) and induces Th1 and Th2 cytokines (Arestides, He, Westlake, et al., 2002). OX40 also regulates the development and suppressive function of Tregs (Piconese, Pittoni, Burocchi, et al., 2010; Ruby, Yates, Hirschhorn-Cymerman, et al., 2009; Takeda, Ine, Killeen, et al., 2004; Valzasina et al., 2005). In murine models, administration of an OX40 mAb exhibited antitumor effects in both immunogenic and poorly immunogenic tumor models as a single agent (Weinberg, Rivera, Prell, et al., 2000). In murine models implanted with prostate or sarcoma tumors, anti-OX40 or anti-CTLA-4 monotherapy had limited effects on tumor growth or survival, but the combination of the two agents significantly prolonged survival (Redmond, Linch, & Kasiewicz, 2014). Intratumoral coinjection of anti-CTLA-4 and anti-OX-40 mAbs led to depletion of tumor-infiltrating Tregs and a systemic tumor response (Marabelle, Kohrt, Sagiv-Barfi, et al., 2013). OX40 agonism was also tested in combination with a PD-1 blocking antibody in a murine ovarian cancer model; while individual anti-PD-1 or Ox40 mAb treatment was ineffective, combination mAb treatment significantly inhibited tumor growth (Guo et al., 2014). Several trials are ongoing now to combine OX-40 agonists (MEDI6383, MOXR0916) with PD-1 or PD-L1 blockade, but clinical results have not yet been reported (NCT02221960, NCT02410512).

3.2 Targeting GITR + PD-1/PD-L1

Glucocorticoid-induced TNFR-related protein (GITR) is a surface protein that is upregulated upon T cell activation and is expressed at constitutively high levels of FOXP3+ Tregs (Schaer, Cohen, & Wolchok, 2010). Murine models have shown that GITR stimulation (with an agonistic antibody or with cognate ligand) promotes effector T cell proliferation, cytokine production (Kanamaru, Youngnak, Hashiguchi, et al., 2004; Ronchetti et al., 2007), resistance to Treg suppression (Ko, Yamazaki, Nakamura, et al., 2005; Mitsui, Nishikawa, Muraoka, et al., 2010; Nishikawa, Kato, Hirayama, et al., 2008; Shimizu, Yamazaki, Takahashi, Ishida, & Sakaguchi, 2002), and inhibition of Treg suppressive function (Valzasina et al., 2005). In *in vivo* models, administration of a GITR agonist antibody is associated with reduction of intratumoral Treg accumulation and potentiation of antitumor CD8+ effector T cell function (Cohen, Schaer, Liu, et al., 2010; Ko et al., 2005; Mitsui et al., 2010), as well as antitumor activity (Cohen et al., 2010; Ko et al., 2005; Turk et al., 2004). When given in combination with PD-1 blockade, increased activity was also seen. For example, anti-GITR and anti-PD-1 administered to mice with ID8 ovarian cancer, 20% of mice were tumor-free after 90 days while either anti-PD-1 or anti-GITR antibody alone exhibited little antitumor effect (Lu et al., 2014). Anti-GITR antibodies are in clinical development (TRX518, MK4166) is being tested in solid tumors as a single agent (NCT01239134) and in combination with PD-1 blockade (NCT0213275).

3.3 Targeting 4-1BB (CD137) + CTLA-4 or PD-1/PD-L1

4-1BB is a surface protein primarily present on activated T cells, NK cells, DCs, and Tregs (Vinay & Kwon, 2011, 2012; Wilcox, Chapoval, Gorski, et al., 2002). Stimulation of 4-1BB by its ligand, 4-1BBL, or by agonist antibodies enhances the activation of several immune cell populations, and 4-1BB signaling can enhance cytotoxic T cell survival and function (Hernandez-Chacon, Li, Wu, et al., 2011; Lee et al., 2002; Starck, Scholz, Dorken, & Daniel, 2005). Preclinical data show antitumor effects of agonist antibodies targeting 4-1BB (Melero, Shuford, Newby, et al., 1997), as well as in combination with coinhibitory checkpoint antibodies. Curran and colleagues demonstrated in a B16 melanoma mice model, the combination of CTLA-4 and 4-1BB antibodies in the context of an Flt3-ligand vaccine prolonged mice survival compared to either antibody alone. Combination therapy also increased infiltration by CD8+ and CD4+ effector T cells when compared to single agent antibody therapy (Curran, Kim, Montalvo, Al-Shamkhani, & Allison, 2011). Urelumab, a human 4-1BB antibody, was tested in the Phase 1 setting in 83 patients with melanoma, RCC, ovarian, and prostate cancer. Three patients with melanoma had partial responses and four patients had stable disease. However, there was also significant liver toxicity at higher doses of therapy, including two deaths (Sznol, Hodi, Margolin, et al., 2008), necessitating dose-reduction which has abated the liver toxicity. Studies of urelumab and another 4-1BB agonist, PF-05082566, in combinations with PD-1 blockade are ongoing now (NCT02534506, NCT02554812).

4. COMBINATIONS WITH OTHER MOLECULES IN THE TUMOR MICROENVIRONMENT

4.1 Targeting IL-2+CTLA-4

High-dose IL-2 was one of the first immunotherapies utilized for the treatment of cancer. In a large retrospective review of 255 patients with metastatic RCC enrolled on seven phase II trials, the objective response rate was 14% with 12 (5%) complete responses and 24 (9%) partial responses (Fyfe et al., 1995). In a similar review of 270 patients with metastatic melanoma, the objective response rate was 16% with 17 (6%) complete responses and 26 (10%) partial responses (Atkins, Lotze, Dutcher, et al., 1999). Unfortunately, treatment with IL-2 was associated with severe treatment-related adverse events in both RCC and melanoma, leading to a treatment-related mortality rate of 2–4%.

As the clinical activity of CTLA-4 blocking antibodies appears to be IL-2 dependent (Shrikant, Khoruts, & Mescher, 1999), there has also been interest in whether IL-2 can augment the effects of CTLA-4 blockade. Indeed, recent data have demonstrated that combination therapy with IL-2 and

CTLA-4 blockade significantly increased tumor-infiltrating CD8 and NK cells as was associated with delayed tumor growth and prolonged survival in mice (Kohlhapp, Broucek, Hughes, et al., 2015). However, this synergy was not evidently seen in a previous phase I/II trial of ipilimumab plus IL-2 therapy. Thirty-six patients with metastatic melanoma were enrolled; all patients received high-dose IL-2 and ipilimumab (24 of whom received the maximum dose of 3 mg/kg). Eight patients (22%) experienced objective tumor regression, which was not significantly different from the response rates seen to ipilimumab alone (Maker, Phan, Attia, et al., 2005).

More recently, Bowen and colleagues have tried an alternative approach with a phase I trial of intratumoral injections of both ipilimumab and IL-2. Twelve patients were treated at three ipilimumab dose levels and 10 patients were evaluable at the time of presentation. Four patients had a systemic objective response (40%) and among those categorized as progressive disease, one was found to be a complete response upon subsequent resection of lesions (Bowen, Meek, Williams, et al., 2015). There is also a clinical trial (NCT02203604) currently enrolling patients which will reexamine the systemic combination of high-dose IL-2 and ipilimumab.

4.2 Targeting IDO+CTLA-4 or PD-1/PD-L1

Indoleamine 2,3-dioxigenase 1 (IDO1) is the first step an enzymatic cascade that catalyzes tryptophan into kynurenine (Munn, Zhou, Attwood, et al., 1998). The dual impact of IDO1 activity on decreasing tryptophan and increasing kynurenine appears to have several immunosuppressive effects (Vacchelli, Aranda, Eggermont, et al., 2014) including inhibition of T cell activation and proliferation (Munn, Sharma, Baban, et al., 2005), decreasing T cell receptor complex expression (Fallarino, Grohmann, You, et al., 2006), and promoting regulatory T cells (Curti, Pandolfi, Valzasina, et al., 2007; Montagnoli, Fallarino, Gaziano, et al., 2006; Sharma, Hou, Baban, et al., 2010). Preclinical data suggest that targeting IDO may result in clinical efficacy, particularly in combination with T cell checkpoint blockade (Holmgaard, Zamarin, Munn, Wolchok, & Allison, 2013). In IDO knockout mice bearing B16 melanomas, anti-CTLA-4 treatment resulted in a significantly increased survival compared to wild-type mice; this benefit was associated with increased accumulation of tumor-infiltrating CD4+ and CD8+ T cells. The antitumor effects of PD-1/PD-L1 blockade and anti-GITR were also enhanced in IDO knockout mice. In wild-type mice, combination therapy with an anti-CTLA-4 antibody and 1MT (a pharmacologic inhibitor of IDO) increased Teff/Treg ratio, delayed tumor growth, and significantly improved survival.

Initial results in humans with melanoma have also been encouraging. INCB024360 (IDO1 inhibitor) was tested in combination with ipilimumab in patients with metastatic melanoma (Gibney, Hamid, Gangadhar, et al., 2014). Notably, at higher doses (eg, 300 mg BID) several patients developed significant alterations in liver function tests. These were reversible with corticosteroids and treatment discontinuation, but prompted dose-reduction in clinical trials. Enrollment was restarted at 25 mg PO BID and only one of eight patients had grade 3 liver function elevation. Initial efficacy was promising with three of eight patients having immune-related partial response (irPR) and three with irSD (Gibney et al., 2014); updated results are still pending. There are now a number of clinical trials of IDO inhibitors (INCB024360, GDC-0919) in combination with PD-1/PD-L1 blockade that are open and enrolling patients (NCT02178722, NCT02318277, NCT02298153, NCT02471846), but no results have yet been reported.

5. OTHER ANTICANCER TREATMENT MODALITIES IN COMBINATION WITH T CELL CHECKPOINT BLOCKADE

5.1 Oncolytic Viruses

Oncolytic viruses offer a potential strategy to synergize with T cell checkpoint blockade by inducing immunogenic cell death and promoting an anti tumor inflammatory response. A recent preclinical study of Newcastle disease virus (NDV) demonstrated the systemic immune effects that can occur even when virus is injected locally (Zamarin, Holmgaard, Subudhi, et al., 2014). In a murine model with B16 melanomas established on the bilateral flanks, direct unilateral intratumoral injection of NDV produced significant increases in tumor-specific CD4 and CD8+ T cells in the contralateral (noninjected) tumor. Moreover, the combination of local NDV and systemic CTLA-4 blockade lead to significant increases in tumor regression systemically.

Talimogene laherparepvec (T-VEC) is an oncolytic virus currently in clinical development. T-VEC is a herpes simplex virus-1-based oncolytic immunotherapy. OPTiM, a phase III trial of T-VEC vs GM-CSF in unresectable stage IIIB-IV melanoma ($n=436$), met the primary endpoint of improved durable response rate in the T-VEC arm, although the comparison arm is no longer considered a standard therapy following the approval of ipilimumab (and PD-1 blockade) (Andtbacka et al., 2015).

T-VEC is now being tested in combination with both CTLA-4 and PD-1 blockade. In a phase 1b trial of T-VEC plus ipilimumab, 18 patients unresectable melanoma were enrolled. Treatment was well tolerated (grade 3–4 immune-related adverse events occurred in only two patients) and efficacy was promising (objective response was seen in 56%, and 33% had complete response) (Puzanov, Milhem, Andtbacka, et al., 2015). There is now a phase II trial of ipilimumab with or without T-VEC enrolling patients (NCT01740297) as well as a phase I/II study of T-VEC combined with PD-1 blockade (NCT02263508) (Ribas, Puzanov, Gajewski, et al., 2015).

5.2 Cytotoxic Chemotherapy

The host immune response may have an impact on response to cytotoxic chemotherapy, as the presence of tumor-infiltrating lymphocytes are associated with improved response to chemotherapy in a variety of tumor types (Denkert, von Minckwitz, Brase, et al., 2015; Galon, Costes, Sanchez-Cabo, et al., 2006; Liu, Zhang, Ye, et al., 2012; Mao et al., 2014; Shinto, Hase, Hashiguchi, et al., 2014). Additionally, some chemotherapies may even improve antitumor immunity, such as low-dose cyclophosphamide (Ghiringhelli, Menard, Puig, et al., 2007; Le & Jaffee, 2012) and gemcitabine (Shevchenko, Karakhanova, Soltek, et al., 2013) via selective depletion of immunosuppressive cells such as Tregs; paclitaxel and 5-fluorouriacil have also been found to inhibit MDSCs and restore antitumor activity of CD8+ T cells (Sevko, Michels, Vrohlings, et al., 2013; Umansky & Sevko, 2012; Vincent, Mignot, Chalmin, et al., 2010). These findings have prompted efforts to combine cytotoxic chemotherapy with T cell checkpoint blockade, although no clear breakthroughs have been identified to date.

In patients with melanoma, ipilimumab plus dacarbazine had a significantly improved overall survival compared to dacarbazine alone (11.2 vs 9.1 months); however, because there was no ipilimumab-alone arm, it is unclear whether there was any incremental benefit for the addition of chemotherapy (Robert et al., et al., 2011). Similarly, in patients with melanoma treated with ipilimumab alone or in combination with dacarbazine or carboplatin/paclitaxel, there was no evident improvement in response (29.4% with Ipi alone, 27.8% with Ipi/dacarbazine, 11.1% with Ipi/carboplatin/paclitaxel) (Weber, Hamid, Amin, et al., 2013).

The combination of ipilimumab and carboplatin/paclitaxel was explored further in patients with NSCLC and extensive-stage SCLC (Lynch,

Bondarenko, Luft, et al., 2012; Reck, Bondarenko, Luft, et al., 2013). Both trials demonstrated a moderate improvement in immune-related progression-free survival (irPFS) for patient receiving "phased" ipilimumab, which consisted of two doses of ipilimumab alone followed by four concurrent doses of ipilimumab plus carboplatin/paclitaxel. There was no irPFS advantage for patients who received ipilimumab plus chemotherapy first.

PD-1/PD-L1 blockade is being investigated in combination with a variety of chemotherapeutic agents in NSCLC. Initial results with nivolumab in combination with platinum doublet chemotherapy showed efficacy that was not substantially different from historical experience with chemotherapy alone (ORR was 33–50% and the 24-week PFS was 36–71%), while toxicity was a bit higher (45% of patients had grade 3 or grade 4 toxicity) (Antonia, Brahmer, Gettinger, et al., 2014). More recent reports have shown more promising efficacy and safety signals in platinum-based chemotherapy combinations with pembrolizumab and atezolizumab, with objective responses in 60–75%, although the sample size of studies to date remains small (Liu, Powderly, Camidge, et al., 2015; Papadimitrakopoulou, Patnaik, Borghaei, et al., 2015). Several Phase 3 studies of chemotherapy in combination with PD-1 or PD-L1 blockade in patients with NSCLC are planned.

5.3 Radiation

The abscopal effect refers to a phenomenon of tumor regression at a site distant from the primary site of radiotherapy (RT) (Mole, 1953). There is evidence that RT in itself has an immunomodulatory effect (Garnett et al., 2004; Lugade et al., 2005; Parker, Jones, Strober, & Knox, 2013; Reits, Hodge, Herberts, et al., 2006; Zeng, Harris, Lim, Drake, & Tran, 2013) and depleting intratumoral CD8+ T cells reduces the antitumor effect of RT in melanoma mice models (Gupta, Probst, Vuong, et al., 2012; Lee, Auh, Wang, et al., 2009). RT may affect antitumor immunity by enhancing antigen presentation by upregulation of major histocompatibility complex class I (MHC-1) expression of malignant cells and upregulation of tumor-associated antigens (Sharma, Bode, Wenger, et al., 2011). Sharabi and colleagues found RT of B16 melanoma significantly increased both CD4+ and CD8+ TILs, and combining RT with PD-1 blockade increased the ratio of Teff/Treg cells (Sharabi, Nirschl, Kochel, et al., 2015). In another mouse model, RT in combination with PD-1 blockade enhanced RT efficacy and reduced the local accumulation of MDSCs (Deng, Liang, Burnette, et al., 2014).

Case reports highlighting the potential synergy between RT and immune checkpoint blockade (Golden, Demaria, Schiff, Chachoua, & Formenti, 2013; Postow, Callahan, Barker, et al., 2012) have prompted larger studies to examine this combination. In an analysis of 21 patients who received radiation after disease progression on ipilimumab (Grimaldi, Simeone, Giannarelli, et al., 2014), 13 patients (62%) had evidence of an abscopal response, including 9 of whom had partial responses (43%) and 2 had stable disease (10%). Median OS for patients with abscopal responses was 22.4 months compared to 8.3 months for patients without. In another phase I trial, 22 patients with metastatic melanoma received radiation to a single index lesion followed by four cycles of ipilimumab (Twyman-Saint Victor, Rech, Maity, et al., 2015). Eighteen percent of patients had a partial response, 18% had stable disease, and 64% had progressive disease. In a phase I trial of MPDL3280A, an anti-PD-L1 monoclonal antibody, combined with RT, all five patients who were treated demonstrated clinical activity (Sagiv-Barfi, Rajapaksa, Czerwinski, et al., 2014).

Of additional note, a recent report examined cell lines derived from the unirradiated tumors of mice that relapsed after radiation to a different lesion and four cycles of ipilimumab (Twyman-Saint Victor et al., 2015). They found that these cell lines demonstrated PD-L1 upregulation and that the addition of PD-L1 blockade improved response to radiation and CTLA-4 blockade. This suggests that "triple therapy" of PD-1/CTLA-4 blockade with radiation therapy may demonstrate significant efficacy in future clinical trials. Several trials are ongoing at this time examining radiation in combination with T cell checkpoint blockade.

6. CONCLUSIONS

The recent successes of T cell checkpoint inhibitors have firmly established the concept that antitumor immunity can be effectively and safely leveraged as a weapon to treat patients with cancer. The durability of responses to T cell checkpoint inhibitors largely exceeds what has previously been clinically possible with other forms of anticancer therapy, and the capacity of these therapies to work in an ever-growing list of malignancies provides new hope for an unprecedentedly broad group of patients. Yet, only a minority of patients will benefit from single-agent T cell checkpoint blockade. Combination immunotherapies represent a critical next step in the progress toward improving responses. The experience with CTLA-4 + PD-1 blockade in patients with metastatic melanoma has highlighted the remarkable power of dual checkpoint blockade, while the toxicity data from this combination have also emphasized the importance of careful dose finding studies to identify safe and effective dosing regimens, which may be different in different diseases. As we continue to move forward, significant work remains to be done to determine both the optimal combination regimens as well as predictors of response in order to better personalize treatment. Building upon the progress we have detailed in this chapter, we believe the field is poised for many more breakthroughs and significant new improvements in the treatment of patients with cancers.

REFERENCES

Agata, Y., Kawasaki, A., Nishimura, H., et al. (1996). Expression of the PD-1 antigen on the surface of stimulated mouse T and B lymphocytes. *International Immunology*, *8*(5), 765–772.

al-Shamkhani, A., Birkeland, M. L., Puklavec, M., Brown, M. H., James, W., & Barclay, A. N. (1996). OX40 is differentially expressed on activated rat and mouse T cells and is the sole receptor for the OX40 ligand. *European Journal of Immunology*, *26*(8), 1695–1699.

Andtbacka, R. H., Kaufman, H. L., Collichio, F., et al. (2015). Talimogene laherparepvec improves durable response rate in patients with advanced melanoma. *Journal of Clinical Oncology*, *33*, 2780–2788.

Ansell, S. M., Lesokhin, A. M., Borrello, I., et al. (2015). PD-1 blockade with nivolumab in relapsed or refractory Hodgkin's lymphoma. *The New England Journal of Medicine*, *372*(4), 311–319.

Antonia, S. J., Bendell, J. C., Taylor, M. H., et al. (2015). Phase I/II study of nivolumab with or without ipilimumab for treatment of recurrent small cell lung cancer (SCLC): CA209-032. *American Society of Clinical Oncology Meeting Abstracts*, *33*(15 suppl), 7503.

Antonia, S. J., Brahmer, J. R., Gettinger, S. N., et al. (2014). Nivolumab (anti-PD-1; BMS-936558, ONO-4538) in combination with platinum-based doublet chemotherapy (PT-DC) in advanced non-small cell lung cancer (NSCLC). *American Society of Clinical Oncology Meeting Abstracts*, *32*(15 suppl), 8113.

Antonia, S. J., Gettinger, S. N., Chow, L. Q. M., et al. (2014). Nivolumab (anti-PD-1; BMS-936558, ONO-4538) and ipilimumab in first-line NSCLC: Interim phase I results. *American Society of Clinical Oncology Meeting Abstracts*, *32*(15 suppl), 8023.

Antonia, S. J., Goldberg, S. B., Balmanoukian, A. S., et al. (2015). Phase Ib study of MEDI4736, a programmed cell death ligand-1 (PD-L1) antibody, in combination with tremelimumab, a cytotoxic T-lymphocyte-associated protein-4 (CTLA-4) antibody, in patients (pts) with advanced NSCLC. *American Society of Clinical Oncology Meeting Abstracts*, *33*(15 suppl), 3014.

Arestides, R. S., He, H., Westlake, R. M., et al. (2002). Costimulatory molecule OX40L is critical for both Th1 and Th2 responses in allergic inflammation. *European Journal of Immunology*, *32*(10), 2874–2880.

Atkins, M. B., Lotze, M. T., Dutcher, J. P., et al. (1999). High-dose recombinant interleukin 2 therapy for patients with metastatic melanoma: Analysis of 270 patients treated between 1985 and 1993. *Journal of Clinical Oncology: Official Journal of the American Society of Clinical Oncology*, *17*(7), 2105–2116.

Baixeras, E., Huard, B., Miossec, C., et al. (1992). Characterization of the lymphocyte activation gene 3-encoded protein. A new ligand for human leukocyte antigen class II antigens. *The Journal of Experimental Medicine*, *176*(2), 327–337.

Bansal-Pakala, P., Halteman, B. S., Cheng, M. H., & Croft, M. (2004). Costimulation of CD8 T cell responses by OX40. *Journal of Immunology*, *172*(8), 4821–4825.

Barber, D. L., Wherry, E. J., Masopust, D., et al. (2006). Restoring function in exhausted CD8 T cells during chronic viral infection. *Nature*, *439*(7077), 682–687.

Baum, P. R., Gayle, R. B., 3rd, Ramsdell, F., et al. (1994). Molecular characterization of murine and human OX40/OX40 ligand systems: Identification of a human OX40 ligand as the HTLV-1-regulated protein gp34. *The EMBO Journal*, *13*(17), 3992–4001.

Bennett, F., Luxenberg, D., Ling, V., et al. (2003). Program death-1 engagement upon TCR activation has distinct effects on costimulation and cytokine-driven proliferation: Attenuation of ICOS, IL-4, and IL-21, but not CD28, IL-7, and IL-15 responses. *Journal of Immunology*, *170*(2), 711–718.

Bowen, R. C., Meek, S., Williams, M., et al. (2015). A phase I study of intratumoral injection of ipilimumab and interleukin-2 in patients with unresectable stage III-IV melanoma. *American Society of Clinical Oncology Meeting Abstracts*, *33*(15 suppl), 3018.

Brahmer, J. R., Drake, C. G., Wollner, I., et al. (2010). Phase I study of single-agent anti-programmed death-1 (MDX-1106) in refractory solid tumors: Safety, clinical activity, pharmacodynamics, and immunologic correlates. *Journal of Clinical Oncology*, *28*(19), 3167–3175.

Brahmer, J., Reckamp, K. L., Baas, P., et al. (2015). Nivolumab versus docetaxel in advanced squamous-cell non-small-cell lung cancer. *The New England Journal of Medicine*, *373*, 123–135.

Brunner, M. C., Chambers, C. A., Chan, F. K., Hanke, J., Winoto, A., & Allison, J. P. (1999). CTLA-4-mediated inhibition of early events of T cell proliferation. *Journal of Immunology*, *162*(10), 5813–5820.

Burnet, M. (1957a). Cancer: A biological approach. I. The processes of control. *British Medical Journal*, *1*(5022), 779–786.

Burnet, M. (1957b). Cancer: A biological approach. III. Viruses associated with neoplastic conditions. IV. Practical applications. *British Medical Journal*, *1*(5023), 841–847.

Carreno, B. M., Bennett, F., Chau, T. A., et al. (2000). CTLA-4 (CD152) can inhibit T cell activation by two different mechanisms depending on its level of cell surface expression. *Journal of Immunology*, *165*(3), 1352–1356.

Chemnitz, J. M., Parry, R. V., Nichols, K. E., June, C. H., & Riley, J. L. (2004). SHP-1 and SHP-2 associate with immunoreceptor tyrosine-based switch motif of programmed death 1 upon primary human T cell stimulation, but only receptor ligation prevents T cell activation. *Journal of Immunology*, *173*(2), 945–954.

Cohen, A. D., Schaer, D. A., Liu, C., et al. (2010). Agonist anti-GITR monoclonal antibody induces melanoma tumor immunity in mice by altering regulatory T cell stability and intra-tumor accumulation. *PLoS One*, *5*(5). e10436.

Coley, W. B. (1893). The treatment of malignant tumors by repeated inoculations of erysipelas. With a report of ten original cases. *Clinical Orthopaedics and Related Research*, *1991*(262), 3–11.

Curran, M. A., Kim, M., Montalvo, W., Al-Shamkhani, A., & Allison, J. P. (2011). Combination CTLA-4 blockade and 4-1BB activation enhances tumor rejection by increasing T-cell infiltration, proliferation, and cytokine production. *PLoS One*, *6*(4). e19499.

Curran, M. A., Montalvo, W., Yagita, H., & Allison, J. P. (2010). PD-1 and CTLA-4 combination blockade expands infiltrating T cells and reduces regulatory T and myeloid cells within B16 melanoma tumors. *Proceedings of the National Academy of Sciences of the United States of America*, *107*(9), 4275–4280.

Curti, A., Pandolfi, S., Valzasina, B., et al. (2007). Modulation of tryptophan catabolism by human leukemic cells results in the conversion of CD25- into CD25 + T regulatory cells. *Blood*, *109*(7), 2871–2877.

Day, C. L., Kaufmann, D. E., Kiepiela, P., et al. (2006). PD-1 expression on HIV-specific T cells is associated with T-cell exhaustion and disease progression. *Nature*, *443*(7109), 350–354.

Deng, L., Liang, H., Burnette, B., et al. (2014). Irradiation and anti-PD-L1 treatment synergistically promote antitumor immunity in mice. *The Journal of Clinical Investigation*, *124*(2), 687–695.

Denkert, C., von Minckwitz, G., Brase, J. C., et al. (2015). Tumor-infiltrating lymphocytes and response to neoadjuvant chemotherapy with or without carboplatin in human epidermal growth factor receptor 2-positive and triple-negative primary breast cancers. *Journal of Clinical Oncology: Official Journal of the American Society of Clinical Oncology*, *33*(9), 983–991.

Dong, H., Zhu, G., Tamada, K., & Chen, L. (1999). B7-H1, a third member of the B7 family, co-stimulates T-cell proliferation and interleukin-10 secretion. *Nature Medicine*, *5*(12), 1365–1369.

Escudier, B., Pluzanska, A., Koralewski, P., et al. (2007). Bevacizumab plus interferon alfa-2a for treatment of metastatic renal cell carcinoma: A randomised, double-blind phase III trial. *Lancet*, *370*(9605), 2103–2111.

Fallarino, F., Grohmann, U., You, S., et al. (2006). The combined effects of tryptophan starvation and tryptophan catabolites down-regulate T cell receptor zeta-chain and induce a regulatory phenotype in naive T cells. *Journal of Immunology*, *176*(11), 6752–6761.

Fourcade, J., Sun, Z., Benallaoua, M., et al. (2010). Upregulation of Tim-3 and PD-1 expression is associated with tumor antigen-specific CD8+ T cell dysfunction in melanoma patients. *The Journal of Experimental Medicine*, *207*(10), 2175–2186.

Freeman, G. J., Borriello, F., Hodes, R. J., et al. (1993). Uncovering of functional alternative CTLA-4 counter-receptor in B7-deficient mice. *Science*, *262*(5135), 907–909.

Freeman, G. J., Long, A. J., Iwai, Y., et al. (2000). Engagement of the PD-1 immunoinhibitory receptor by a novel B7 family member leads to negative regulation of lymphocyte activation. *The Journal of Experimental Medicine*, *192*(7), 1027–1034.

Fyfe, G., Fisher, R. I., Rosenberg, S. A., Sznol, M., Parkinson, D. R., & Louie, A. C. (1995). Results of treatment of 255 patients with metastatic renal cell carcinoma who received high-dose recombinant interleukin-2 therapy. *Journal of Clinical Oncology: Official Journal of the American Society of Clinical Oncology*, *13*(3), 688–696.

Galon, J., Costes, A., Sanchez-Cabo, F., et al. (2006). Type, density, and location of immune cells within human colorectal tumors predict clinical outcome. *Science*, *313*(5795), 1960–1964.

Garnett, C. T., Palena, C., Chakraborty, M., Tsang, K. Y., Schlom, J., & Hodge, J. W. (2004). Sublethal irradiation of human tumor cells modulates phenotype resulting in enhanced killing by cytotoxic T lymphocytes. *Cancer Research*, *64*(21), 7985–7994.

Garon, E. B., Rizvi, N. A., Hui, R., et al. (2015). Pembrolizumab for the treatment of non-small-cell lung cancer. *The New England Journal of Medicine*, *372*(21), 2018–2028.

Ghiringhelli, F., Menard, C., Puig, P. E., et al. (2007). Metronomic cyclophosphamide regimen selectively depletes CD4+CD25+ regulatory T cells and restores T and NK effector functions in end stage cancer patients. *Cancer Immunology, Immunotherapy: CII*, *56*(5), 641–648.

Gibney, G. T., Hamid, O., Gangadhar, T. C., et al. (2014). Preliminary results from a phase 1/2 study of INCB024360 combined with ipilimumab (ipi) in patients (pts) with melanoma. *American Society of Clinical Oncology Meeting Abstracts*, *32*(15 suppl), 3010.

Golden, E. B., Demaria, S., Schiff, P. B., Chachoua, A., & Formenti, S. C. (2013). An abscopal response to radiation and ipilimumab in a patient with metastatic non-small cell lung cancer. *Cancer Immunology Research*, *1*(6), 365–372.

Greene, J. L., Leytze, G. M., Emswiler, J., et al. (1996). Covalent dimerization of CD28/CTLA-4 and oligomerization of CD80/CD86 regulate T cell costimulatory interactions. *The Journal of Biological Chemistry*, *271*(43), 26762–26771.

Grimaldi, A. M., Simeone, E., Giannarelli, D., et al. (2014). Abscopal effects of radiotherapy on advanced melanoma patients who progressed after ipilimumab immunotherapy. *Oncoimmunology, 3*. e28780.

Grosso, J. F., Goldberg, M. V., Getnet, D., et al. (2009). Functionally distinct LAG-3 and PD-1 subsets on activated and chronically stimulated CD8 T cells. *Journal of Immunology, 182*(11), 6659–6669.

Grosso, J. F., Kelleher, C. C., Harris, T. J., et al. (2007). LAG-3 regulates CD8+ T cell accumulation and effector function in murine self- and tumor-tolerance systems. *The Journal of Clinical Investigation, 117*(11), 3383–3392.

Guo, Z., Wang, X., Cheng, D., Xia, Z., Luan, M., & Zhang, S. (2014). PD-1 blockade and OX40 triggering synergistically protects against tumor growth in a murine model of ovarian cancer. *PLoS One, 9*(2). e89350.

Gupta, A., Probst, H. C., Vuong, V., et al. (2012). Radiotherapy promotes tumor-specific effector CD8+ T cells via dendritic cell activation. *Journal of Immunology, 189*(2), 558–566.

Hamid, O., Robert, C., Daud, A., et al. (2013). Safety and tumor responses with lambrolizumab (anti-PD-1) in melanoma. *The New England Journal of Medicine, 369*(2), 134–144.

Hammers, H. J., Plimack, E. R., Infante, J. R., et al. (2014). Phase I study of nivolumab in combination with ipilimumab in metastatic renal cell carcinoma (mRCC). *American Society of Clinical Oncology Meeting Abstracts, 32*(15 suppl), 4504.

Herbst, R. S., Soria, J. C., Kowanetz, M., et al. (2014). Predictive correlates of response to the anti-PD-L1 antibody MPDL3280A in cancer patients. *Nature, 515*(7528), 563–567.

Hernandez-Chacon, J. A., Li, Y., Wu, R. C., et al. (2011). Costimulation through the CD137/4-1BB pathway protects human melanoma tumor-infiltrating lymphocytes from activation-induced cell death and enhances antitumor effector function. *Journal of Immunotherapy, 34*(3), 236–250.

Hodi, F. S., O'Day, S. J., McDermott, D. F., et al. (2010). Improved survival with ipilimumab in patients with metastatic melanoma. *The New England Journal of Medicine, 363*(8), 711–723.

Holmgaard, R. B., Zamarin, D., Munn, D. H., Wolchok, J. D., & Allison, J. P. (2013). Indoleamine 2,3-dioxygenase is a critical resistance mechanism in antitumor T cell immunotherapy targeting CTLA-4. *The Journal of Experimental Medicine, 210*(7), 1389–1402.

Huang, C. T., Workman, C. J., Flies, D., et al. (2004). Role of LAG-3 in regulatory T cells. *Immunity, 21*(4), 503–513.

Huard, B., Tournier, M., Hercend, T., Triebel, F., & Faure, F. (1994). Lymphocyte-activation gene 3/major histocompatibility complex class II interaction modulates the antigenic response of CD4+ T lymphocytes. *European Journal of Immunology, 24*(12), 3216–3221.

Ishida, Y., Agata, Y., Shibahara, K., & Honjo, T. (1992). Induced expression of PD-1, a novel member of the immunoglobulin gene superfamily, upon programmed cell death. *The EMBO Journal, 11*(11), 3887–3895.

Jin, H. T., Anderson, A. C., Tan, W. G., et al. (2010). Cooperation of Tim-3 and PD-1 in CD8 T-cell exhaustion during chronic viral infection. *Proceedings of the National Academy of Sciences of the United States of America, 107*(33), 14733–14738.

Kanamaru, F., Youngnak, P., Hashiguchi, M., et al. (2004). Costimulation via glucocorticoid-induced TNF receptor in both conventional and CD25+ regulatory CD4+ T cells. *Journal of Immunology, 172*(12), 7306–7314.

Ko, K., Yamazaki, S., Nakamura, K., et al. (2005). Treatment of advanced tumors with agonistic anti-GITR mAb and its effects on tumor-infiltrating Foxp3+CD25+CD4+ regulatory T cells. *The Journal of Experimental Medicine, 202*(7), 885–891.

Kohlhapp, F., Broucek, J., Hughes, T., et al. (2015). NK cells and CD8+ T cells cooperate to improve therapeutic responses in melanoma treated with interleukin-2 (IL-2) and CTLA-4 blockade. *Journal for Immunotherapy of Cancer, 3*(1), 18.

Krummel, M. F., & Allison, J. P. (1995). CD28 and CTLA-4 have opposing effects on the response of T cells to stimulation. *The Journal of Experimental Medicine, 182*(2), 459–465.

Krummel, M. F., & Allison, J. P. (1996). CTLA-4 engagement inhibits IL-2 accumulation and cell cycle progression upon activation of resting T cells. *The Journal of Experimental Medicine, 183*(6), 2533–2540.

Larkin, J., Chiarion-Sileni, V., Gonzalez, R., et al. (2015). Combined nivolumab and ipilimumab or monotherapy in untreated melanoma. *The New England Journal of Medicine, 373*(1), 23–34.

Latchman, Y., Wood, C. R., Chernova, T., et al. (2001). PD-L2 is a second ligand for PD-1 and inhibits T cell activation. *Nature Immunology, 2*(3), 261–268.

Le, D. T., & Jaffee, E. M. (2012). Regulatory T-cell modulation using cyclophosphamide in vaccine approaches: A current perspective. *Cancer Research, 72*(14), 3439–3444.

Le, D. T., Wang-Gillam, A., Picozzi, V., et al. (2015). Safety and survival with GVAX pancreas prime and Listeria monocytogenes-expressing mesothelin (CRS-207) boost vaccines for metastatic pancreatic cancer. *Journal of Clinical Oncology, 33*(12), 1325–1333.

Lee, Y., Auh, S. L., Wang, Y., et al. (2009). Therapeutic effects of ablative radiation on local tumor require CD8+ T cells: Changing strategies for cancer treatment. *Blood, 114*(3), 589–595.

Lee, H. W., Park, S. J., Choi, B. K., Kim, H. H., Nam, K. O., & Kwon, B. S. (2002). 4-1BB promotes the survival of CD8+ T lymphocytes by increasing expression of Bcl-xL and Bfl-1. *Journal of Immunology, 169*(9), 4882–4888.

Linsley, P. S., Bradshaw, J., Greene, J., Peach, R., Bennett, K. L., & Mittler, R. S. (1996). Intracellular trafficking of CTLA-4 and focal localization towards sites of TCR engagement. *Immunity, 4*(6), 535–543.

Linsley, P. S., Brady, W., Urnes, M., Grosmaire, L. S., Damle, N. K., & Ledbetter, J. A. (1991). CTLA-4 is a second receptor for the B cell activation antigen B7. *The Journal of Experimental Medicine, 174*(3), 561–569.

Linsley, P. S., Greene, J. L., Brady, W., Bajorath, J., Ledbetter, J. A., & Peach, R. (1994). Human B7-1 (CD80) and B7-2 (CD86) bind with similar avidities but distinct kinetics to CD28 and CTLA-4 receptors. *Immunity, 1*(9), 793–801.

Liu, S. V., Powderly, J. D., Camidge, D. R., et al. (2015). Safety and efficacy of MPDL3280A (anti-PDL1) in combination with platinum-based doublet chemotherapy in patients with advanced non-small cell lung cancer (NSCLC). *American Society of Clinical Oncology Meeting Abstracts, 33*(15 suppl), 8030.

Liu, H., Zhang, T., Ye, J., et al. (2012). Tumor-infiltrating lymphocytes predict response to chemotherapy in patients with advance non-small cell lung cancer. *Cancer Immunology, Immunotherapy: CII, 61*(10), 1849–1856.

Lu, L., Xu, X., Zhang, B., Zhang, R., Ji, H., & Wang, X. (2014). Combined PD-1 blockade and GITR triggering induce a potent antitumor immunity in murine cancer models and synergizes with chemotherapeutic drugs. *Journal of Translational Medicine, 12*, 36.

Lugade, A. A., Moran, J. P., Gerber, S. A., Rose, R. C., Frelinger, J. G., & Lord, E. M. (2005). Local radiation therapy of B16 melanoma tumors increases the generation of tumor antigen-specific effector cells that traffic to the tumor. *Journal of Immunology, 174*(12), 7516–7523.

Lynch, T. J., Bondarenko, I., Luft, A., et al. (2012). Ipilimumab in combination with paclitaxel and carboplatin as first-line treatment in stage IIIB/IV non-small-cell lung cancer: Results from a randomized, double-blind, multicenter phase II study. *Journal of Clinical Oncology, 30*(17), 2046–2054.

Maker, A. V., Phan, G. Q., Attia, P., et al. (2005). Tumor regression and autoimmunity in patients treated with cytotoxic T lymphocyte-associated antigen 4 blockade and interleukin 2: A phase I/II study. *Annals of Surgical Oncology, 12*(12), 1005–1016.

Mallett, S., Fossum, S., & Barclay, A. N. (1990). Characterization of the MRC OX40 antigen of activated CD4 positive T lymphocytes—A molecule related to nerve growth factor receptor. *The EMBO Journal, 9*(4), 1063–1068.

Mao, Y., Qu, Q., Zhang, Y., Liu, J., Chen, X., & Shen, K. (2014). The value of tumor infiltrating lymphocytes (TILs) for predicting response to neoadjuvant chemotherapy in breast cancer: A systematic review and meta-analysis. *PLoS One, 9*(12). e115103.

Marabelle, A., Kohrt, H., Sagiv-Barfi, I., et al. (2013). Depleting tumor-specific Tregs at a single site eradicates disseminated tumors. *The Journal of Clinical Investigation, 123*(6), 2447–2463.

McMahan, R. H., Golden-Mason, L., Nishimura, M. I., et al. (2010). Tim-3 expression on PD-1 + HCV-specific human CTLs is associated with viral persistence, and its blockade restores hepatocyte-directed in vitro cytotoxicity. *The Journal of Clinical Investigation, 120*(12), 4546–4557.

Melero, I., Shuford, W. W., Newby, S. A., et al. (1997). Monoclonal antibodies against the 4-1BB T-cell activation molecule eradicate established tumors. *Nature Medicine, 3*(6), 682–685.

Mitsui, J., Nishikawa, H., Muraoka, D., et al. (2010). Two distinct mechanisms of augmented antitumor activity by modulation of immunostimulatory/inhibitory signals. *Clinical Cancer Research: An Official Journal of the American Association for Cancer Research, 16*(10), 2781–2791.

Miyazaki, T., Dierich, A., Benoist, C., & Mathis, D. (1996). LAG-3 is not responsible for selecting T helper cells in CD4-deficient mice. *International Immunology, 8*(5), 725–729.

Mole, R. H. (1953). Whole body irradiation: Radiobiology or medicine? *The British Journal of Radiology, 26*(305), 234–241.

Monney, L., Sabatos, C. A., Gaglia, J. L., et al. (2002). Th1-specific cell surface protein Tim-3 regulates macrophage activation and severity of an autoimmune disease. *Nature, 415*(6871), 536–541.

Montagnoli, C., Fallarino, F., Gaziano, R., et al. (2006). Immunity and tolerance to Aspergillus involve functionally distinct regulatory T cells and tryptophan catabolism. *Journal of Immunology, 176*(3), 1712–1723.

Morales, A., Eidinger, D., & Bruce, A. W. (1976). Intracavitary Bacillus Calmette-Guerin in the treatment of superficial bladder tumors. *The Journal of Urology, 116*(2), 180–183.

Motzer, R. J., Rini, B. I., McDermott, D. F., et al. (2015). Nivolumab for metastatic renal cell carcinoma: Results of a randomized phase II trial. *Journal of Clinical Oncology, 33*(13), 1430–1437.

Munn, D. H., Sharma, M. D., Baban, B., et al. (2005). GCN2 kinase in T cells mediates proliferative arrest and anergy induction in response to indoleamine 2,3-dioxygenase. *Immunity, 22*(5), 633–642.

Munn, D. H., Zhou, M., Attwood, J. T., et al. (1998). Prevention of allogeneic fetal rejection by tryptophan catabolism. *Science, 281*(5380), 1191–1193.

Ngiow, S. F., von Scheidt, B., Akiba, H., Yagita, H., Teng, M. W., & Smyth, M. J. (2011). Anti-TIM3 antibody promotes T cell IFN-gamma-mediated antitumor immunity and suppresses established tumors. *Cancer Research, 71*(10), 3540–3551.

Nishikawa, H., Kato, T., Hirayama, M., et al. (2008). Regulatory T cell-resistant CD8+ T cells induced by glucocorticoid-induced tumor necrosis factor receptor signaling. *Cancer Research, 68*(14), 5948–5954.

Nishimura, H., Nose, M., Hiai, H., Minato, N., & Honjo, T. (1999). Development of lupus-like autoimmune diseases by disruption of the PD-1 gene encoding an ITIM motif-carrying immunoreceptor. *Immunity, 11*(2), 141–151.

Nishimura, H., Okazaki, T., Tanaka, Y., et al. (2001). Autoimmune dilated cardiomyopathy in PD-1 receptor-deficient mice. *Science*, *291*(5502), 319–322.

Okazaki, T., Okazaki, I. M., Wang, J., et al. (2011). PD-1 and LAG-3 inhibitory co-receptors act synergistically to prevent autoimmunity in mice. *The Journal of Experimental Medicine*, *208*(2), 395–407.

Okazaki, T., Tanaka, Y., Nishio, R., et al. (2003). Autoantibodies against cardiac troponin I are responsible for dilated cardiomyopathy in PD-1-deficient mice. *Nature Medicine*, *9*(12), 1477–1483.

Papadimitrakopoulou, V., Patnaik, A., Borghaei, H., et al. (2015). Pembrolizumab (pembro; MK-3475) plus platinum doublet chemotherapy (PDC) as front-line therapy for advanced non-small cell lung cancer (NSCLC): KEYNOTE-021 Cohorts A and C. *American Society of Clinical Oncology Meeting Abstracts*, *33*(15 suppl), 8031.

Pardoll, D. M. (2012). The blockade of immune checkpoints in cancer immunotherapy. *Nature Reviews. Cancer*, *12*(4), 252–264.

Parker, J. J., Jones, J. C., Strober, S., & Knox, S. J. (2013). Characterization of direct radiation-induced immune function and molecular signaling changes in an antigen presenting cell line. *Clinical Immunology*, *148*(1), 44–55.

Parry, R. V., Chemnitz, J. M., Frauwirth, K. A., et al. (2005). CTLA-4 and PD-1 receptors inhibit T-cell activation by distinct mechanisms. *Molecular and Cellular Biology*, *25*(21), 9543–9553.

Piconese, S., Pittoni, P., Burocchi, A., et al. (2010). A non-redundant role for OX40 in the competitive fitness of Treg in response to IL-2. *European Journal of Immunology*, *40*(10), 2902–2913.

Postow, M. A., Callahan, M. K., Barker, C. A., et al. (2012). Immunologic correlates of the abscopal effect in a patient with melanoma. *The New England Journal of Medicine*, *366*(10), 925–931.

Postow, M. A., Chesney, J., Pavlick, A. C., et al. (2015). Nivolumab and ipilimumab versus ipilimumab in untreated melanoma. *The New England Journal of Medicine*, *372*, 2006–2017.

Powles, T., Eder, J. P., Fine, G. D., et al. (2014). MPDL3280A (anti-PD-L1) treatment leads to clinical activity in metastatic bladder cancer. *Nature*, *515*(7528), 558–562.

Puzanov, I., Milhem, M. M., Andtbacka, R. H. I., et al. (2015). Survival, safety, and response patterns in a phase 1b multicenter trial of talimogene laherparepvec (T-VEC) and ipilimumab (ipi) in previously untreated, unresected stage IIIB-IV melanoma. *American Society of Clinical Oncology Meeting Abstracts*, *33*(15 suppl), 9063.

Reck, M., Bondarenko, I., Luft, A., et al. (2013). Ipilimumab in combination with paclitaxel and carboplatin as first-line therapy in extensive-disease-small-cell lung cancer: Results from a randomized, double-blind, multicenter phase 2 trial. *Annals of Oncology: Official Journal of the European Society for Medical Oncology/ESMO*, *24*(1), 75–83.

Redmond, W. L., Linch, S. N., & Kasiewicz, M. J. (2014). Combined targeting of costimulatory (OX40) and coinhibitory (CTLA-4) pathways elicits potent effector T cells capable of driving robust antitumor immunity. *Cancer Immunology Research*, *2*(2), 142–153.

Reits, E. A., Hodge, J. W., Herberts, C. A., et al. (2006). Radiation modulates the peptide repertoire, enhances MHC class I expression, and induces successful antitumor immunotherapy. *The Journal of Experimental Medicine*, *203*(5), 1259–1271.

Ribas, A. (2012). Tumor immunotherapy directed at PD-1. *The New England Journal of Medicine*, *366*(26), 2517–2519.

Ribas, A., Puzanov, I., Gajewski, T., et al. (2015). A multicenter, open-label trial of talimogene laherparepvec (T-VEC) plus pembrolizumab vs pembrolizumab monotherapy in previously untreated, unresected, stage IIIB-IV melanoma. *American Society of Clinical Oncology Meeting Abstracts*, *33*(15 suppl). TPS9081.

Rizvi, N. A., Gettinger, S. N., Goldman, J. W., et al. (2015). Safety and efficacy of first-line nivolumab (NIVO; anti-programmed death-1 [PD-1]) and ipilimumab in non-small cell lung cancer (NSCLC). *Journal of Thoracic Oncology: Official Publication of the International Association for the Study of Lung Cancer*, (Abstract #786).

Rizvi, N. A., Mazieres, J., Planchard, D., et al. (2015). Activity and safety of nivolumab, an anti-PD-1 immune checkpoint inhibitor, for patients with advanced, refractory squamous non-small-cell lung cancer (CheckMate 063): A phase 2, single-arm trial. *The Lancet Oncology*, *16*(3), 257–265.

Robert, C., Ribas, A., Wolchok, J. D., et al. (2014). Anti-programmed-death-receptor-1 treatment with pembrolizumab in ipilimumab-refractory advanced melanoma: A randomised dose-comparison cohort of a phase 1 trial. *Lancet*, *384*(9948), 1109–1117.

Robert, C., Schachter, J., Long, G. V., et al. (2015). Pembrolizumab versus ipilimumab in advanced melanoma. *The New England Journal of Medicine*, *372*(26), 2521–2532.

Robert, C., Thomas, L., Bondarenko, I., et al. (2011). Ipilimumab plus dacarbazine for previously untreated metastatic melanoma. *The New England Journal of Medicine*, *364*(26), 2517–2526.

Ronchetti, S., Nocentini, G., Bianchini, R., Krausz, L. T., Migliorati, G., & Riccardi, C. (2007). Glucocorticoid-induced TNFR-related protein lowers the threshold of CD28 costimulation in CD8+ T cells. *Journal of Immunology*, *179*(9), 5916–5926.

Rosenberg, S. A., Yang, J. C., Sherry, R. M., et al. (2011). Durable complete responses in heavily pretreated patients with metastatic melanoma using T-cell transfer immunotherapy. *Clinical Cancer Research: An Official Journal of the American Association for Cancer Research*, *17*(13), 4550–4557.

Rosenberg, S. A., Yang, J. C., Topalian, S. L., et al. (1994). Treatment of 283 consecutive patients with metastatic melanoma or renal cell cancer using high-dose bolus interleukin 2. *JAMA : The Journal of the American Medical Association*, *271*(12), 907–913.

Ruby, C. E., Yates, M. A., Hirschhorn-Cymerman, D., et al. (2009). Cutting Edge: OX40 agonists can drive regulatory T cell expansion if the cytokine milieu is right. *Journal of Immunology*, *183*(8), 4853–4857.

Sagiv-Barfi, I., Rajapaksa, A., Czerwinski, D., et al. (2014). Abstract 2941: Local tumor irradiation combined with α-PDL-1 immune checkpoint inhibition results in local and systemic anti-tumor responses: Successful translation of a mouse model to a human case series. *Cancer Research*, *74*(19 Supplement), 2941.

Sakuishi, K., Apetoh, L., Sullivan, J. M., Blazar, B. R., Kuchroo, V. K., & Anderson, A. C. (2010). Targeting Tim-3 and PD-1 pathways to reverse T cell exhaustion and restore anti-tumor immunity. *The Journal of Experimental Medicine*, *207*(10), 2187–2194.

Saunders, P. A., Hendrycks, V. R., Lidinsky, W. A., & Woods, M. L. (2005). PD-L2:PD-1 involvement in T cell proliferation, cytokine production, and integrin-mediated adhesion. *European Journal of Immunology*, *35*(12), 3561–3569.

Schaer, D. A., Cohen, A. D., & Wolchok, J. D. (2010). Anti-GITR antibodies—Potential clinical applications for tumor immunotherapy. *Current Opinion in Investigational Drugs*, *11*(12), 1378–1386.

Schneider, H., Downey, J., Smith, A., et al. (2006). Reversal of the TCR stop signal by CTLA-4. *Science*, *313*(5795), 1972–1975.

Schreiber, R. D., Old, L. J., & Smyth, M. J. (2011). Cancer immunoediting: Integrating immunity's roles in cancer suppression and promotion. *Science*, *331*(6024), 1565–1570.

Sevko, A., Michels, T., Vrohlings, M., et al. (2013). Antitumor effect of paclitaxel is mediated by inhibition of myeloid-derived suppressor cells and chronic inflammation in the spontaneous melanoma model. *Journal of Immunology*, *190*(5), 2464–2471.

Sharabi, A. B., Nirschl, C. J., Kochel, C. M., et al. (2015). Stereotactic radiation therapy augments antigen-specific PD-1-mediated antitumor immune responses via cross-presentation of tumor antigen. *Cancer Immunology Research*, *3*(4), 345–355.

Sharma, A., Bode, B., Wenger, R. H., et al. (2011). γ-Radiation promotes immunological recognition of cancer cells through increased expression of cancer-testis antigens in vitro and in vivo. *PLoS One*, *6*(11): e28217.

Sharma, M. D., Hou, D. Y., Baban, B., et al. (2010). Reprogrammed foxp3(+) regulatory T cells provide essential help to support cross-presentation and CD8(+) T cell priming in naive mice. *Immunity*, *33*(6), 942–954.

Shevchenko, I., Karakhanova, S., Soltek, S., et al. (2013). Low-dose gemcitabine depletes regulatory T cells and improves survival in the orthotopic Panc02 model of pancreatic cancer. *International Journal of Cancer*, *133*(1), 98–107.

Shimizu, J., Yamazaki, S., Takahashi, T., Ishida, Y., & Sakaguchi, S. (2002). Stimulation of CD25(+)CD4(+) regulatory T cells through GITR breaks immunological self-tolerance. *Nature Immunology*, *3*(2), 135–142.

Shinto, E., Hase, K., Hashiguchi, Y., et al. (2014). CD8+ and FOXP3+ tumor-infiltrating T cells before and after chemoradiotherapy for rectal cancer. *Annals of Surgical Oncology*, *21*(Suppl 3), S414–S421.

Shrikant, P., Khoruts, A., & Mescher, M. F. (1999). CTLA-4 blockade reverses CD8+ T cell tolerance to tumor by a CD4+ T cell- and IL-2-dependent mechanism. *Immunity*, *11*(4), 483–493.

Sierro, S., Romero, P., & Speiser, D. E. (2011). The CD4-like molecule LAG-3, biology and therapeutic applications. *Expert Opinion on Therapeutic Targets*, *15*(1), 91–101.

Starck, L., Scholz, C., Dorken, B., & Daniel, P. T. (2005). Costimulation by CD137/4-1BB inhibits T cell apoptosis and induces Bcl-xL and c-FLIP(short) via phosphatidylinositol 3-kinase and AKT/protein kinase B. *European Journal of Immunology*, *35*(4), 1257–1266.

Strebhardt, K., & Ullrich, A. (2008). Paul Ehrlich's magic bullet concept: 100 years of progress. *Nature Reviews. Cancer*, *8*(6), 473–480.

Sznol, M., Hodi, F. S., Margolin, K., et al. (2008). Phase I study of BMS-663513, a fully human anti-CD137 agonist monoclonal antibody, in patients (pts) with advanced cancer (CA). *American Society of Clinical Oncology Meeting Abstracts*, *26*(15 suppl), 3007.

Sznol, M., Kluger, H. M., Callahan, M. K., et al. (2014). Survival, response duration, and activity by BRAF mutation (MT) status of nivolumab (NIVO, anti-PD-1, BMS-936558, ONO-4538) and ipilimumab (IPI) concurrent therapy in advanced melanoma (MEL). *American Society of Clinical Oncology Meeting Abstracts*, *32*(15 suppl). LBA9003.

Takeda, I., Ine, S., Killeen, N., et al. (2004). Distinct roles for the OX40-OX40 ligand interaction in regulatory and nonregulatory T cells. *Journal of Immunology*, *172*(6), 3580–3589.

Taube, J. M., Klein, A., Brahmer, J. R., et al. (2014). Association of PD-1, PD-1 ligands, and other features of the tumor immune microenvironment with response to anti-PD-1 therapy. *Clinical Cancer Research: An Official Journal of the American Association for Cancer Research*, *20*(19), 5064–5074.

Thomas, L. (1959). Delayed hypersensitivity in health and disease. In H. S. Lawrence (Ed.), *Cellular and humoral aspects of the hypersensitive states* (pp. 529–532). New York, NY: Hoeber-Harper.

Tivol, E. A., Borriello, F., Schweitzer, A. N., Lynch, W. P., Bluestone, J. A., & Sharpe, A. H. (1995). Loss of CTLA-4 leads to massive lymphoproliferation and fatal multiorgan tissue destruction, revealing a critical negative regulatory role of CTLA-4. *Immunity*, *3*(5), 541–547.

Topalian, S. L., Hodi, F. S., Brahmer, J. R., et al. (2012). Safety, activity, and immune correlates of anti-PD-1 antibody in cancer. *The New England Journal of Medicine*, *366*(26), 2443–2454.

Triebel, F., Jitsukawa, S., Baixeras, E., et al. (1990). LAG-3, a novel lymphocyte activation gene closely related to CD4. *The Journal of Experimental Medicine*, *171*(5), 1393–1405.

Tumeh, P. C., Harview, C. L., Yearley, J. H., et al. (2014). PD-1 blockade induces responses by inhibiting adaptive immune resistance. *Nature*, *515*(7528), 568–571.

Turk, M. J., Guevara-Patino, J. A., Rizzuto, G. A., Engelhorn, M. E., Sakaguchi, S., & Houghton, A. N. (2004). Concomitant tumor immunity to a poorly immunogenic melanoma is prevented by regulatory T cells. *The Journal of Experimental Medicine*, *200*(6), 771–782.

Twyman-Saint Victor, C., Rech, A. J., Maity, A., et al. (2015). Radiation and dual checkpoint blockade activate non-redundant immune mechanisms in cancer. *Nature*, *520*(7547), 373–377.

Umansky, V., & Sevko, A. (2012). Overcoming immunosuppression in the melanoma microenvironment induced by chronic inflammation. *Cancer Immunology, Immunotherapy: CII*, *61*(2), 275–282.

Vacchelli, E., Aranda, F., Eggermont, A., et al. (2014). Trial watch: IDO inhibitors in cancer therapy. *Oncoimmunology*, *3*(10). e957994.

Valzasina, B., Guiducci, C., Dislich, H., Killeen, N., Weinberg, A. D., & Colombo, M. P. (2005). Triggering of OX40 (CD134) on CD4(+)CD25 + T cells blocks their inhibitory activity: A novel regulatory role for OX40 and its comparison with GITR. *Blood*, *105*(7), 2845–2851.

Vinay, D. S., & Kwon, B. S. (2011). 4-1BB signaling beyond T cells. *Cellular & Molecular Immunology*, *8*(4), 281–284.

Vinay, D. S., & Kwon, B. S. (2012). Immunotherapy of cancer with 4-1BB. *Molecular Cancer Therapeutics*, *11*(5), 1062–1070.

Vincent, J., Mignot, G., Chalmin, F., et al. (2010). 5-Fluorouracil selectively kills tumor-associated myeloid-derived suppressor cells resulting in enhanced T cell-dependent antitumor immunity. *Cancer Research*, *70*(8), 3052–3061.

Walunas, T. L., Bakker, C. Y., & Bluestone, J. A. (1996). CTLA-4 ligation blocks CD28-dependent T cell activation. *The Journal of Experimental Medicine*, *183*(6), 2541–2550.

Walunas, T. L., Lenschow, D. J., Bakker, C. Y., et al. (1994). CTLA-4 can function as a negative regulator of T cell activation. *Immunity*, *1*(5), 405–413.

Waterhouse, P., Penninger, J. M., Timms, E., et al. (1995). Lymphoproliferative disorders with early lethality in mice deficient in Ctla-4. *Science*, *270*(5238), 985–988.

Weber, J. S., D'Angelo, S. P., Minor, D., et al. (2015). Nivolumab versus chemotherapy in patients with advanced melanoma who progressed after anti-CTLA-4 treatment (CheckMate 037): A randomised, controlled, open-label, phase 3 trial. *The Lancet Oncology*, *16*(4), 375–384.

Weber, J., Hamid, O., Amin, A., et al. (2013). Randomized phase I pharmacokinetic study of ipilimumab with or without one of two different chemotherapy regimens in patients with untreated advanced melanoma. *Cancer Immunity*, *13*, 7.

Weinberg, A. D., Rivera, M. M., Prell, R., et al. (2000). Engagement of the OX-40 receptor in vivo enhances antitumor immunity. *Journal of Immunology*, *164*(4), 2160–2169.

Wilcox, R. A., Chapoval, A. I., Gorski, K. S., et al. (2002). Cutting edge: Expression of functional CD137 receptor by dendritic cells. *Journal of Immunology*, *168*(9), 4262–4267.

Wolchok, J. D., Kluger, H., Callahan, M. K., et al. (2013). Nivolumab plus ipilimumab in advanced melanoma. *The New England Journal of Medicine*, *369*(2), 122–133.

Wolchok, J. D., Neyns, B., Linette, G., et al. (2010). Ipilimumab monotherapy in patients with pretreated advanced melanoma: A randomised, double-blind, multicentre, phase 2, dose-ranging study. *The Lancet Oncology*, *11*(2), 155–164.

Woo, S. R., Turnis, M. E., Goldberg, M. V., et al. (2012). Immune inhibitory molecules LAG-3 and PD-1 synergistically regulate T-cell function to promote tumoral immune escape. *Cancer Research*, *72*(4), 917–927.

Workman, C. J., & Vignali, D. A. (2003). The CD4-related molecule, LAG-3 (CD223), regulates the expansion of activated T cells. *European Journal of Immunology, 33*(4), 970–979.

Zamarin, D., Holmgaard, R. B., Subudhi, S. K., et al. (2014). Localized oncolytic virotherapy overcomes systemic tumor resistance to immune checkpoint blockade immunotherapy. *Science Translational Medicine, 6*(226), 226ra232.

Zeng, J., Harris, T. J., Lim, M., Drake, C. G., & Tran, P. T. (2013). Immune modulation and stereotactic radiation: Improving local and abscopal responses. *BioMed Research International, 2013*, 658126.

CHAPTER SEVEN

Adoptive T-Cell Therapy for Cancer

James C. Yang, Steven A. Rosenberg[1]
Center for Clinical Research, National Cancer Institute, Bethesda, MD, United States
[1]Corresponding author: e-mail address: sar@nih.gov

Contents

Abstract

Recent developments have demonstrated that immunotherapies are capable of achieving durable antitumor responses in patients with metastatic cancer. One modality that has been able to induce durable complete regressions in patients with melanoma has been adoptive cell therapy (ACT). This has slowly been expanded to other cancer types using new approaches such as genetically engineered T-cells and other methods of antigen targeting. It now appears that immune targeting of mutated "neoantigens" plays a major role in successful ACT, as well as in other immunotherapies such as checkpoint inhibitors. This realization presents not only new challenges to ACT but also new opportunities in that all tumors now may have potential antigens to attack that can be revealed by tumor genomic sequencing. There are a variety of exciting approaches to translate these new findings into clinical trials applying ACT to the majority of cancer types.

Only recently has the goal of immunologically rejecting a metastatic cancer in a patient been realized in a consistent fashion. The concept has been advanced for over a century, but was only seen sporadically or anecdotally. Two major developments that have led to this success have been the demonstration that many cancer patients have a T-cell repertoire capable of recognizing their cancer and the realization that the tumor microenvironment

Advances in Immunology, Volume 130
ISSN 0065-2776
http://dx.doi.org/10.1016/bs.ai.2015.12.006

is inhibitory to the function of this repertoire. A detailed molecular understanding of these phenomena has not only led to immunotherapies that consistently lead to tumor rejection but also point the way to achieving it in many different types of human cancer. Other chapters in this volume will detail the mechanisms and successes of checkpoint blockade, particularly in patients with melanoma, lung cancer, and other cancers, while this chapter will focus on the T-cell repertoire and methods of augmenting it to better achieve complete and lasting tumor eradication.

1. IDENTIFYING TUMOR-ASSOCIATED ANTIGENS RECOGNIZED BY T-CELLS

Prior to a molecular understanding of tumor rejection antigens, there was evidence that some T-cells from patients with cancer could immunologically recognize their cancer. The clearest and most consistent examples were from patients with melanoma, where tumor infiltrating lymphocytes (TIL) or peripheral blood lymphocytes (PBL) stimulated in vitro with tumor was shown to lyse their autologous tumor line in an major histocompatability complex (MHC)-specific fashion (Herin et al., 1987; Knuth, Danowski, Oettgen, & Old, 1984; Muul, Spiess, Director, & Rosenberg, 1987). One major advantage of studying melanoma was that patient-specific tumor lines could be readily established from surgical specimens in many patients and were available for molecular studies and immunological assays. In addition, this autologous tumor recognition by T-cells appeared to be much more prevalent in melanoma patients for reasons not understood at the time. When TIL, simply expanded in recombinant interleukin-2 (rIL-2), were cocultured with their autologous tumor (either cryopreserved fresh tumor, enzymatically digested single cell suspensions, or in vitro-maintained tumor cell lines) cytolysis in short-term chromium release assay or cytokine release could be demonstrated for most melanoma patients studied (Topalian, Solomon, & Rosenberg, 1989). These tumor-reactive T-cells were then concurrently investigated both clinically and immunologically.

Laboratory studies of tumor-reactive T-cells led to a key discovery in our molecular understanding of the T-cell response to human cancer in 1991 when van der Bruggen et al. cloned the antigen recognized by a tumor-reactive T-cell clone from a patient with melanoma (van der Bruggen et al., 1991). Following this achievement, many antigens were cloned using

T-cell clones often obtained from patients with melanoma (Boon, Cerottini, Van den Eynde, van der Bruggen, & Van, 1994; Kawakami et al., 1994). TIL were frequently the source of these clones and the initial antigens discovered were dominated by tissue differentiation antigens shared by melanomas and melanocytes (Rosenberg, Kawakami, Robbins, & Wang, 1996). These proteins are synthesized in huge amounts by cells in the melanocytic lineage, and the high frequency of their representation in tumor cDNA libraries may account for their early dominance in antigen identification. In addition, the attractiveness of targeting shared antigens which could potentially be attacked using "off-the-shelf" reagents focused attention on these types of targets. The other major class of shared target antigens was the tumor-germline antigens (initially termed tumor-testis antigens based on limited expression data). These were proteins, mostly encoded by genes on the X-chromosome, that were expressed during embryogenesis, but seldom on adult tissues other than germline or placental tissues. These were found to be immunogenic both by high-avidity humoral responses and by generating T-cell responses in patients with cancer (Jager et al., 1998; Scanlan et al., 2002). As more has been learned about the tumor-germline antigens, their main drawback seems to be a low prevalence of expression by the most common tumor histologic types, and inconsistent expression within individual patients and tumors (Park et al., 2016). Another class of shared antigens is normal proteins that are highly overexpressed by tumors, but expressed on some normal tissues. These included proteins such as CEA, mesothelin, or hTERT. Initial efforts to enhance T-cell repertoires against these self-antigens focused on vaccine strategies. Using dozens of approaches and enrolling thousands of patients, these vaccine strategies have only induced anecdotal objective responses in patients with measurable disease (Klebanoff, Acquavella, Yu, & Restifo, 2011). Clinical testing of vaccines has migrated more toward use in the adjuvant setting, but such studies require large randomized studies with years of follow-up, and have been largely negative as well. During this time, a large body of evidence has been developed which shows that there were many immunosuppressive influences in the tumor microenvironment that can suppress T-cell activation, expansion, or function. These included T-regulatory cells, inhibitory "checkpoint" receptors on T-cells, myeloid-derived suppressor cells (MDSC) and enzymes, and cytokines that impair T-cell function (Bronte & Mocellin, 2009; Gajewski et al., 2006). One approach to circumventing this local immunosuppression was to generate and expand a tumor-reactive T-cell repertoire ex vivo and adoptively transfer it to the patient along with other reagents that could support these cells or perhaps block local immunosuppressive factors.

2. CELL THERAPY WITH GENETICALLY ENGINEERED T-CELLS

A major advance that allowed this to be developed in concert with antigen discovery work was the ability to gene engineer mature human peripheral blood T-cells with high efficiency and safety (Morgan et al., 2006). Current techniques using gamma retroviruses or lentiviruses can introduce stable genetic changes with high frequency in human PBL without requiring selection and such cells have been administered to hundreds of patients without complications from gene modification after many years of follow-up. On cloning T-cell receptors (TCR) which recognized defined tumor-associated antigens, these TCR were genetically engineered into the PBL of HLA-appropriate patients and the expanded cells given in adoptive transfer protocols (Morgan et al., 2006). Another approach used a novel chimeric antigen receptor (CAR) where the ligand-binding domain was derived from the single chain variable fragment of a monoclonal antibody which bound a cell surface molecule on tumors (Eshhar, Waks, & Gross, 2014; Eshhar, Waks, Gross, & Schindler, 1993). This was covalently linked in tandem via a transmembrane domain to the CD3-zeta activating moiety from T-cells to trigger activation on ligation of the CAR in an MHC-independent fashion (often with interposed costimulatory domains to enhance activation or cell survival) (Brentjens et al., 2007). The design and use of CAR in gene-engineered T-cells is dealt with in greater detail elsewhere in this publication.

Trials of such genetically retargeted T-cells showed that some could achieve tumor regression but also established the risks associated with antigen selection. Clinical trials of receptor-engineered T-cells targeting unmutated, shared tumor-associated antigens showed these antigens could divided into ones that: (1) were associated with unacceptable autoimmunity, (2) induced acceptable or tolerable autoimmunity, and (3) those that showed no apparent autoimmunity. As an example of the first outcome, T-cells transduced with TCRs against the melanoma/melanocyte differentiation antigens such as MART-1 or gp100 caused tumor regression but also attacked the skin, eyes, and inner ear due to the presence of melanocytes at those locations (Johnson et al., 2009). Rashes, vitiligo, and decreases in vision and hearing limited the utility of these reagents. On targeting CEA with a high-avidity TCR generated in an HLA-A2 transgenic mouse, some early evidence of antitumor activity was seen, but all patients treated developed severe, life-threatening colitis (Parkhurst et al., 2011). On the other hand, when the B-cell marker, CD19 (present on both benign and malignant T-cells) was targeted with CARs, impressive responses in patients with B-cell lymphomas and leukemias were seen (Goff et al., 2010; Kochenderfer et al., 2010; Maude et al., 2014). The accompanying B-cell aplasia that also resulted was manageable with surveillance for hypogammaglobulinemia and infection and appropriate supportive care. In view of the dramatic responses seen, the risk-benefit ratio of this anti-B-cell autoimmunity was considered acceptable. The best current example of a self-antigen effectively targeted with receptor-engineered T-cells with no apparent autoimmunity has been adoptive cell therapy (ACT) against the tumor-germline antigen, NY-ESO-1. Originally discovered by screening patient sera for antitumor IgG responses using the SEREX technique, this member of the tumor-germline family of antigens was found expressed by a patient's esophageal cancer (Chen et al., 1997). It is also expressed by a small subset of melanomas and by 80% of synovial sarcomas due to a common translocation in this tumor. Autologous PBL from patients were transduced for clinical testing with an anti-NY-ESO-1 TCR that was HLA-A0201 restricted (Robbins et al., 2011). As is common with T-cell transfers, the recipient was immunosuppressed with a single cycle of nonmyeloablative chemotherapy prior to cell transfer. This transiently eliminates host T-regulatory cells and MDSC, induces the homeostatic cytokines IL-7 and IL-15, and depletes resident lymphocytes competing for these T-cell growth factors, all shown to enhance ACT or survival of T-cells after transfer in mouse models (Klebanoff, Khong, Antony, Palmer, & Restifo, 2005). When HLA-A0201 + patients with measurable metastatic melanoma or synovial sarcoma were given anti-NY-ESO-1 receptor transduced T-cells (with a brief course of supportive systemic IL-2), objective response rates by RECIST criteria of 55% and 61%, respectively, were seen, with some complete and durable responses seen (Fig. 1; Robbins et al., 2015). No treatment-related autoimmunity was evident in these patients. A wide variety of other tumor-associated target antigens are being studied as targets for receptor-engineered T-cells, but have yet to be demonstrated to mediate major tumor regressions and to be safe.

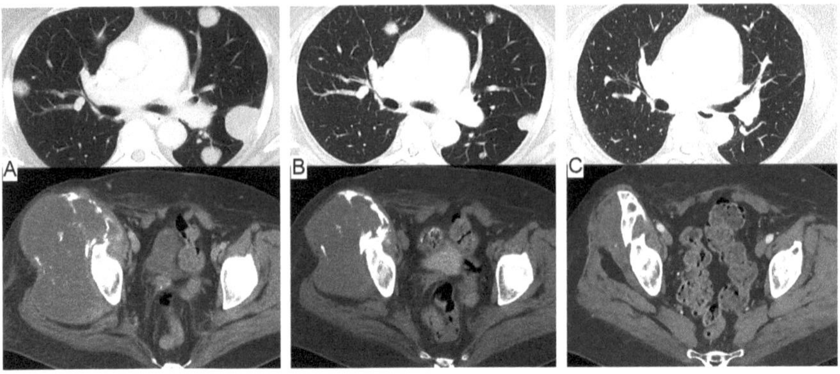

Figure 1 Patient with an NY-ESO-1-expressing metastatic synovial cell sarcoma showing ongoing regression of lung metastases and a large pelvic primary tumor after a single adoptive transfer of autologous T-cells genetically engineered with a T-cell receptor recognizing NY-ESO-1. (A) Pretreatment scans, (B) 5 months posttreatment, and (C) 5 years posttreatment.

3. CELL THERAPY WITH TIL

Even prior to conducting laboratory studies on antigen identification and clinical translation protocols using receptor-engineered T-cells, empirical trials using melanoma TIL were also undertaken. The availability of such autologous T-cells with consistent antitumor reactivity allowed the development of the principles and practices of adoptive lymphocyte transfer in patients. Over two-thirds of T-cells grown from resected melanoma metastases show autologous tumor recognition after being cultured in IL-2 (Goff et al., 2010). These can be readily expanded in vitro to large numbers, and used for adoptive T-cell transfer. This was initially done in patients with measurable metastatic melanoma with the patients receiving just TIL (grown in bulk from enzymatically dispersed tumor and administered without regard to in vitro reactivity) and systemic IL-2 (Rosenberg et al., 1988, 1994). Some patients also received a single dose of cyclophosphamide (at 25 mg/kg) prior to cell administration. Overall the objective response rate was 34% with neither differences between patients given or not given cyclophosphamide nor between patients who had or not had prior systemic IL-2 therapy alone. Despite giving a median of nearly 2×10^{11} cells per patient, many of these responses proved to be of short duration, with only 7% persisting for a year after treatment. Earlier studies of gene-marked TIL showed that in vivo survival after administration was very brief with most patients having no TIL detectable by RT-PCR a month later (Rosenberg et al., 1990). After this experience and with supportive data from murine models, transient host lymphodepletion using cyclophosphamide and fludarabine was introduced just prior to cell infusion. In addition, changes in the methods of growing and selecting TIL for administration were made. When TIL were grown from individual fragments of a resected metastasis, a high degree of heterogeneity was seen between the different fragment cultures. When fresh or cultured autologous tumor was available, these TIL fragment cultures could be shown to often differ in their tumor recognition. Therefore, for many patients, one was able to select subcultures with greater tumor recognition to expand and administer. In addition, the final expansion of these select fragment cultures was also made much faster and more consistent using a method incorporating anti-CD3 antibody and irradiated feeder cells in addition to IL-2 (Riddell & Greenberg, 1990). Ultimately, when all of these principles were employed in treating a group of patients with melanoma, an objective response rate of 54% was seen (using three different lymphodepleting regimens that did not confer significant differences in outcome) with 20% of these patients maintaining durable complete responses after 5–8 years of follow-up (Rosenberg et al., 2011; Table 1). The same high response rates were seen in patients previously unresponsive

Table 1 Response Rate and Response Durations of 93 Patients with Measurable Metastatic Melanoma Treated with Preparative Lymphodepletion, TIL, and Interleukin-2 (Three Different Lymphodepleting Regimens Were Used but Showed No Significant Differences in Outcome so They Are Considered Together)

	PR	**CR**	**ORR**
Total Pts	**Number of Patients (Duration in Months)**		
93	32 (34%)	20 (22%)	52 (56%)
	84, 36, 29, 28, 21	137+, 135+, 134+, 124+	
	14, 14, 13, 12, 11	120+, 120+, 116+, 113+	
	9, 8, 7, 7, 7	104+, 101+, 100+, 95+	
	6, 6, 6, 6, 6	94+, 94+, 94+,93+	
	5, 5, 4, 4, 4	82+, 64+, 63+, 19	
	3, 3, 3, 2, 2		
	2, 2		

Most patients had progressed after prior IL-2 therapy and only two patients received more than one TIL transfer. The data are updated with a median follow-up of 9.8 years.

to IL-2 alone, demonstrating that the cell transfer was responsible for the regressions. High levels of persistent T-cell survival were also seen which were associated with a higher likelihood of response. More recently, this experience has been reproduced in another 101 patients with melanoma, where the overall and complete response rates were 53% and 24%, respectively (manuscript in preparation). These trials perhaps best illustrate the concept that adoptive T-cell transfer is capable of causing and sustaining complete regressions of disseminated cancer in patients.

4. CELL THERAPY TARGETING MUTATED "NEOANTIGENS"

An interesting observation during these trials with melanoma-derived TIL was that these TIL cultures often contained reactivity against melanoma/melanocyte associated differentiation antigens but infrequently induced vitiligo and rarely were associated with eye, ear, or any autoimmune toxicities. This suggested that the relevant target antigens might be something else. The (low)prevalence of reactivity against the known tumor-germline antigens also could not account for the high rates of response (Kvistborg et al., 2012). In some TIL cultures, clear autologous tumor recognition could be demonstrated but no recognition of HLA-matched melanomas was seen suggesting an antigen repertoire with private specificity was responsible. Several prior studies using expression cloning to identify tumor antigens had demonstrated that these mutated "neoantigens" were indeed targeted by some T-cells (Coulie et al., 1995; Robbins et al., 1996; Wolfel et al., 1995). The need to procure autologous tumor as a reagent and the difficulties of translating such reactivities into patient-specific therapeutics had blunted enthusiasm for this class of tumor antigen. Now, the advent of rapid and relatively inexpensive whole exomic sequencing (WES) greatly increased access to the genetic information needed to pursue this avenue of investigation. So studies were undertaken to explore tumor-specific mutated proteins as T-cell targets. A series of three melanomas with tumor-reactive TIL (of known HLA restriction) were fully sequenced and all nonsynonymous exomic mutations identified (Robbins et al., 2013). As most epitopes presented by MHC Class I are 9 or 10 amino acids in length, 19-amino acid sequences consisting of each mutant amino acid and its 9 flanking wild-type residues on either side were analyzed for every mutation and epitopes within them rank ordered using HLA-binding algorithms and the known HLA restriction element. Although there were thousands of potential "neoepitopes" for each tumor, when only the 40 candidate epitopes with the highest predicted affinity were synthesized and tested for recognition, multiple mutated epitopes were recognized by all three TIL. The three patient TIL recognized the 5th, 18th, 19th, and 38th best-binding peptides for patient #1; the 2nd, 17th, and 23rd for patient #2; and the 2nd, 4th, 24th, and 36th for patient #3. These experiments indicated that multiple high-avidity epitopes created by tumor-specific mutations in melanoma were frequently being processed and presented and were generating T-cell responses. Others also found that mutated antigens were being recognized by T-cell responses in patients responding to other immunotherapies such as checkpoint inhibitors.

Coincident with these laboratory discoveries, compelling circumstantial evidence was accumulating that immune recognition of mutated antigens was playing a significant role in other effective clinical immunotherapies. It has long been known that melanoma was a tumor histology that was often one of the most responsive to immunotherapies such as interleukin-2 and anti-CTLA4 and can rarely undergo some degree of spontaneous regression. When large numbers of human tumors were subjected to DNA sequencing, melanoma was found to be one of the most highly mutated of all human tumors, presumably due to UV mutagenesis (Lawrence et al., 2013). Interestingly, uveal and mucosal melanomas with no connection to UV irradiation were much less mutated and appeared much less responsive to the same immunotherapies (Krauthammer et al., 2012). As newer agents such as anti-PD1 or anti-PDL1 antibodies were tested, again melanoma was the most responsive histology (Topalian et al., 2012). Yet other tumors such as lung, bladder, and head and neck cancers with high rates of mutation (mostly related to cigarette carcinogens) also showed significant response rates (American Association for Cancer Research, 2015; Brahmer et al., 2015). Perhaps most compelling was the finding that most colon cancers, which had low rates of mutation, did not respond to these agents, but the subset with DNA mismatch repair defects (and much higher rates of mutation) was highly responsive to anti-PD1 as were a set of noncolorectal tumors with mismatch repair defects (Table 2; Le et al., 2015). Yet validation of the hypothesis that mutation-driven immune responses were involved in tumor rejection required that specific mutation-reactive T-cells be shown to cause tumor rejection when adoptively transferred to a patient. In 2014, Tran et al. studied a patient with chemotherapy refractory metastatic cholangiocarcinoma (Tran et al., 2014). A lung metastasis was resected, WES performed and multiple TIL cultures grown from individual fragments

Table 2 Impact of Mismatch Repair Deficiency on Mutation Frequency and Response to Pembrolizumab

	Mismatch Repair-Deficient CRC	**Mismatch Repair Proficient CRC**	**Mismatch Repair-Deficient Non-CRC**
Pts treated	10	18	7
Complete responses	0	0	1
Partial responses	4	0	4
Response rate	40%	0%	71%
Pt tumors sequenced	7	6	2
Mean # somatic mutations	1875	73	1455

CRC, colorectal cancer.

from the tumor. Initially, the patient was treated with Cy-Flu followed by 4×10^{10} bulk TIL from multiple fragments (without reactivity testing as there was no autologous tumor available) along with four doses of IL-2. The patient had a minor response but relapsed after 7 months. Her tumor was found to harbor 26 nonsynonymous mutations and minigenes encoding each mutated amino acids and the 12 wild-type residues on either side (to encompass not only short Class I presented epitopes, but all possible Class II presented epitopes) were synthesized and concatenated into three tandem minigenes. Autologous dendritic cells were transfected with RNA from these minigenes and each TIL fragment culture independently tested for recognition of any of the products of the tandem minigenes. One was reactive with one of the tandem minigenes and the individual minigene within it that conferred TIL recognition proved to encode ERBB2 interacting protein (ERBB2IP) containing an E805G mutation which was within the epitope presented by the Class II molecule HLA-DQB1*0601. One TIL culture was 95% pure for a VB22+ T-cell clone reactive with this peptide and it was grown to large numbers for retreatment. The patient then received another TIL infusion of 12×10^{10} cells, 95% of which were this CD4+ clone (retrospective analysis showed the first infusion contained 1/12th the amount of this clone), again with Cy-Flu and four doses of IL-2. Nearly 2 years later, she has had a near complete regression of her tumors which ongoing shrinkage (Fig. 2). Since then, investigation of multiple other tumors of various types has confirmed that mutation-reactive TIL can be found in most gastrointestinal cancers (Tran et al., 2014) as well as lung cancer, ovarian cancer,

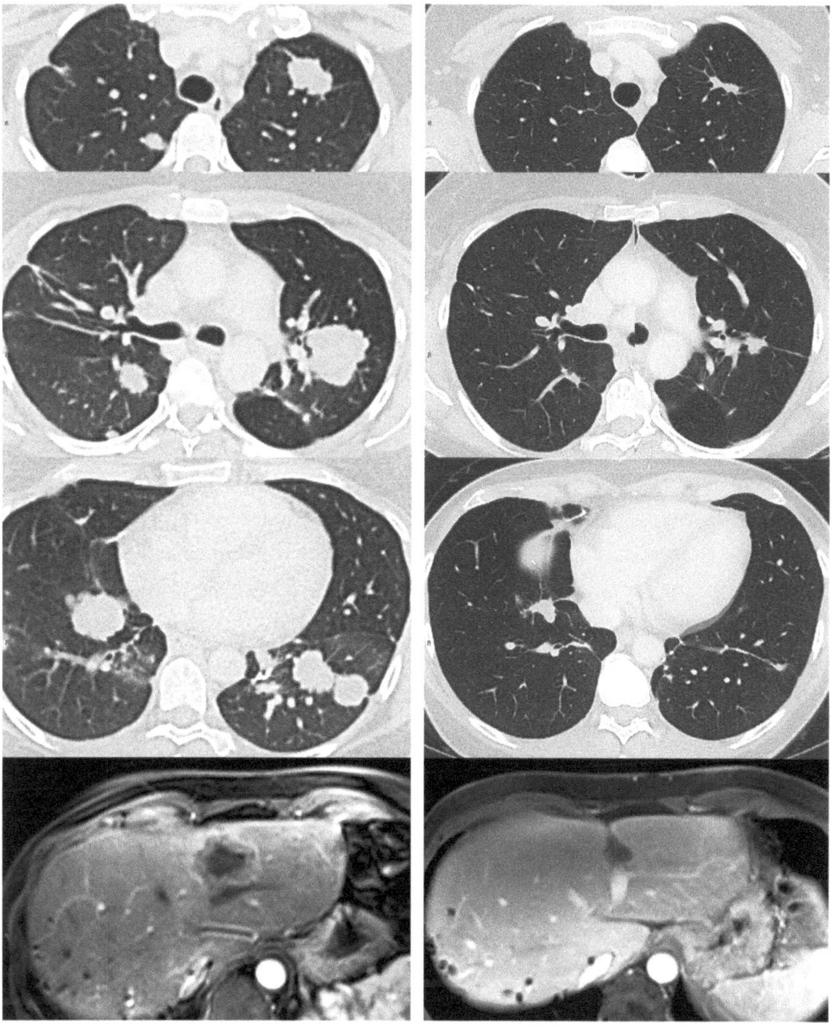

Figure 2 Patient with metastatic cholangiocarcinoma after adoptive T-cell transfer of tumor infiltrating lymphocytes highly selected (95% clonal) for a CD4+ clone reactive with a mutated epitope in ERBB2 interactive protein, expressed by her cancer. Pretreatment scans of lung and liver (left) and matching scans 20 months later (right) are shown.

breast cancer, and pancreatic cancer. Current efforts are directed at determining if these T-cells can be effective in adoptive immunotherapy and optimizing their activity using combinations of other immunotherapeutic agents. The presence of these "nonself" antigens resulting from somatic mutations in human tumors and the potential for having high-affinity TCRs in the T-cell repertoire against them that have not been edited by central thymic deletion opens the door to finding endogenous antitumor responses to be exploited against most cancers types. Such endogenous reactivity against neoantigens may in fact be the common pathway responsible for the efficacy of currently approved immunotherapies such as interleukin-2 and checkpoint inhibitors (Van Allen et al., 2015).

The ability to genetically modify human T-cells with high efficiency is not confined to only redirecting specificity. It also allows the function of T-cells to be altered to better effect tumor rejection. One recent clinical experience introducing IL-12 secretion into melanoma TIL to improve antigen processing and paracrine cytokine secretion in the tumor microenvironment resulted in a 63% objective response rate in a group of 16 patients given much smaller numbers of TIL than usual with just lymphodepletion but no IL-2 (Zhang et al., 2015). Although encouraging, the responses proved mostly short-lived and toxic IL-12 levels were seen in one patient. Other pharmacologic and methodological approaches applied during T-cell growth or in vivo are also being investigated to modify effector functions, and circumvent local immunosuppression or control the state of differentiation of transferred T-cells is being investigated based on preclinical data (Gattinoni et al., 2009; Hinrichs et al., 2008). Here again, the adoptive transfer of in vitro expanded T-cells represents the best opportunity to precisely "sculpt" the T-cell response using genetic and somatic modifications to achieve tumor rejection.

5. THE FUTURE OF ACT

To consistently achieve complete and durable tumor rejections with T-cell transfer, several problems need to be overcome. It appears that enhancement of the antitumor T-cell repertoire must be combined with measures to overcome the immunosuppressive local tumor microenvironment, so immunotherapeutic combinations will likely be necessary. Removing the tumor-reactive T-cells and growing them in vitro provides a potent way to rapidly increase their number and separates them from potentially detrimental measures applied to suppress the hostile tumor microenvironment. Logistically, the hurdle of growing very large numbers of T-cells on a patient-specific basis also needs to be addressed with either better engineering or more potent T-cells. Only a limited number of normal self-antigens have been safely and successfully targeted by adoptive T-cell therapy, but these are largely absent in the most common solid tumors. This exposes our lack of good shared target antigens and the paucity of off-the-shelf reagents that can be used against the most common human cancer types. The advent of efficient methods to genetically analyze individuals' tumors has made the pursuit of tumor-specific mutated antigens feasible and attractive. Yet the obstacle of mutational heterogeneity within tumors must be addressed and new means of enriching for the correct T-cells are needed. Still, this concept of "engrafting" a patient with a greatly enhanced T-cell repertoire against their cancer through adoptive transfer may accomplish the first critical step leading to successful tumor rejection.

REFERENCES

American Association for Cancer Research. (2015). Expanding the reach of anti-PD-1 therapy. *Cancer Discovery*, *5*, 684–685.

Boon, T., Cerottini, J. C., Van den Eynde, B., van der Bruggen, P., & Van, P. A. (1994). Tumor antigens recognized by T lymphocytes. *Annual Review of Immunology*, *12*, 337–365.

Brahmer, J., Reckamp, K. L., Baas, P., Crino, L., Eberhardt, W. E., Poddubskaya, E., et al. (2015). Nivolumab versus docetaxel in advanced squamous-cell non-small-cell lung cancer. *The New England Journal of Medicine*, *373*, 123–135.

Brentjens, R. J., Santos, E., Nikhamin, Y., Yeh, R., Matsushita, M., La, P. K., et al. (2007). Genetically targeted T cells eradicate systemic acute lymphoblastic leukemia xenografts. *Clinical Cancer Research*, *13*, 5426–5435.

Bronte, V., & Mocellin, S. (2009). Suppressive influences in the immune response to cancer. *Journal of Immunotherapy*, *32*, 1–11.

Chen, Y. T., Scanlan, M. J., Sahin, U., Tureci, O., Gure, A. O., Tsang, S., et al. (1997). A testicular antigen aberrantly expressed in human cancers detected by autologous antibody screening. *Proceedings of the National Academy of Sciences of the United States of America*, *94*, 1914–1918.

Coulie, P. G., Lehmann, F., Lethe, B., Herman, J., Lurquin, C., Andrawiss, M., et al. (1995). A mutated intron sequence codes for an antigenic peptide recognized by cytolytic T lymphocytes on a human melanoma. *Proceedings of the National Academy of Sciences of the United States of America*, *92*, 7976–7980.

Eshhar, Z., Waks, T., & Gross, G. (2014). The emergence of T-bodies/CAR T cells. *Cancer Journal*, *20*, 123–126.

Eshhar, Z., Waks, T., Gross, G., & Schindler, D. G. (1993). Specific activation and targeting of cytotoxic lymphocytes through chimeric single chains consisting of antibody-binding domains and the gamma or zeta subunits of the immunoglobulin and T-cell receptors. *Proceedings of the National Academy of Sciences of the United States of America*, *90*, 720–724.

Gajewski, T. F., Meng, Y., Blank, C., Brown, I., Kacha, A., Kline, J., et al. (2006). Immune resistance orchestrated by the tumor microenvironment. *Immunological Reviews*, *213*, 131–145.

Gattinoni, L., Zhong, X. S., Palmer, D. C., Ji, Y., Hinrichs, C. S., Yu, Z., et al. (2009). Wnt signaling arrests effector T cell differentiation and generates CD8+ memory stem cells. *Nature Medicine*, *15*, 808–813.

Goff, S. L., Smith, F. O., Klapper, J. A., Sherry, R., Wunderlich, J. R., Steinberg, S. M., et al. (2010). Tumor infiltrating lymphocyte therapy for metastatic melanoma: Analysis of tumors resected for TIL. *Journal of Immunotherapy*, *33*, 840–847.

Herin, M., Lemoine, C., Weynants, P., Vessiere, F., Van, P. A., Knuth, A., et al. (1987). Production of stable cytolytic T-cell clones directed against autologous human melanoma. *International Journal of Cancer*, *39*, 390–396.

Hinrichs, C. S., Spolski, R., Paulos, C. M., Gattinoni, L., Kerstann, K. W., Palmer, D. C., et al. (2008). IL-2 and IL-21 confer opposing differentiation programs to CD8+ T cells for adoptive immunotherapy. *Blood*, *111*, 5326–5333.

Jager, E., Chen, Y. T., Drijfhout, J. W., Karbach, J., Ringhoffer, M., Jager, D., et al. (1998). Simultaneous humoral and cellular immune response against cancer-testis antigen NY-ESO-1: Definition of human histocompatibility leukocyte antigen (HLA)-A2-binding peptide epitopes. *The Journal of Experimental Medicine*, *187*, 265–270.

Johnson, L. A., Morgan, R. A., Dudley, M. E., Cassard, L., Yang, J. C., Hughes, M. S., et al. (2009). Gene therapy with human and mouse T-cell receptors mediates cancer regression and targets normal tissues expressing cognate antigen. *Blood*, *114*, 535–546.

Kawakami, Y., Eliyahu, S., Delgado, C. H., Robbins, P. F., Rivoltini, L., Topalian, S. L., et al. (1994). Cloning of the gene coding for a shared human melanoma antigen recognized by autologous T cells infiltrating into tumor. *Proceedings of the National Academy of Sciences of the United States of America*, *91*, 3515–3519.

Klebanoff, C. A., Acquavella, N., Yu, Z., & Restifo, N. P. (2011). Therapeutic cancer vaccines: Are we there yet? *Immunological Reviews*, *239*, 27–44.

Klebanoff, C. A., Khong, H. T., Antony, P. A., Palmer, D. C., & Restifo, N. P. (2005). Sinks, suppressors and antigen presenters: How lymphodepletion enhances T cell-mediated tumor immunotherapy. *Trends in Immunology*, *26*, 111–117.

Knuth, A., Danowski, B., Oettgen, H. F., & Old, L. J. (1984). T-cell-mediated cytotoxicity against autologous malignant melanoma: Analysis with interleukin 2-dependent T-cell cultures. *Proceedings of the National Academy of Sciences of the United States of America*, *81*, 3511–3515.

Kochenderfer, J. N., Wilson, W. H., Janik, J. E., Dudley, M. E., Stetler-Stevenson, M., Feldman, S. A., et al. (2010). Eradication of B-lineage cells and regression of lymphoma in a patient treated with autologous T cells genetically engineered to recognize CD19. *Blood*, *116*, 4099–4102.

Krauthammer, M., Kong, Y., Ha, B. H., Evans, P., Bacchiocchi, A., McCusker, J. P., et al. (2012). Exome sequencing identifies recurrent somatic RAC1 mutations in melanoma. *Nature Genetics*, *44*, 1006–1014.

Kvistborg, P., Shu, C. J., Heemskerk, B., Fankhauser, M., Thrue, C. A., Toebes, M., et al. (2012). TIL therapy broadens the tumor-reactive CD8(+) T cell compartment in melanoma patients. *Oncoimmunology*, *1*, 409–418.

Lawrence, M. S., Stojanov, P., Polak, P., Kryukov, G. V., Cibulskis, K., Sivachenko, A., et al. (2013). Mutational heterogeneity in cancer and the search for new cancer-associated genes. *Nature*, *499*, 214–218.

Le, D. T., Uram, J. N., Wang, H., Bartlett, B. R., Kemberling, H., Eyring, A. D., et al. (2015). PD-1 blockade in tumors with mismatch-repair deficiency. *The New England Journal of Medicine*, *372*, 2509–2520.

Maude, S. L., Frey, N., Shaw, P. A., Aplenc, R., Barrett, D. M., Bunin, N. J., et al. (2014). Chimeric antigen receptor T cells for sustained remissions in leukemia. *The New England Journal of Medicine*, *371*, 1507–1517.

Morgan, R. A., Dudley, M. E., Wunderlich, J. R., Hughes, M. S., Yang, J. C., Sherry, R. M., et al. (2006). Cancer regression in patients after transfer of genetically engineered lymphocytes. *Science*, *314*, 126–129.

Muul, L. M., Spiess, P. J., Director, E. P., & Rosenberg, S. A. (1987). Identification of specific cytolytic immune responses against autologous tumor in humans bearing malignant melanoma. *Journal of Immunology*, *138*, 989–995.

Park, T. S., Groh, E. M., Patel, K., Kerkar, S. P., Lee, C. C., & Rosenberg, S. A. (2016). Expression of MAGE-A and NY-ESO-1 in primary and metastatic cancers. *Journal of Immunotherapy*, *39*(1), 1–7.

Parkhurst, M. R., Yang, J. C., Langan, R. C., Dudley, M. E., Nathan, D. A., Feldman, S. A., et al. (2011). T cells targeting carcinoembryonic antigen can mediate regression of metastatic colorectal cancer but induce severe transient colitis. *Molecular Therapy*, *19*, 620–626.

Riddell, S. R., & Greenberg, P. D. (1990). The use of anti-CD3 and anti-CD28 monoclonal antibodies to clone and expand human antigen-specific T cells. *Journal of Immunological Methods*, *128*, 189–201.

Robbins, P. F., El-Gamil, M., Li, Y. F., Kawakami, Y., Loftus, D., Appella, E., et al. (1996). A mutated beta-catenin gene encodes a melanoma-specific antigen recognized by tumor infiltrating lymphocytes. *The Journal of Experimental Medicine*, *183*, 1185–1192.

Robbins, P. F., Kassim, S. H., Tran, T. L., Crystal, J. S., Morgan, R. A., Feldman, S. A., et al. (2015). A pilot trial using lymphocytes genetically engineered with an NY-ESO-1-reactive T-cell receptor: Long-term follow-up and correlates with response. *Clinical Cancer Research*, *21*, 1019–1027.

Robbins, P. F., Lu, Y. C., El-Gamil, M., Li, Y. F., Gross, C., Gartner, J., et al. (2013). Mining exomic sequencing data to identify mutated antigens recognized by adoptively transferred tumor-reactive T cells. *Nature Medicine*, *19*, 747–752.

Robbins, P. F., Morgan, R. A., Feldman, S. A., Yang, J. C., Sherry, R. M., Dudley, M. E., et al. (2011). Tumor regression in patients with metastatic synovial cell sarcoma and melanoma using genetically engineered lymphocytes reactive with NY-ESO-1. *Journal of Clinical Oncology*, *29*, 917–924.

Rosenberg, S. A., Aebersold, P., Cornetta, K., Kasid, A., Morgan, R. A., Moen, R., et al. (1990). Gene transfer into humans—Immunotherapy of patients with advanced melanoma, using tumor-infiltrating lymphocytes modified by retroviral gene transduction. *The New England Journal of Medicine*, *323*, 570–578.

Rosenberg, S. A., Kawakami, Y., Robbins, P. F., & Wang, R. (1996). Identification of the genes encoding cancer antigens: Implications for cancer immunotherapy. *Advances in Cancer Research*, *70*, 145–177.

Rosenberg, S. A., Packard, B. S., Aebersold, P. M., Solomon, D., Topalian, S. L., Toy, S. T., et al. (1988). Use of tumor-infiltrating lymphocytes and interleukin-2 in the immunotherapy of patients with metastatic melanoma. A preliminary report. *The New England Journal of Medicine*, *319*, 1676–1680.

Rosenberg, S. A., Yang, J. C., Sherry, R. M., Kammula, U. S., Hughes, M. S., Phan, G. Q., et al. (2011). Durable complete responses in heavily pretreated patients with metastatic melanoma using T-cell transfer immunotherapy. *Clinical Cancer Research*, *17*, 4550–4557.

Rosenberg, S. A., Yannelli, J. R., Yang, J. C., Topalian, S. L., Schwartzentruber, D. J., Weber, J. S., et al. (1994). Treatment of patients with metastatic melanoma with autologous tumor-infiltrating lymphocytes and interleukin 2. *Journal of the National Cancer Institute*, *86*, 1159–1166.

Scanlan, M. J., Welt, S., Gordon, C. M., Chen, Y. T., Gure, A. O., Stockert, E., et al. (2002). Cancer-related serological recognition of human colon cancer: Identification of potential diagnostic and immunotherapeutic targets. *Cancer Research*, *62*, 4041–4047.

Topalian, S. L., Hodi, F. S., Brahmer, J. R., Gettinger, S. N., Smith, D. C., McDermott, D. F., et al. (2012). Safety, activity, and immune correlates of anti-PD-1 antibody in cancer. *The New England Journal of Medicine*, *366*, 2443–2454.

Topalian, S. L., Solomon, D., & Rosenberg, S. A. (1989). Tumor-specific cytolysis by lymphocytes infiltrating human melanomas. *Journal of Immunology*, *142*, 3714–3725.

Tran, E., Turcotte, S., Gros, A., Robbins, P. F., Lu, Y. C., Dudley, M. E., et al. (2014). Cancer immunotherapy based on mutation-specific CD4+ T cells in a patient with epithelial cancer. *Science*, *344*, 641–645.

Van Allen, E. M., Miao, D., Schilling, B., Shukla, S. A., Blank, C., Zimmer, L., et al. (2015). Genomic correlates of response to CTLA-4 blockade in metastatic melanoma. *Science*, *350*, 207–211.

van der Bruggen, P., Traversari, C., Chomez, P., Lurquin, C., De, P. E., Van den Eynde, B., et al. (1991). A gene encoding an antigen recognized by cytolytic T lymphocytes on a human melanoma. *Science*, *254*, 1643–1647.

Wolfel, T., Hauer, M., Schneider, J., Serrano, M., Wolfel, C., Klehmann-Hieb, E., et al. (1995). A p16INK4a-insensitive CDK4 mutant targeted by cytolytic T lymphocytes in a human melanoma. *Science*, *269*, 1281–1284.

Zhang, L., Morgan, R. A., Beane, J. D., Zheng, Z., Dudley, M. E., Kassim, S. H., et al. (2015). Tumor-infiltrating lymphocytes genetically engineered with an inducible gene encoding interleukin-12 for the immunotherapy of metastatic melanoma. *Clinical Cancer Research*, *21*, 2278–2288.

INDEX

Note: Page numbers followed by "*f*" indicate figures, and "*t*" indicate tables.

D

E

G

H

I

M

R

S

T

V

CONTENTS OF RECENT VOLUMES

Volume 88

Volume 89

Volume 90

Volume 91

Volume 92

Volume 93

Volume 94

Volume 95

Volume 96

Volume 97

Volume 98

Volume 99

Volume 100

Volume 101

Volume 102

Volume 103

Volume 104

Volume 105

Volume 106

Volume 107

Volume 108

Volume 109

Volume 110

Volume 111

Volume 112

Volume 113

Volume 114

Volume 115

Volume 116

Volume 117

Volume 118

Volume 119

Volume 120

Volume 121

Volume 122

Volume 123

Volume 124

Volume 125

Volume 126

Volume 127

Volume 128

Volume 129

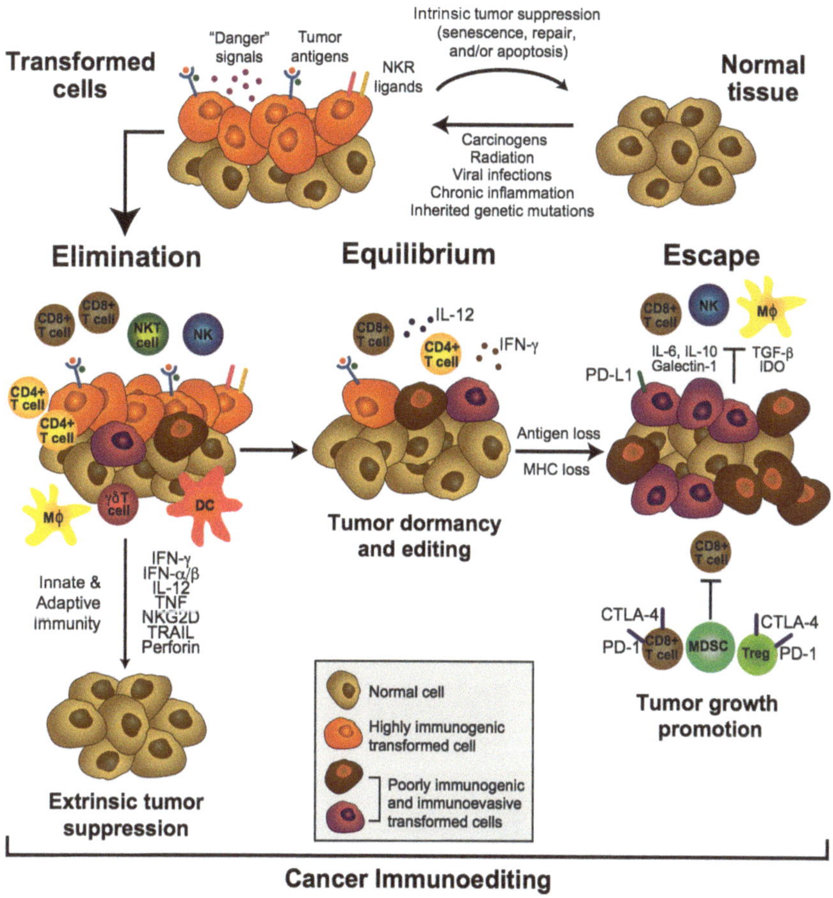

Chapter 2, Figure 1 (See page 31 of this volume.)

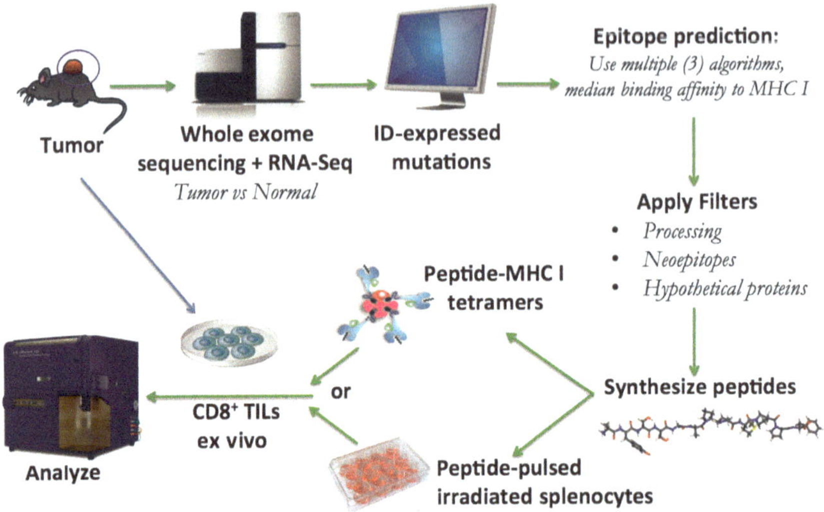

Chapter 2, Figure 2 (See page 49 of this volume.)

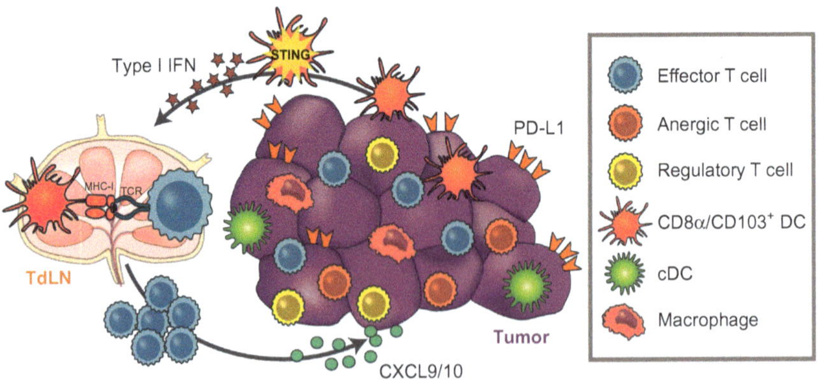

Chapter 3, Figure 1 (See page 78 of this volume.)

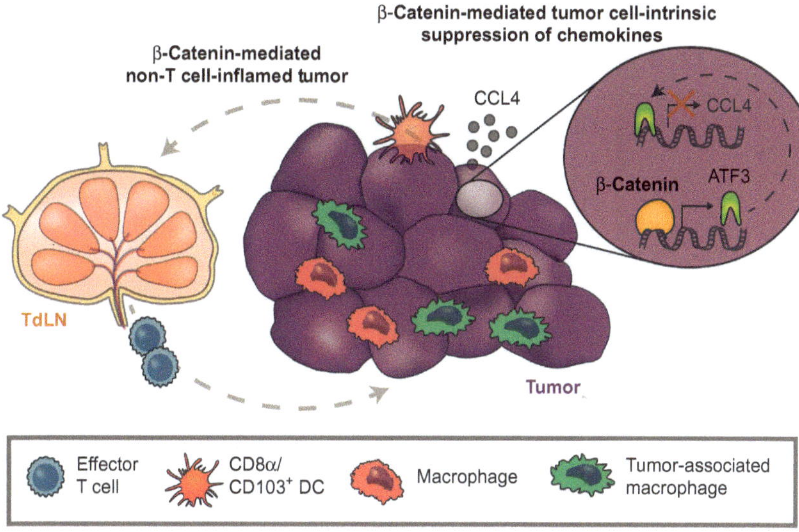

Chapter 3, Figure 2 (See page 81 of this volume.)

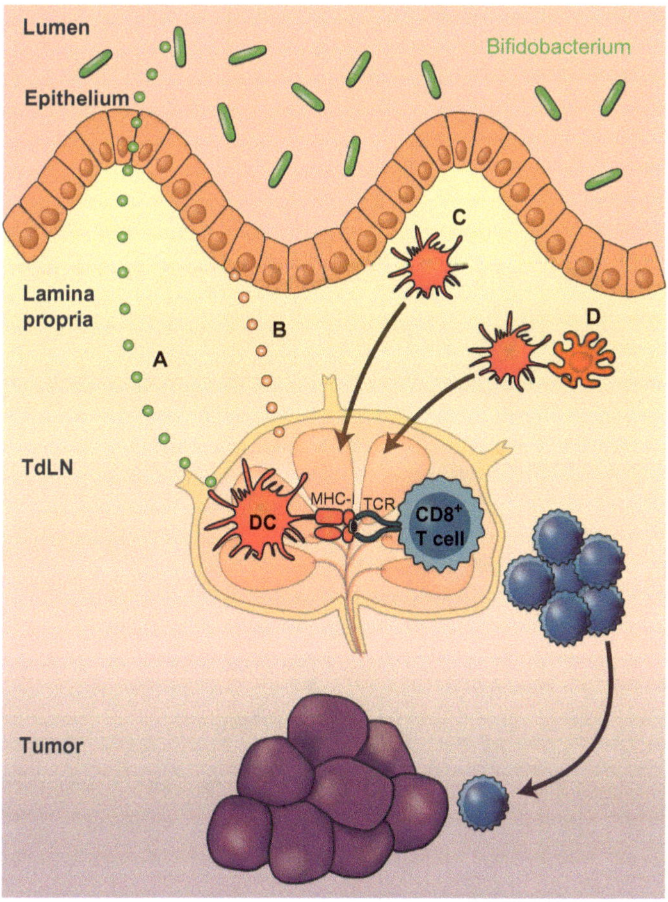

Chapter 3, Figure 3 (See page 87 of this volume.)

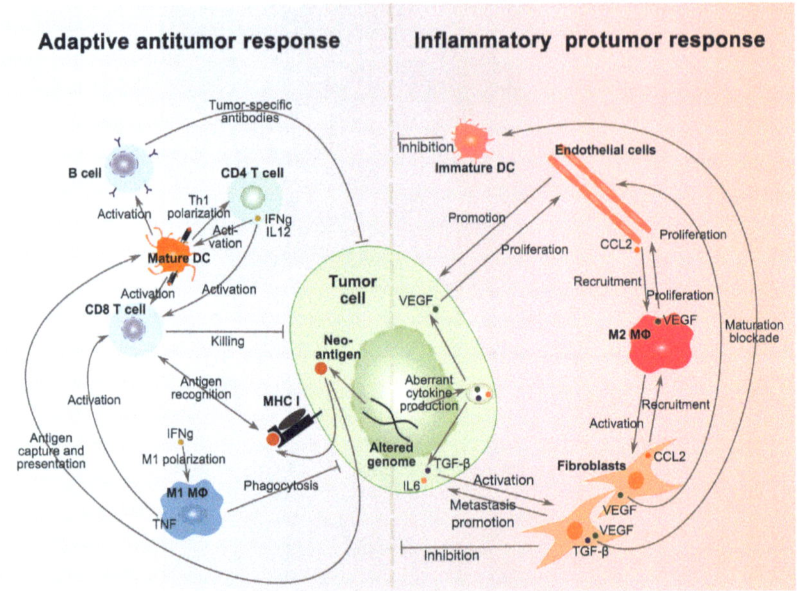

Chapter 4, Figure 1 (See page 100 of this volume.)

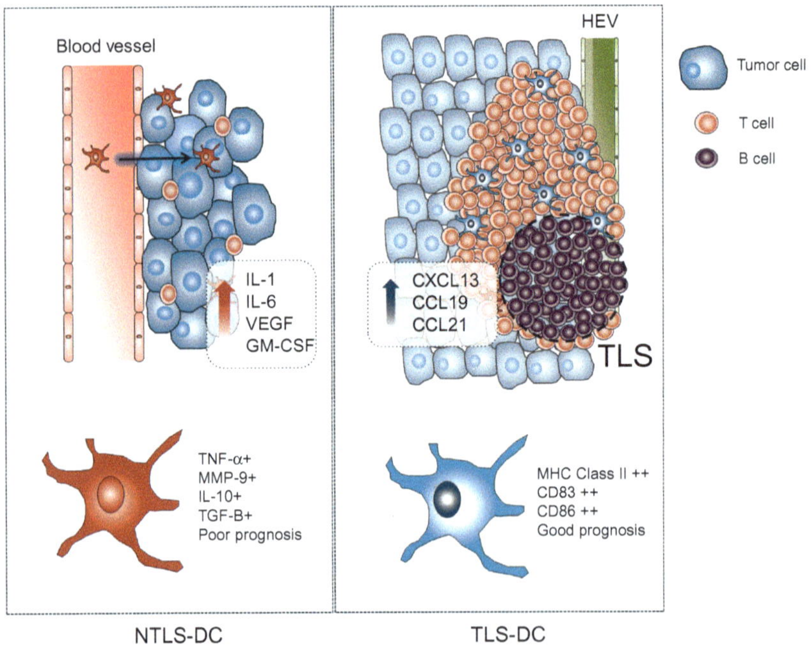

Chapter 4, Figure 4 (See page 136 of this volume.)

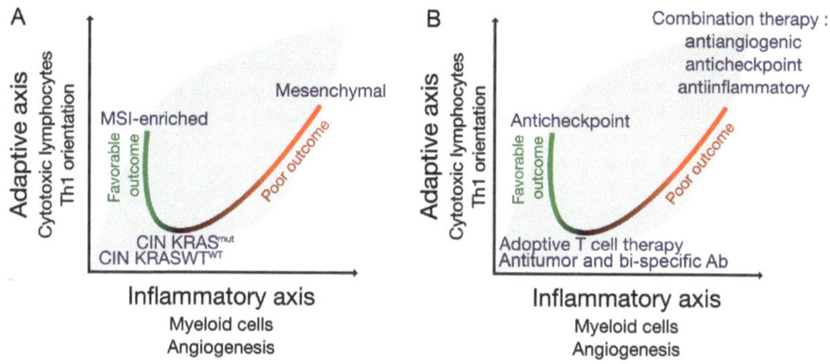

Chapter 4, Figure 5 (See page 153 of this volume.)

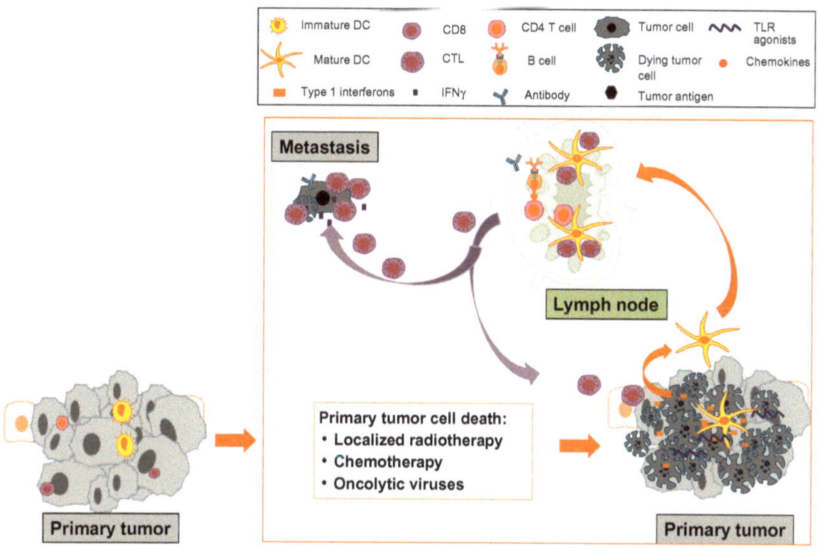

Chapter 5, Figure 1 (See page 221 of this volume.)

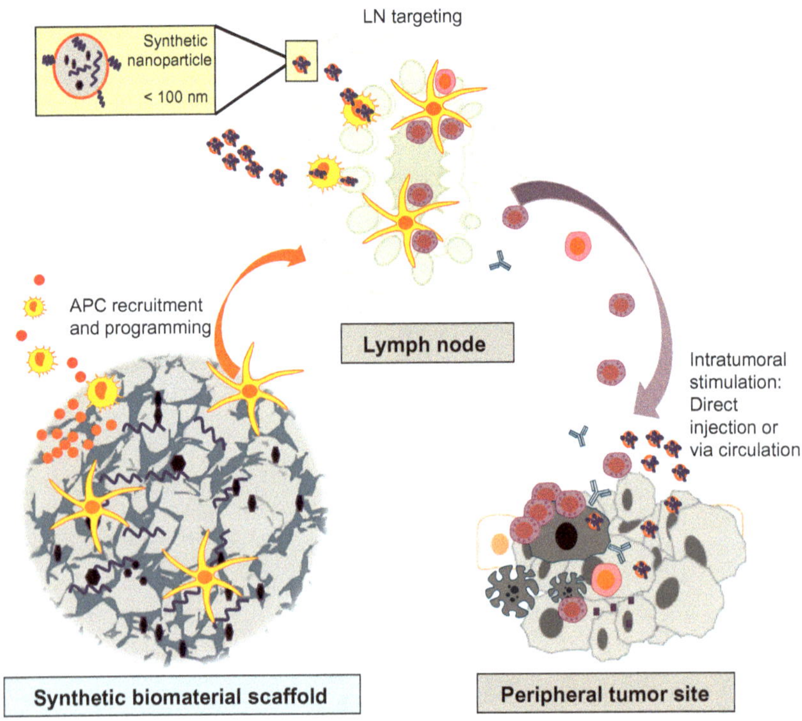

Chapter 5, Figure 2 (See page 223 of this volume.)

FSC
www.fsc.org
MIX
Papier | Fördert gute Waldnutzung
FSC® C083411

Kolibri 360 GmbH
Ferdinand-Jühlke-Straße 7
99095 Erfurt, Deutschland
produktsicherheit@kolibri360.de